AF572422

# *Prospects for Future Climate*

## *A Special US/USSR Report on Climate and Climate Change*


Edited by

Michael C. MacCracken
Alan D. Hecht
Mikhail I. Budyko
Yuri A. Izrael


*Prepared under the auspices of the US/USSR Agreement on Protection of the Environment*

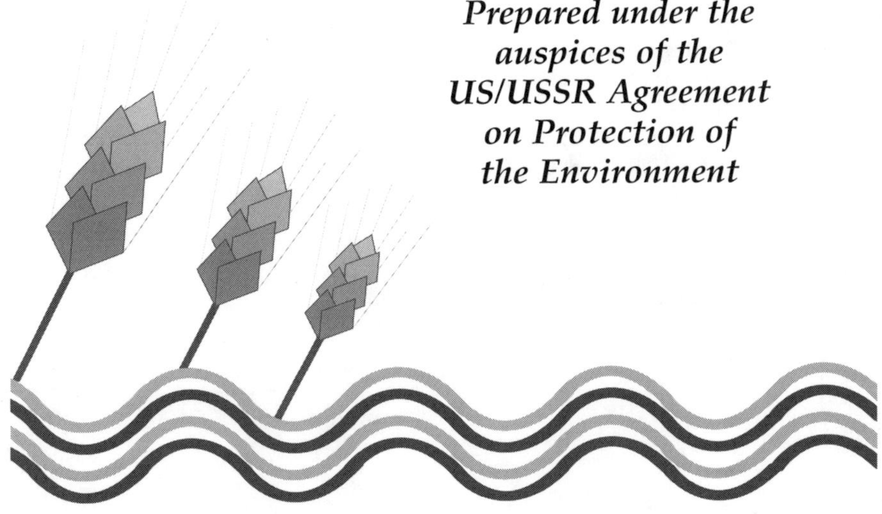

This report was prepared jointly by authors from the US and USSR, each supported by their institutional and sponsoring agencies. The editing and preparation of the English version of the report was carried out under the auspices of the US Department of Energy by the Lawrence Livermore National Laboratory under contract W-7405-Eng-48 with support in part from the US National Oceanic and Atmospheric Administration.

**Library of Congress Cataloging in Publication Data**

Prospects for future climate: a special US/USSR report on climate and climate change / edited by Michael C. MacCracken ... [et al.].

p. cm.

"US/USSR agreement on protection of the environment".

Includes bibliographical references (p. 235).

ISBN 0-87371-440-7

1. Climatic changes. I. Climatology. II. MacCracken, Michael C.

QC981.8.C5P759 1990

# TRANSMITTAL LETTER

The Honorable William Reilly
US Chairman

Council of Ministers
of the USSR

## Joint US/USSR Commission on Protection of the Environment

The composition of the global atmosphere is being altered as a consequence of emissions from activities ranging from energy generation to expansion of agriculture. Both empirical and theoretical studies demonstrate that the changes in composition will lead to global warming and other changes in climate. These climate changes will in turn affect important agricultural, hydrologic, and other activities on which global society depends.

Under the auspices of Working Group VIII (The Influence of Environmental Changes on Climate) of the US/USSR Agreement on Protection of the Environment signed in 1972, scientists from both countries have been collaborating for about fifteen years on research projects to better understand the nature of climate and climate change. This collaboration has involved many scientific symposia, exchange visits of scientists, joint projects, activities, publications, and exchanges of important climatic data sets. This report presents results arising primarily from our joint studies and, although generally consistent with the findings of the broader scientific community, is not intended as a detailed scientific review of all research on the greenhouse question. Taken together with the analyses and findings of other international reviews, however, this report contributes significantly to scientific understanding of this issue.

Together, important insight has been gained from our scientific collaboration, and much more can be gained from continued and expanded cooperation in the future. The imminence of the prospective chemical, climatic, agricultural, hydrological, and other impacts necessi-tates that we continue our very high level of scientific cooperation to further advance scientific understanding and to build an even stronger basis for policy development and evaluation.

Mikhail I. Budyko
USSR Co-chairman
Working Group VIII

Alan D. Hecht
US Co-chairman
Working Group VIII

# AUTHORS

This report of Working Group VIII (The Influence of Environmental Changes on Climate) under the US/USSR Agreement on Protection of the Environment represents the collective assessment of the authors from both countries.


USSR Authors

Irina I. Borzenkova
Mikhail I. Budyko
Eleonora K. Byutner
George S. Golitsyn
Yuri A. Izrael
Igor L. Karol
Kira I. Kobak
Vladimir M. Kotlyakov
Gennadi V. Menzhulin
Igor A. Shiklomanov
Andre A. Velichko
Konstantin Ya. Vinnikov

US Authors

Raymond S. Bradley
Robert Etkins
Peter H. Gleick
Martin I. Hoffert
Thomas R. Karl
John E. Kutzbach
Michael C. MacCracken
Cynthia E. Rosenzweig
Michael E. Schlesinger
Donald J. Wuebbles

## ACKNOWLEDGMENTS

The valuable series of joint symposia, workshops, exchanges, and projects extending back about 15 years has provided the foundation for this special report. The impetus for assembling this assessment of our understanding came from the call for this special report by President Reagan and President Gorbachev in the joint communique issued at the conclusion of their summit meeting in December 1987.

In the two years since that meeting, an outline was agreed upon, lead chapter authors from each country were selected, draft versions of the chapters were prepared, and the authors met to integrate and improve their contributions. The manuscript has then been reviewed by scientists from each country, and final changes have been incorporated by the editors.

We want to acknowledge the dedicated efforts by the authors from both countries to work toward agreement and understanding, identifying both what we know and what more we need to know. The support of the many participating and supporting organizations in both countries is also gratefully appreciated.

The editors also want to express their gratitude to Pam Drumtra, Coralyn K. McGregor, Penny Keenan, and Catherine M. Williams of the Lawrence Livermore National Laboratory, US and to Irina Turchinovich, Yuliya Monakhova, and Valentina Yanuta of the State Hydrological Institute, USSR for their many efforts to prepare the materials for publication.

The Editors

# CONTENTS

# SUMMARY

This special joint report describes the considerable progress that has been made in understanding the nature and extent of past and present climate changes on the Earth. Much of what we have learned about climate and about the potential effects of climate change on agriculture and hydrology has resulted from research undertaken over the past 10 years. The prospect of human-induced climate change resulting from the rapid increase in the concentrations of greenhouse gases in the atmosphere (and other anthropogenic changes) makes prediction of future climate and climate impacts an essential international research priority.

Since the 1970s, Soviet and US scientists have recognized that global warming would be inevitable as a consequence of the on-going perturbations to atmospheric composition. Analyses of observations by scientists of both nations indicate that the global average temperature has increased by 0.4 to 0.5°C over the past hundred years. The challenge, based on current knowledge, is to predict specific changes in global climate and their impact on the environment, especially over the next 50 years. To build the foundation on which to base projections of future conditions, cooperative US and USSR workshops, symposia, and research studies sponsored by Working Group VIII have resulted in:

(1) Improved documentation of climate changes in the extratropical Northern Hemisphere during past warm periods extending back several million years,
(2) Increased availability and detailed analyses of data sets on global-scale changes in temperature and precipitation over the past 100 years,
(3) Measurements and geological estimates of carbon dioxide changes over past geologic periods,
(4) Intensified reconstructions and analyses of paleoclimatic conditions, using both empirical and numerical modeling approaches, and
(5) Development of improved theoretical capabilities (models) for studying climate change and for projecting future climate conditions.

Even with these many scientific advances, the challenge of projecting future climatic conditions remains difficult for many reasons. The physical and chemical mechanisms responsible for changes in atmospheric composition and climate are complicated, particularly with respect to feedbacks involving ocean, cloud, and chemistry processes. Projecting the extent of climate change requires projecting economic growth, especially energy use and technology change, into the future. Such projections can never be certain.

Despite the difficulty of the task, scientists from the US and USSR have pursued two important approaches to understanding and predicting future climates.

The first approach, which has been carried out over the past 10 years, was an ambitious effort focusing on the empirical analysis of climate. An extensive data base of modern and paleoclimatic conditions has been compiled that covers much of the Northern Hemisphere and some areas in the Southern Hemisphere. As a result of these activities, there is growing recognition that changes in the atmospheric composition played a major role in contributing to past periods of global warmth, just as we expect that future warming will result from the increasing concentrations of $CO_2$ and other greenhouse gases in the atmosphere. Consequently, past intervals of warm climate, may provide important insights into the characteristics of the future climate.

However, in some ways climates of these past periods differ from the unique conditions that we project for the future. For example, we are in a period of rapidly changing atmospheric composition, so that the slower warming of the oceans compared to the continental areas may complicate these projections. In addition, because climatic patterns of the past were also affected by other factors, further study and consideration of such complications is necessary if we are to gain full benefit from the insights provided by this approach.

A second approach to climate prediction has been the development of climate models that explicitly represent the dynamics, thermodynamics, and hydrology of the atmosphere based on the fundamental laws of physics. The models are constantly being improved and accurately represent many aspects of the global climate, but they remain generally limited by their consideration of equilibrium rather than time-dependent conditions. The model projections also remain limited by their relatively coarse geographic resolution and by inadequate understanding of important atmospheric processes. Both numerical models and empirical techniques are further limited in their ability to project changes in the variability and frequency of extreme events.

However, it is possible to combine insights from both numerical models and empirical methods to provide some projections of general climate trends for the future. *For the concentrations of $CO_2$ and other greenhouse gases roughly equivalent to those that are expected by the beginning of the twenty-first century, both empirical and modeling approaches suggest that the global warming over the next quarter century will be about equivalent to that over the last century.* The actual warming is subject to some moderation or amplification about this rising mean temperature as a result of fluctuations introduced by natural variability, but temperatures are likely to remain above pre-industrial values. Temperature increases are expected to be greater in high than in low latitudes and to be larger in winter than in summer. Both approaches also suggest that the US and USSR will generally receive more precipitation, particularly in winter. In some regions, there may be more frequent intervals of increased stress on summer soil moisture, especially in midcontinental regions which, because of higher evaporation, would tend to become drier than the current average conditions.

For the second quarter of the next century, results are less certain, in part because it is difficult to forecast the economic and agricultural development that leads to increased emissions and resulting changes in atmospheric composition. *For plausible increases in the concentrations of greenhouse gases, both modeling and empirical approaches suggest that the warming will continue to 2050 and beyond.* Global precipitation and evaporation will also increase, precipitation being more in winter than summer and evaporation more in summer than winter. Some regions will experience increases and some decreases in soil moisture, although details are still uncertain.

The potential consequences of these global trends for agriculture and water management and for society in general may be extremely serious. Concerted international attention must continue to be focused on further increasing our understanding of potential impacts and on exploring possible responses to the expected changes and impacts.

For the past 17 years, the steadily intensifying, enriching, and rewarding interactions of Working Group VIII have contributed to improving our understanding of this important issue. This report is based on the research stemming from many of these interactions. The joint participation in this analysis by scientists from the two nations provides insight and broadens overall understanding by reliance on both modeling and empirical perspectives.

The challenge ahead is at least as great as the accomplishments so far achieved. The positive feedback generated by our increasing interactions in recent years under these auspices offers great hope for fulfilling the important requirements that our nations and peoples place on scientists to provide the information that can ensure environmental protection in the years ahead.

# CHAPTER 1. INTRODUCTION

The US-USSR Agreement on Protection of the Environment, signed in 1972, has provided a framework for ten Working Groups to cooperate in joint research on atmospheric and environmental problems. From the very beginning, issues of climate prediction, reconstructions of past climates, and human influence on the atmosphere have been given special consideration, with Working Group VIII (The Influence of Environmental Changes on Climate) focusing specifically on these topics. Initial studies under this Agreement concentrated on the physical characteristics of climate and on climate modeling. The program has grown over the years to include cooperative efforts on carbon cycle modeling, atmospheric chemistry including ozone depletion, radiation and cloud dynamics, and exchanges of climatic data sets.

In 1976, United States and Soviet scientists began a special effort to reconstruct descriptions of the past climatic conditions of both countries. A 7-year effort of data collection and analysis resulted in a two-part special report, *Late Quaternary Environments of the Soviet Union,* (Velichko et al., 1984) and *Late Quaternary Environments of the United States* (Porter and Wright, 1984). These studies produced an extensive data base for use in validating model simulations of past climate change. The preparation of this report stimulated considerable effort in reconstructing the climatic conditions of preceding geological intervals, especially past warm periods that might provide insights into future climatic conditions.

Similarly, in 1975 United States and Soviet scientists began special joint efforts to advance our abilities to model the global climate. A series of extended exchange visits and exchanges of data sets contributed to initial model developments. Over the past several years, testing and intercomparison of the results of these models, both with the results of other models and with newly available satellite-derived data sets, have led to further improvements.

What has evolved from these joint efforts is an innovative approach to projecting future climate change based on the hypothesis that we can combine the strengths of what we have learned about past warm periods and what we can simulate with our models. Thus, we are combining understanding gained from studies of the climatic patterns of warmer geological periods such as the Mid-Pliocene climatic optimum (from about 4.3 to 3.3 million years ago), which was probably characterized by generally higher levels of carbon dioxide ($CO_2$), and from model simulations of the changes in climate to be expected with the higher $CO_2$

concentrations expected in the future to suggest the characteristics of the climate changes that may occur during the first half of the next century.

A major portion of the US-USSR collaborative efforts over the past several years has been to gather data, to test, and to evaluate this two-pronged approach (US-USSR, 1982; Budyko and MacCracken, 1987). This has occurred at the same time that it has become recognized that substantial climate change may result from human activities, especially as a consequence of the increasing emissions of $CO_2$ and other trace gases. Human-induced climate change has been the focus of several major joint workshops and is now the subject of extensive international activities.

At the conclusion of their 1987 summit meeting in Washington, DC, President Reagan and President Gorbachev signed a joint communique emphasizing their support for the environmental agreement and their encouragement for cooperative studies of climate change. In particular, both sides agreed to prepare a joint report on "future climates" aimed at providing both nations with an assessment of their cooperative work and a statement on the likelihood and nature of future climate change, especially as it may be affected by human factors.

This report, preparation of which began in the autumn of 1988, takes an important step in furthering understanding of climate change, both in a bilateral and multilateral sense.[1] Specifically, this report

(1) Draws primarily on cooperative research by United States and Soviet scientists,
(2) Identifies areas of scientific consensus,
(3) Identifies areas of scientific uncertainty where additional work is required, and
(4) Recommends areas for continued cooperation.

The chapters are organized to reflect the broad areas of cooperative research between the United States and the Soviet Union, particularly of those aspects related to the potential warming from the increasing concentrations of $CO_2$ and other greenhouse gases. Our cooperation has allowed both nations to pursue a common objective with sometimes different, but always complementary, approaches. This cooperation has been strengthened by our joint commitment to exchange and share ideas and to work toward a common understanding about the prospects for future climate change. Where our scientific views of the future climate

---

[1] Since this evaluation was begun, the international community has organized the Intergovernmental Panel on Climate Change (IPCC) under the auspices of the World Meteorological Organization (WMO) and the United Nations Environment Program (UNEP) to provide a forum for international debate on the causes of and responses to future climate change. Key elements of the IPCC process are the preparation of an assessment of current knowledge of (1) the causes of climate change and possible means by which future climates can be predicted, and (2) the impacts of climate change on human activities and the environment. This report addresses both of these topics, but not the third IPCC responsibility of reviewing response strategies.

differ as a result of our different approaches or interpretations of data, we have agreed to consider these differences of opinion as the basis for cooperative research projects and activities.

We agree on many important conclusions. The burning of fossil fuels and widespread deforestation have resulted in a steady rise of the concentration of $CO_2$ in the atmosphere. The concentrations of other gases present in trace amounts, such as methane and chlorofluorocarbons, are also known to be increasing. Collectively, the rising concentrations of $CO_2$ and these other trace gases will significantly enhance the Earth's natural greenhouse effect. Both paleoclimatic evidence and modeling studies agree that such changes will lead to major climate change. These changes in the natural climate may have important influences on the agricultural, hydrologic, and other resources upon which our societies depend.

The US and the USSR, as major producers of $CO_2$ and the other trace gases, face the difficult task of planning and managing future agricultural, hydrologic, and energy systems, taking into account the prospect of future climate change. The highest level of scientific cooperation and assessment is essential for developing appropriate and realistic national and international policies for addressing the challenge of global climate change. This report builds upon the many significant accomplishments of our collaboration in the past and strengthens the foundation for continuing research and cooperation.

# *CHAPTER 2. PAST CHANGES IN GLOBAL CLIMATE*

## 2.1. INTRODUCTION

The Earth's climate has always been changing and will no doubt continue to change. Until fairly recently, climates of the past, if not unknown, have at best been perceived in only general and qualitative terms. This situation is now changing. New observational techniques, accurate radiometric dating methods, and comprehensive sampling strategies have evolved. As a result, there has been a dramatic increase in knowledge of the early evolution of the atmosphere and ocean, of the shifting shapes and locations of continents, and of the waxing and waning of ice sheets, forests, lakes, and deserts. This knowledge has in turn stimulated both qualitative and quantitative inquiries into the extent of paleoclimatic change and the factors contributing to these changes.

Paleoclimatic studies are important for several reasons:

(1) To gain a physical understanding of the causes of climate change.

(2) To determine the spatial and temporal distributions of climate change so that the magnitude and rate of natural climate variability can be compared to those expected from climate changes that may be induced by human activities.

(3) To identify the features of warm climates that may provide insights into possible future climatic conditions.

(4) To test the accuracy of paleoclimatic simulations made with climate models. These data provide a means of testing climate models using changes in climate comparable in magnitude to those projected for the next century.

The purpose of this chapter is to review the status of our work in these various areas. Many of these topics have been subjects of joint Soviet-American studies since 1976 (US/USSR, 1977). The first several years were devoted to discussing methods for making quantitative estimates of past climates. More recently, research has emphasized exchanges of data, analysis of climatic histories, and publishing of monographs in both countries. We are now entering a third stage, that of developing global analyses and investigating the causes of climate change.

## 2.2. CAUSES OF CLIMATE CHANGE OVER THE GEOLOGIC PAST

Using information on past climates to aid in projecting the future climate requires that we understand the degree to which past changes in climate were forced by changes in particular factors and the extent to which the changes were simply random. Over long periods, the reasonably persistent and regular changes in climate suggest that the random factor may be relatively small compared with the large changes that have been observed. In a general sense, there is scientific agreement on the factors that have probably had the largest influences on the evolution of climate over the history of the Earth (Budyko, 1974):

(1) Changes in the solar radiation related to solar output;
(2) Changes in the land/ocean distribution associated with tectonic plate movements and consequent changes of orography, ocean circulation, and sea-level;
(3) Changes in atmospheric composition, particularly in the concentrations of carbon dioxide ($CO_2$) and methane ($CH_4$);
(4) Changes in the albedo of the Earth's surface;
(5) Changes in the Earth's orbital parameters; and
(6) Changes of a catastrophic nature, such as meteor impacts, extended series of volcanic eruptions, etc.

The relative importance of these factors on different time scales, however, is a very active area of scientific inquiry.

Restricting discussion to roughly the last 250 million years, for which we can reconstruct many aspects of the climate changes that have occurred, provides a wide range of interesting examples. About 250 million years ago, the solar flux may have been about 1% less than at present (having increased more or less linearly to the present). The continents were then all joined together; they have since split apart and moved gradually to their present locations. These changes have influenced not only land and ocean temperature and rainfall distributions, but also ocean currents, temperature, and salinity. The locations and heights of plateaus and mountains have changed, also causing changes in temperature and rainfall patterns. Changes in ocean geometry have redirected ocean currents, as for example when the rise of the isthmus of Panama about 3.5 to 3.0 million years ago blocked the flows between the Atlantic and Pacific Oceans.

The idea that variations in $CO_2$ had induced climatic fluctuations in the geological past was first proposed by Arrhenius (1896, 1903), who believed that the past warm epochs could be attributed to increased $CO_2$ in the atmosphere. Since that time, studies of geologic records, particularly in ice cores, have documented that large changes have indeed occurred in the concentration of atmospheric $CO_2$, which has a strong influence on the greenhouse effect of the atmosphere. With the prospect that the $CO_2$ concentration will continue to increase over the next century, understanding the past role of changes in $CO_2$ concentration has the potential to provide insights into the nature of the future climatic response. This requires developing quantitative estimates of cause-effect relationships.

The challenge of determining the $CO_2$ concentration for different epochs in the past has been considered in several recent studies (Budyko, 1974; Budyko and Ronov, 1979; Berner et al., 1983; Sundquist, 1985; and Budyko, et al., 1985). Geological evidence suggests that the $CO_2$ concentration was on the order of 5 to 10 times greater than the present level for much of the Mesozoic (the geological interval roughly 225 to 67 million years ago). Such a change alone could explain a significant portion of the estimated warmth of that period. Estimates by various groups agree that there has been an overall decrease in the $CO_2$ concentration over the past 100 million years, only to be reversed by human activities over the past 200 years. This decrease, compounded by continental movements and regional processes of uplift, helps to explain the cooling that has culminated in the development of large ice sheets in Antarctica and Greenland and glacial-interglacial cycles on the other continents.

Table 2.1, expanded from Budyko et al. (1985), presents estimates of changes in absolute and relative $CO_2$ amounts in the atmosphere over about the last 100 million years. (The record for the last few hundred years is discussed in Chapter 4.) These estimates clearly illustrate the general decreasing trend in the $CO_2$ concentration, but the accuracy of such estimates is limited (e.g., Berner et al., 1983; Budyko et al., 1985; Sundquist, 1985). This generally decreasing trend deserves particular attention because of its importance to understanding the natural conditions prevailing before human influence.

Figure 2.1 compares estimates of the $CO_2$ concentration and of variations in volcanicity for this time interval (Budyko et al., 1985). Although the temporal resolution of the available data is very limited, it is possible that, following the $CO_2$ maximum in the Late Cretaceous, there were two more relative maxima during the Eocene and Miocene superimposed on the generally decreasing trend in the $CO_2$ concentration. As shown in Fig. 2.1, the variations in atmospheric $CO_2$ generally correlate with the changes in volcanic activity. This association is an important part of the model used by Budyko et al. (1985) to estimate past changes in the $CO_2$ concentration.

In addition to variations in $CO_2$, the mean global temperature over the time interval in question has been affected by changes in land-ocean distribution, ocean circulation, orography, and surface albedo (glacial effects, vegetation effects, etc.) and, to a limited degree, by the gradual increase in solar radiation. Geological data indicate that the

**Table 2.1a.** Definitions of geological intervals during which values of atmospheric $CO_2$ have changed (from Budyko et al., 1985).

| Geological intervals | | | | Boundaries and duration of the interval before present (millions of years) |
|---|---|---|---|---|
| Era | Period | Epoch | Symbol | |
| Mesozoic | Cretaceous | Late Cretaceous | $K_2$ | 101 – 67 = 34 |
| Cenozoic | Tertiary | Paleocene | $P_1$ | 67 – 58 = 9 |
| | | Eocene | $P_2$ | 58 – 37 = 21 |
| | | Oligocene | $P_3$ | 37 – 25 = 12 |
| | | Miocene | $N_1$ | 25 – 9[a] = 16 |
| | | Pliocene | $N_2$ | 9[a] – 2 = 7 |
| | Quaternary | Pleistocene[b] | Q | 2 – 0.01 = 2 |
| | | Holocene | | 0.01 to pre-industrial |

[a]Although newer evidence places the Pliocene-Miocene boundary at 5 million years, the estimate of $CO_2$ concentration by Budyko et al. (1985) is based on the boundary being at 9 million years.

[b]This epoch includes the last interglacial (the Eemian) that peaked about 125,000 years before present. The estimated $CO_2$ concentration during this interglacial was about 280 to 300 ppmv.

Late Cretaceous (from about 101 to about 67 million years ago) could have been as much as 10°C warmer than the present (Budyko et al., 1985). During the warmest part of the Pliocene (referred to as the Mid-Pliocene climatic optimum and extending from about 4.3 to 3.3 million years ago), data suggest that it was 3 to 4°C warmer than present. Such a tempera-ture elevation corresponds approximately to estimates of the equilibrium warming to be expected if the $CO_2$ concentration doubles as compared with its level during the pre-industrial period (see Chapter 5). Since the Mid-Pliocene, the climate has been cooler, with strong glacial/ interglacial fluctuations over the past 1 to 2 million years, apparently paced by changes in the Earth's orbital parameters (Hays et al., 1976; Imbrie and Imbrie, 1979; Crowley, 1983).

**Table 2.1b.** Estimates of absolute and relative values of atmospheric $CO_2$ for various geological intervals (expanded from Budyko et al., 1985).

| Epoch | Mass of carbon in the atmosphere ($10^{15}$g = PgC) | Atmospheric concentration (ppmv)[a] |
|---|---|---|
| Late Cretaceous | 3700 | 1800 |
| Paleocene | 1600 | 760 |
| Eocene | 2500 | 1200 |
| Oligocene | 700 | 320 |
| Miocene | 1600 | 760 |
| Pliocene | 950 | 450 |
| Pleistocene | 400 to 600 | 200 to 300[b] |
| Holocene | 600 | 280 |

[a]ppmv = parts per million by volume.

[b]Direct measurements are available only for about the last 160,000 years (Lorius et al., 1985; Barnola et al., 1987).

This chapter will seek insights primarily from the climates of the past several million years for which evidence is most detailed and dating is most accurate. Throughout this period, the effects of the generally decreasing atmospheric $CO_2$ concentration and the effects of mountain and plateau uplift (Markov and Velichko, 1967; Ruddiman and Kutzbach, 1989) and continuing movement of the continents need to be considered (see Sections 2.3 and 2.5). In the later part of this period (the last million years), the effects of orbital changes and of changes in the concentrations of greenhouse gases and albedo of the surface are especially important (see Sections 2.4 and 2.5). The glacial-interglacial fluctuations in high and middle latitudes and the monsoon fluctuations in low latitudes (Prell and Kutzbach, 1987) are clearly depicted in the climatic records of the last few hundred thousand years. There is statistical evidence that orbital changes "pace" or "trigger" these changes in climate. There is also emerging evidence that large changes in the $CO_2$ concentration are an important contributor to these changes in climate.

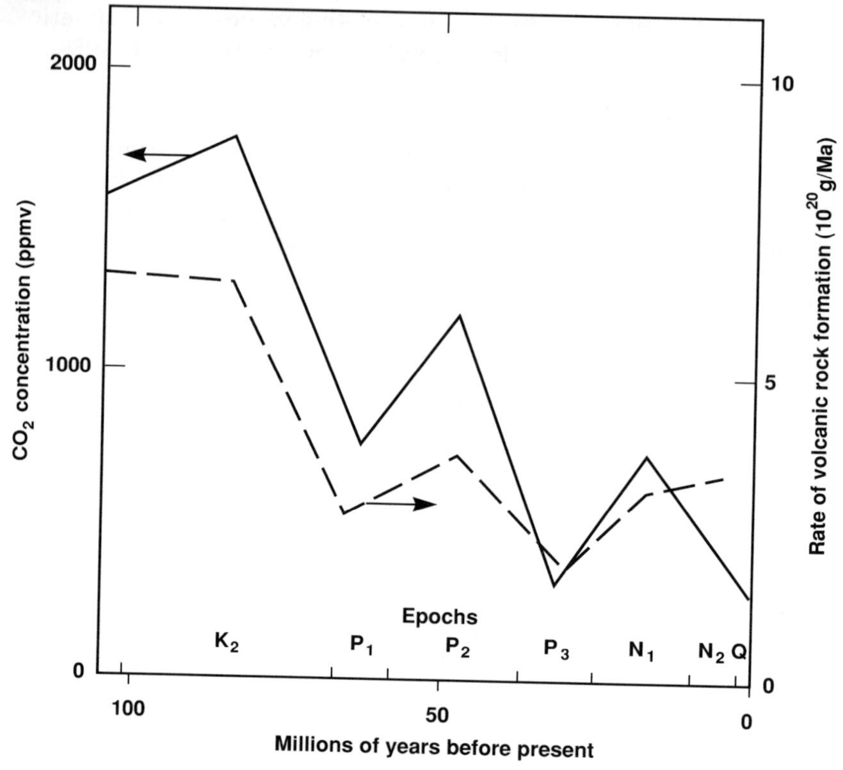


**Figure 2.1.** Estimates of the changes in $CO_2$ concentration and rate of volcanic rock formation over the past 100 million years (from Budyko et al., 1985). Designations for the epochs are given in Table 2.1.

## 2.3. CLIMATES OF THE CENOZOIC ERA

The hypothesis of a global cooling trend during the Cenozoic era, the geological interval including the Tertiary and Quaternary periods (i.e., during the last 67 million years), was put forward more than 20 years ago. Sinitsyn (1967) considered this process to have been uneven, occurring with rhythmic fluctuations that were most significant in the Pleistocene epoch. Emiliani (1966) applied oxygen-isotope techniques to study deep-sea sediments from the Caribbean Sea. He discovered that the bottom-water temperature of the world ocean changed considerably during the Cenozoic, decreasing from about 15°C at the Mesozoic-Cenozoic boundary to values almost equal to modern by the end of the Pliocene. Subsequently, extensive efforts have been made to reconstruct climate changes over the last 100 million years by isotopic and other means (e.g., Bowen, 1966; Lowenstam, 1968; Savin et al., 1975; Savin, 1982; Hecht, 1985; Yasamanov, 1985; Golbert, 1987; and others).

Figure 2.2 presents reconstructions of surface- and bottom-water temperatures for the last 130 million years based on oxygen isotope data from different regions of the Pacific Ocean (Savin et al., 1975;

Borzenkova, 1981; Savin, 1982; etc.). Assuming that data on benthic (bottom-living) fauna reflect water temperature changes in the high latitudes, the record shown in curve 2 indicates that global cooling began in the Late Cretaceous. This cooling could very likely have been associated with a sharp decrease in the atmospheric $CO_2$ concentration from about 2000 ppmv in the Santonian-Coniacian optimum (from 88 to 84 million years ago) to about 900 ppmv in the Maestrichtian (from 70 to 66 million years ago).

Lithological and paleobotanic evidence provides an opportunity to reconstruct temperature changes for land areas (e.g., Ronov and Balukhovsky, 1981; Wolfe, 1981; Axelrod, 1984; Barron, 1987; Krasilov, 1987; Mein, 1987; Velichko, 1987; and Vakhrameev, 1988). As shown by comparing curve 4 with curves 1 and 2 in Fig. 2.2, there is generally fair agreement between the land record and estimated temperature changes

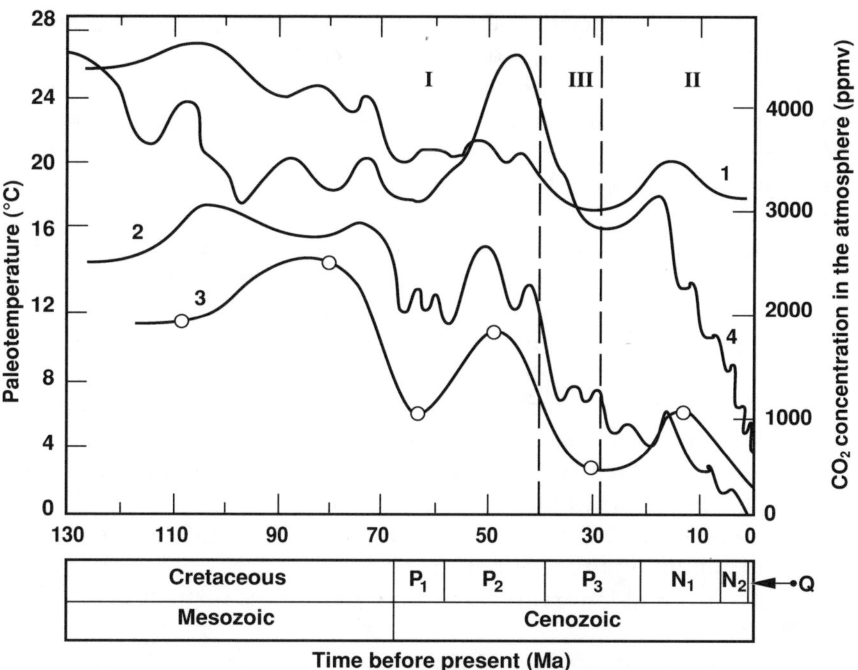


**Figure 2.2.** Estimates of ocean surface temperatures in low (curve 1) and high (curve 2) latitudes (Savin et al., 1975; Borzenkova, 1981; Savin, 1982; ), of atmospheric $CO_2$ concentration (curve 3) based on Budyko et al. (1985), and of surface air temperature over the Russian plains (curve 4) based on Velichko (1987) for the last 120 million years. The curve showing water temperature changes in low latitudes (curve 1) was developed using data from nine sites that were verified with data for the Atlantic Ocean. The time periods are divided into a nonglacial regime (I), a glacial regime (II), and a transition regime (III) from Zubakov (1986). The symbols representing epochs during the Cenozoic are given in Table 2.1.

for the ocean. This comparative analysis indicates that, from about 90 to about 65 million years ago, the temperature decreased by 8 to 10°C in the high northern latitudes, by 4 to 5°C in the middle latitudes, and by 1 to 2°C in the subtropics and tropics. As indicated in Table 2.1, this cooling was followed by a warming in the Paleocene, reaching a local maximum in the Middle Eocene (often called the "Eocene temperature optimum") that was apparently associated with an increase of the atmospheric $CO_2$ concentration (e.g., Borzenkova, 1981; Kulkova, 1987).

In the second half of the Eocene, this evidence indicates that the temperature began falling gradually, ending in a sharp drop at the Eocene-Oligocene boundary about 37 million years ago. This cooling appears to have been associated with a decrease in the atmospheric $CO_2$ concentration, an increase in planetary albedo as the land emerged and the sea receded, and the development of the first ice sheets (marine ice sheets and mountain glaciation) in the Southern Hemisphere. The Eocene-Oligocene boundary thus divides two climatic regimes—the first was predominantly nonglacial, which was typical for the Earth about 200 million years ago (the Mesozoic and Early Cenozoic), and the second was the colder regime more typical of the past few million years (Frakes, 1979; Zubakov, 1986).

The earlier nonglacial climatic regime was characterized by the following features:

(1) Carbon dioxide concentrations in the atmosphere, as estimated from geological evidence, of more than five times the pre-industrial level. Reconstructions indicate that the temperature in high northern latitudes was about 10 to 15°C above the present value.

(2) A general absence (or very limited presence) both of ice sheets on the continents and cold bottom waters in the oceans.

(3) A Northern Hemisphere pole-equator temperature gradient considerably less than the modern one.

(4) An apparent absence of deserts, indicating a high level of moistening on the continents.

(5) Climatic- and landscape-zone boundaries less distinct than at present.

Over the last 40 million years, the climate has cooled, and there has been a transition from the nonglacial to the glacial-interglacial climate regime of the Quaternary (the last 2 million years). This transition was uneven and multi-state ("stepwise"), initially having comparatively small intervals of coolings (from the geological point of view) alternating with periods of relatively stable and warm climate. Thus, we had the onset of the Antarctic ice sheet about 38 million years ago, warming during the early Miocene from about 19.5 to 16.5 million years ago, and further cooling during the Miocene from about 10 to 9 million years ago.

Recent evidence suggests that there has been considerable uplift of the Tibetan Plateau and the Colorado Plateau in the past 5 to 10 million years (Ruddiman and Kutzbach, 1989). Simulation experiments with climate models show that the effects of increased orography could produce some of the cooling and drying that distinguishes these earlier climatic regimes from the present climate.

The strong climate fluctuations that characterized the Pleistocene began about 2 to 3 million years ago when the first Northern Hemisphere ice sheets spread over Greenland, Iceland, and much of North America and when permafrost areas developed in northeastern Asia. About a million years ago, sharp, systematic fluctuations (glacial-interglacial cycles) of the climate began and year-round pack ice covered much of the Arctic Ocean.[1]

Detailed time-series analyses of both marine records estimating changes in glacial ice volume and astronomical records estimating changes in solar radiation produced by small changes in the Earth's orbital parameters suggest that orbital changes may have "paced" or helped fix the timing of these glacial-interglacial cycles (Hays et al., 1976).

Most of the evidence cited in this section is based upon regional studies. There is a need to extend these studies to new areas and to develop more quantitative means of interpreting the data in climatic terms.

## 2.4. PLEISTOCENE ICE CORE DATA AND CLIMATE

Over the past decade, exciting new studies of cores from polar ice sheets have provided important advances in our understanding of climates of the late Pleistocene and Holocene. Analyses of the composition and isotopic ratios of the ice and trapped air bubbles provide an important tool for determining local temperature and global atmospheric composition ($CO_2$, $CH_4$, and aerosols) for the past 160,000 years or more. Data covering the last glacial-interglacial cycle have resulted from the detailed processing of such ice cores, mainly from an especially deep one at the Soviet station "Vostok" in Antarctica.

Analyses of air extracted from Greenland and Antarctic ice cores show that the atmospheric $CO_2$ content during the last glacial maximum about 18,000 years ago was about 190 to 220 ppmv, increasing during the Holocene (the last 10,000 years) to about 270 to 280 ppmv (Neftel et al., 1988). The same general features are apparent in the records of the $CH_4$ concentration, which increased from about 0.35 ppmv at the end of the glacial epoch to approximately 0.65 ppmv during the Holocene (Raynaud et al., 1988; Stauffer et al., 1988).

The Vostok ice core has been analyzed to a depth of 2200 meters, corresponding to a time period from the present back to 165,000 years

[1] Clark (1979) argues that sea ice has been present on the Arctic Ocean for several million years and that present sea ice conditions are "more open" than at anytime in the last few million years. Herman and Hopkins (1980), however, argue that seasonal ice developed 2 to 4 million years ago and year-round ice developed about 700 to 900 thousand years ago.

ago. This interval of time covers entirely the most recent glacial-interglacial cycle. The isotopic profile obtained by Barnola et al. (1987) shows two interglacials: the Holocene, which apparently peaked 6000 to 5000 years ago, and the Eemian, which peaked about 125,000 years ago (see Fig. 2.3). The estimate of local temperature increase during the Eemian interglacial, which is based on variations in the isotopic ratio of deuterium δD, gives a value up to about 2°C above the temperature of the Holocene in the Antarctic. The maximum cooling during the peak of the last glacial was about 10°C locally, even though the global average surface temperature decrease was only about 5°C.

Studies of the Vostok ice core have considerably expanded our knowledge about past atmospheric fluctuations of the concentration of $CO_2$ (Barnola et al., 1987). As shown in Fig. 2.3, there were two major extremes in the $CO_2$ concentration: a low of 190 to 200 ppmv and a high

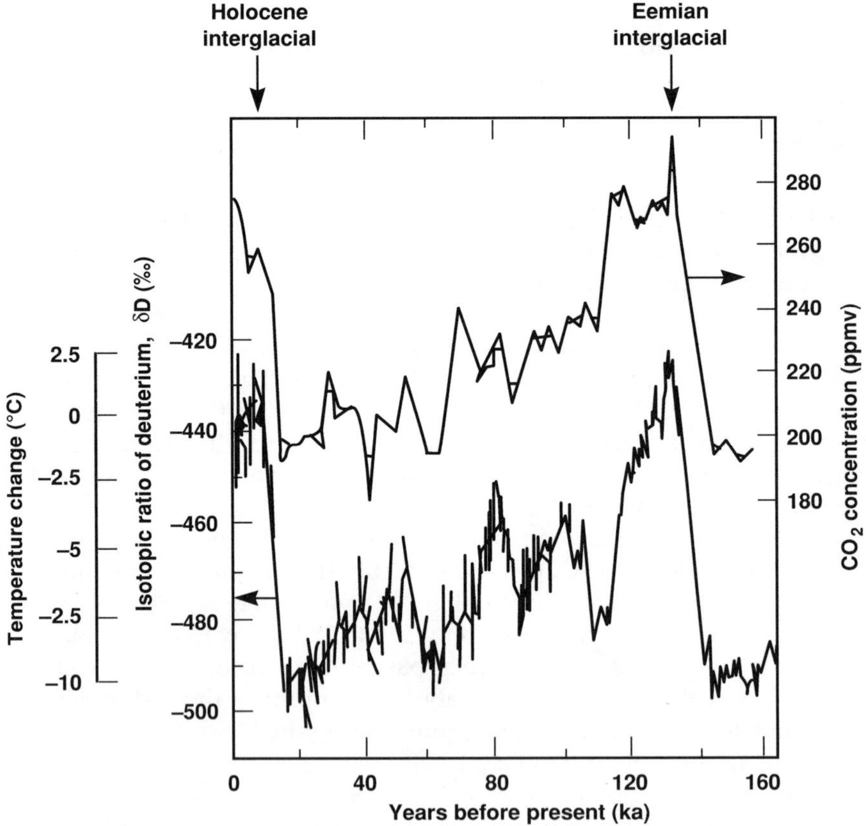


**Figure 2.3.** Carbon dioxide content of the atmosphere (ppmv) and temperature (based on the deuterium isotope ratio δD) as derived from the Vostok ice core for the last 160 thousand years (Barnola et al., 1987). The short horizontal lines associated with the $CO_2$ curve indicate time averages; the vertical lines associated with the δD curve indicate error limits of the measurements.

of 260 to 280–300 ppmv. The lower extreme occurred during glacial maxima and the higher extreme during the warm periods (i.e., during the Holocene and Eemian interglacials). The upper extreme of past $CO_2$ concentration is consistent with estimates of the pre-industrial level of $CO_2$ determined by other means. There are some indications that the $CO_2$ and temperature changes are in phase during the warming; however, during cooling, temperature change precedes the $CO_2$ decrease (Genthon et al., 1987; Jouzel et al., 1987). More work is needed to determine the absolute time scales and relative phase leads or lags more accurately.

There is not yet a clear explanation for the relatively rapid glacial/interglacial changes in the $CO_2$ concentration. It is likely that the ocean played an important role in changing the $CO_2$ concentration via the combination of changes in marine productivity and in ocean circulation (Jouzel et al., 1988), as well as via the influence of variations in sea ice extent. The role of the land biota is less certain. Quite probably, different mechanisms acted at different climatic stages of the glacial-interglacial cycle (Genthon et al., 1987).

The most recent analyses of the core from Vostok (Raynaud et al., 1988) show a good correlation between the $CH_4$ concentration and global temperature (Fig. 2.4). The methane concentration apparently increased by about 80% from the end of the previous ice age to the Eemian (i.e., from about 160 to 115 thousand years ago). This result and other data indicate that the concentrations of these greenhouse gases and perhaps

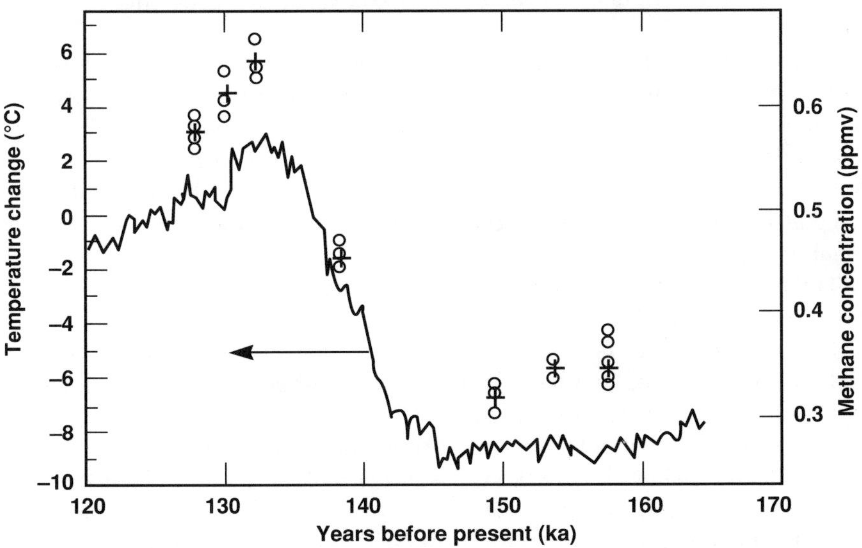


**Figure 2.4.** Methane concentration (ppmv) and the temperature history derived from the deuterium isotope ratio (δD) based on results from the Vostok borehole ice core for the period 120 to 165 thousand years ago (Raynaud et al., 1988). The circles indicate data points and the crosses average values for a particular layer.

other chemical components in the atmosphere (e.g., aerosols) have been closely connected with the climate throughout the entire glacial-interglacial cycle (Stauffer et al., 1988).

## 2.5. PALEOCLIMATE RECONSTRUCTIONS FOR THREE PERIODS OF RELATIVE GLOBAL WARMTH

With the prospect of future warming, the climatic conditions of three geologic periods in the past have received particular attention because of evidence that their climate was relatively warm compared to the present climate. These periods are the Mid-Holocene, the Eemian interglacial, and the Pliocene optimum. These are also periods that have been the subject of discussions at previous US-USSR workshops and in joint publications, for example in the volumes *Late Quaternary Environments of the United States* (Porter and Wright, 1984) and *Late Quaternary Environments of the Soviet Union* (Velichko, 1984).

### 2.5.1. The Mid-Holocene (6000 to 5000 years ago)

Holocene climates have been studied in considerable detail. The continental data sets for much of this period are near-global in distribution, time control is good, and the accuracy of the paleoclimatic estimates is relatively good (Hecht, 1985). Analyses of these data sets generally agree that warmer conditions in northern middle and polar latitudes peaked about 6000 to 5000 years ago during the growing season.

Paleobotanical data provide the most comprehensive information about this period. The availability of many series of radiocarbon datings from numerous sections and on the correspondence of palynological (pollen-spore) spectra with the standard schemes for identifying geological intervals of the Holocene together provide a double control that makes it possible to correlate data from different regions with reasonable reliability.

The reconstructions shown in Fig. 2.5a–c are based on data from over 300 continental sites. Data for North America come from Ager (1983), Baker (1983), Davis (1983), Delcourt and Delcourt (1983), Webb et al. (1983), Webb (1985), COHMAP (1988), and others; data for western Europe come from Leroi-Gourhan (1968), Wijmstra (1969), Zagwijn (1975), Renault-Miskovsky (1976), and others; data for western Asia come from Van Zeist et al. (1975), Weinstein (1976), and Van Zeist and Bottema (1982); data for southeastern Asia come from Duan et al. (1980); and data for northern Eurasia come from Grichuk (1969), Khotinskiy (1977), Yelina (1981), Nikolskaya (1982), and Bezusko and Klimanov (1988).

These authors used several techniques to extract palynological information. For North America, the regional paleoclimatic reconstructions were based partially on regression analysis (Webb et al., 1983;

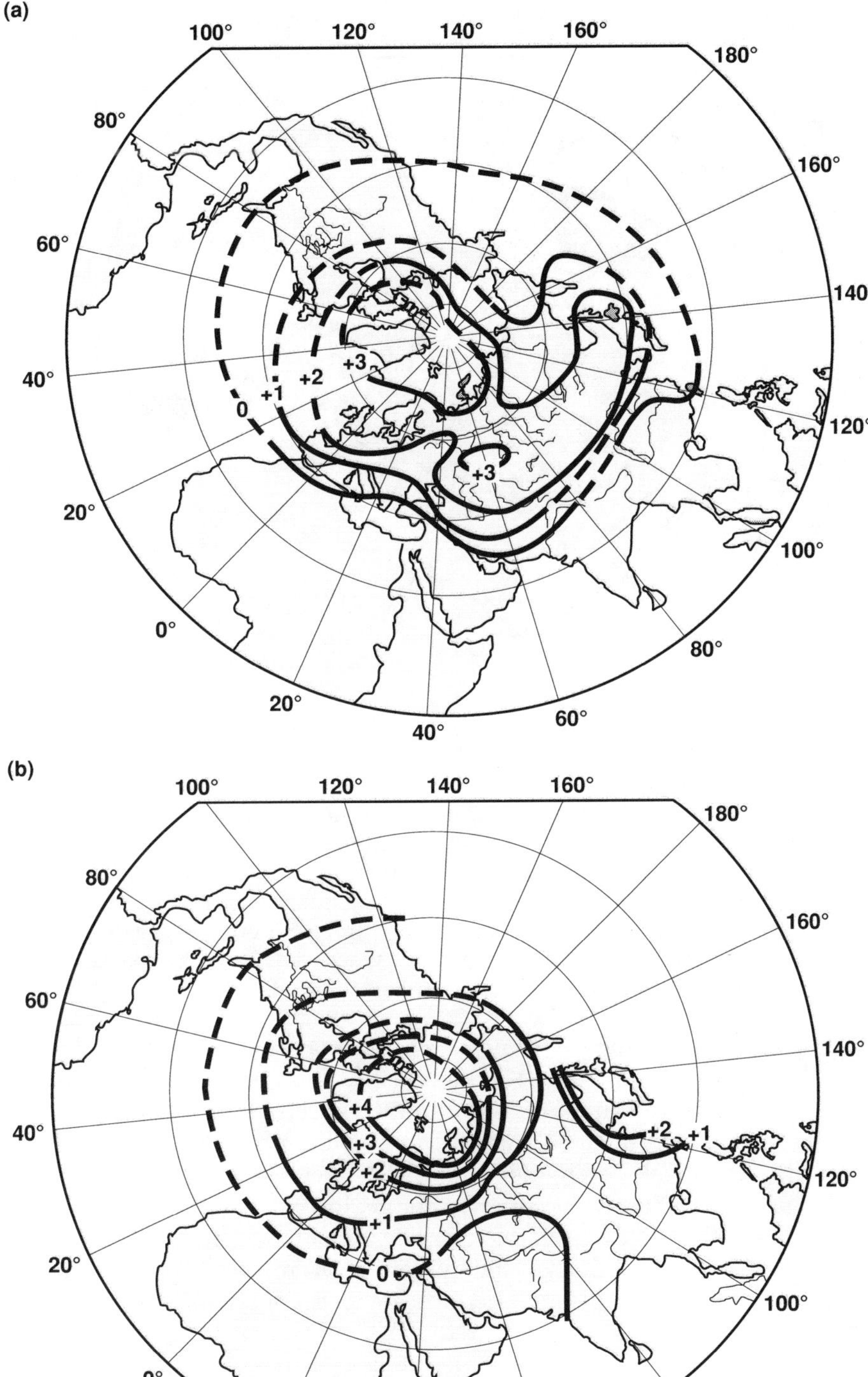
(a)
100°
120°
140°
160°
180°
80°
160°
60°
140°
40°
120°
20°
100°
0°
80°
20°
40°
60°
0
+1
+2
+3
+3
(b)
100°
120°
140°
160°
180°
80°
160°
60°
140°
40°
120°
20°
100°
0°
80°
20°
40°
60°
+4
+3
+2
+1
0
+2
+1

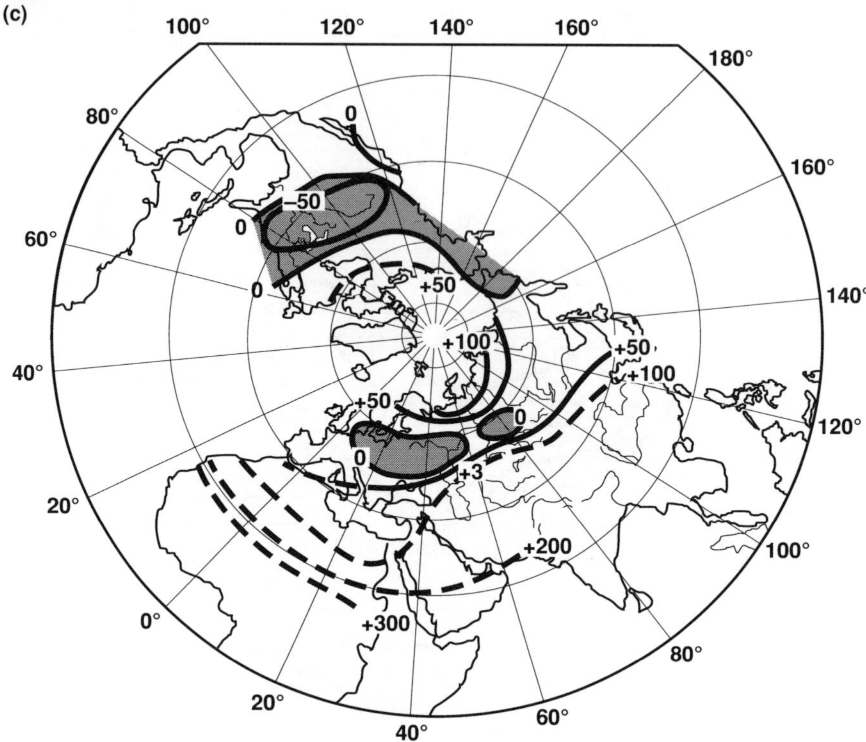


**Figure 2.5.** Paleoclimatic reconstructions for the Mid-Holocene optimum (6000 to 5000 years ago) showing departures from values characteristic of the second half of the nineteenth century for (a) surface air temperature (°C) for January–February, (b) surface air temperature for July–August, and (c) annual precipitation (mm/yr). Maps are based on data from Klimanov (1982), Velichko (1984), and (except for January–February temperature) from Borzenkova and Zubakov (1984). Contours are dashed in regions where data are most sparse and are darkly shaded in regions where precipitation is reduced.

Webb, 1985). For most regions of northern Eurasia, the information-statistical method developed by Klimanov (1982) was used. This method is based on the relationship of the spore-pollen spectra to climatic conditions for the late nineteenth and early twentieth centuries. The relationships are developed by determining the percent content of pollen of each plant species in a sediment or other geological sample and assigning to each sample a probable climatic index based on the region of its formation. Reconstructions of the modern reference records of specific regions assembled using this method indicate an accuracy of ±1°C for temperature and ±25 to 50 mm/yr for the long-term total annual precipitation (Klimanov, 1981; Grichuk, 1985; Velichko, 1985).

For other regions having sufficient palynological information to describe vegetation during this period, the zonal method developed by Khotinsky and Savina (1985) was used. This method, which is less accurate than the information-statistical method, relies on correlations between the present climate and present vegetation patterns. Assuming that the relationships based on the current distributions of climate and species remain valid under altered climatic regimes, the accuracy of the above methods when applied to long-term average temperature is about ±1 to 1.5°C (Velichko, 1985).

For estimating past levels of precipitation, the situation is more complex. For most continental areas, the precipitation estimates depend on there being a consistent empirical relationship in time with soil moisture. In low latitudes, reconstructions of lake levels and river runoff were used to derive the estimates of precipitation (Kutzbach, 1983; Street-Perrott and Harrison, 1985; Borzenkova, 1987; COHMAP, 1988; and others). These methods for reconstructing long-term average annual precipitation typically have an accuracy of about ±25 to 50 mm/yr based on the ability of the methods to reproduce average precipitation for reference periods (e.g., Grichuk, 1985; Velichko, 1985; Winkler et al., 1986); their accuracy for paleoclimatic periods cannot be separately tested.

To reconstruct upper ocean temperatures, analyses of foraminifera (ocean-living protozoa) distributions and of oxygen-isotope variations have been used (see, for example, CLIMAP, 1984; Hecht, 1985; Barash, 1985). The accuracy of ocean temperature estimates based on these techniques is about ±1.5 to 2.0°C.

The estimated increase of Northern Hemisphere temperature resulted mainly from the temperature increase in the high latitudes. The reconstructions for winter and summer show that temperatures in coastal regions near the Arctic Ocean exceeded reference values characteristic of the second half of the nineteenth century by 3°C and more (Andrews et al., 1981; Klimanov, 1982; Borzenkova and Zubakov, 1984; and Khotinsky and Savina, 1985). In subpolar regions of North America and Eurasia, temperatures during the Mid-Holocene exceeded the reference values by about 2°C.

In most midlatitude regions, however, the reconstructions indicate that temperatures exceeded the reference values by only about 1°C. In the interior continental areas in the tropics (and also in some regions of North America and Africa), temperatures even decreased, although generally by less than about 1°C. In southeastern Asia, temperatures exceeded the reference value by about 1°C. The warming reached 2°C and more in the seaside regions of China and in Japan as a result of warm-air transport from the ocean to the continent due to an increased intensity of the monsoon climate. The reconstructions for January show that temperature was higher in Central Asia compared to pre-industrial values and that the winter temperature drop was not as large as it is now. In many areas of Central Asia equatorward of 35 to 40°N, however, temperature changes could not be determined with sufficient accuracy to determine sign or magnitude. Data from the Southern Hemisphere are so limited that it is not clear whether the warming was global in extent.

The reconstructions of annual precipitation for the Mid-Holocene optimum indicate complicated changes in the regional precipitation patterns. Although the Northern Hemisphere temperature increase is estimated to have been about 1°C, the changes in precipitation distribution were apparently much more varied than those occurring during the earlier geologic periods of considerably greater temperature increases. Changes in the seasonal distribution of heat due to changes in the Earth's orbital parameters probably played an important role in causing these variations. The most significant precipitation increase occurred in the tropical and subtropical areas. There is substantial geological evidence of a significant strengthening of the Northern Hemisphere monsoon during the period around 6000 years ago (Street-Perrott and Harrison, 1985; Prell and Kutzbach, 1987). Studies of lake levels, archaeological sites, and paleobotanic remains indicate that annual precipitation, even in the central Sahara, exceeded modern values by 200 to 300 mm/yr. Increased precipitation also occurred in the arid regions of Arabia and the Thar desert. Farther to the north, precipitation also exceeded modern values, although the increases were smaller. In the Mediterranean regions, including the Middle East, precipitation is estimated to have been higher by 50 mm/yr, and in the now arid Aral-Caspian region and in most regions of China and Mongolia by even greater amounts. In North America, southward of 35°N, precipitation is estimated to have exceeded the modern value.

Changes in the precipitation distribution were most varied in the middle latitudes. In northern Europe, precipitation was about equal to the reference values; in the southern part, adjoining the Mediterranean, precipitation was apparently greater than the reference value. In the eastern, midlatitude regions of Eurasia, precipitation exceeded the reference value by about 50 mm/yr. There were also areas where long-term average precipitation decreased modestly, such as in the central Russian plain and southwestern Siberia. In North America, areas in the central and eastern regions are estimated to have experienced the most significant reduction, with rainfall decreasing by up to 50 mm/yr and more (Bartlein et al., 1984; Borzenkova, 1987). The tendency toward reduced rainfall was also evident in the paleoreconstructions along the California coast; in the areas reaching northward as far as Alaska, the departures were very small. In the Arctic latitudes of North America and Eurasia, precipitation again tended to exceed reference values, but the departure values were much lower than in the low latitudes. A detailed discussion of estimated paleoclimatic conditions in North America and western Europe and in the monsoon regions of North Africa and south and east Asia can be found in COHMAP (1988) and associated references.

Physical understanding of important aspects of the Mid-Holocene climate involves the role of orbital parameter changes in producing an enhanced seasonal cycle of solar radiation in the Northern Hemisphere (Kutzbach and Guetter, 1986; COHMAP, 1988; Mitchell et al., 1988;

Kutzbach and Gallimore, 1988; Crowley, 1989; Gallimore and Kutzbach, 1989). The simulations of the climate of this period with climate models show warmer northern summers, colder northern winters over much of North America and warmer winters over much of Eurasia, increased northern summer monsoons, drier northern continental interiors, and reduced Arctic sea ice. Although the mean-annual temperature of the Northern Hemisphere is estimated to have been higher, model-based estimates of the change in global average temperature are near zero.

### 2.5.2. The Eemian Interglacial (125 thousand years ago)

Northern Hemisphere temperature and precipitation anomalies have also been reconstructed for the Eemian interglacial, which is centered about 125 thousand years ago (see Fig. 2.6). This warm interglacial is also referred to as the Sangamon or Mikulino in the continental record and as stage 5e in the ocean sediment record (Velichko, 1982; CLIMAP, 1984; Barash, 1985; Hecht, 1985).

In selecting data for use in reconstructing the climatic conditions of this period, ensuring correspondence of the ages of the land and ocean records is achieved in two ways. The chronostratigraphic method is used to correlate paleontological data with the occurrence of glacial sediments for the Middle and Later Pleistocene intervals, including identification of the facial relationships among interglacial Mikulino fossil soil, riverine alluvium deposits, and marine terraces. Particular succession features of palynological diagrams can also be used to identify the interglacials in different regions, thereby providing a key for correlation of the paleobotanical spectra and synchronization of the paleoclimatic reconstructions for the peak of the last (Eemian) interglacial [zone f–g from Jessen and Milthers (1928); $M_4$, $M_5$, and $M_6$ from Grichuk (1961); and $K_3$ from Gurtovaya (1987)].

To obtain temperature and precipitation values for the continents, changes in pollen and spore counts and distributions were used (with plant species determined). These palynological spectra have been interpreted by using relationships between plant species and climates and regions[2] that are evident in the recent record (Grichuk, 1982, 1985; Bradley, 1985). Analyses at 60 locations have been used for the reconstruction by Velichko et al. (1984). These points, which are not evenly distributed, are situated mainly in Eurasian midlatitude and polar areas. The estimated accuracy in reconstructing the long-term averages of Northern Hemisphere summer and winter temperatures is about ±1 to 1.5°C and in reconstructing long-term average annual precipitation is about ±25 to 50 mm/yr. Surface ocean temperatures were again determined by statistical analysis of foraminifera distributions (CLIMAP, 1984) with an accuracy of about 1 to 1.5°C (see Molfino et al., 1982).

---

[2]A method described as "climagrams and arealograms" in Soviet literature (e.g., Grichuk 1982).

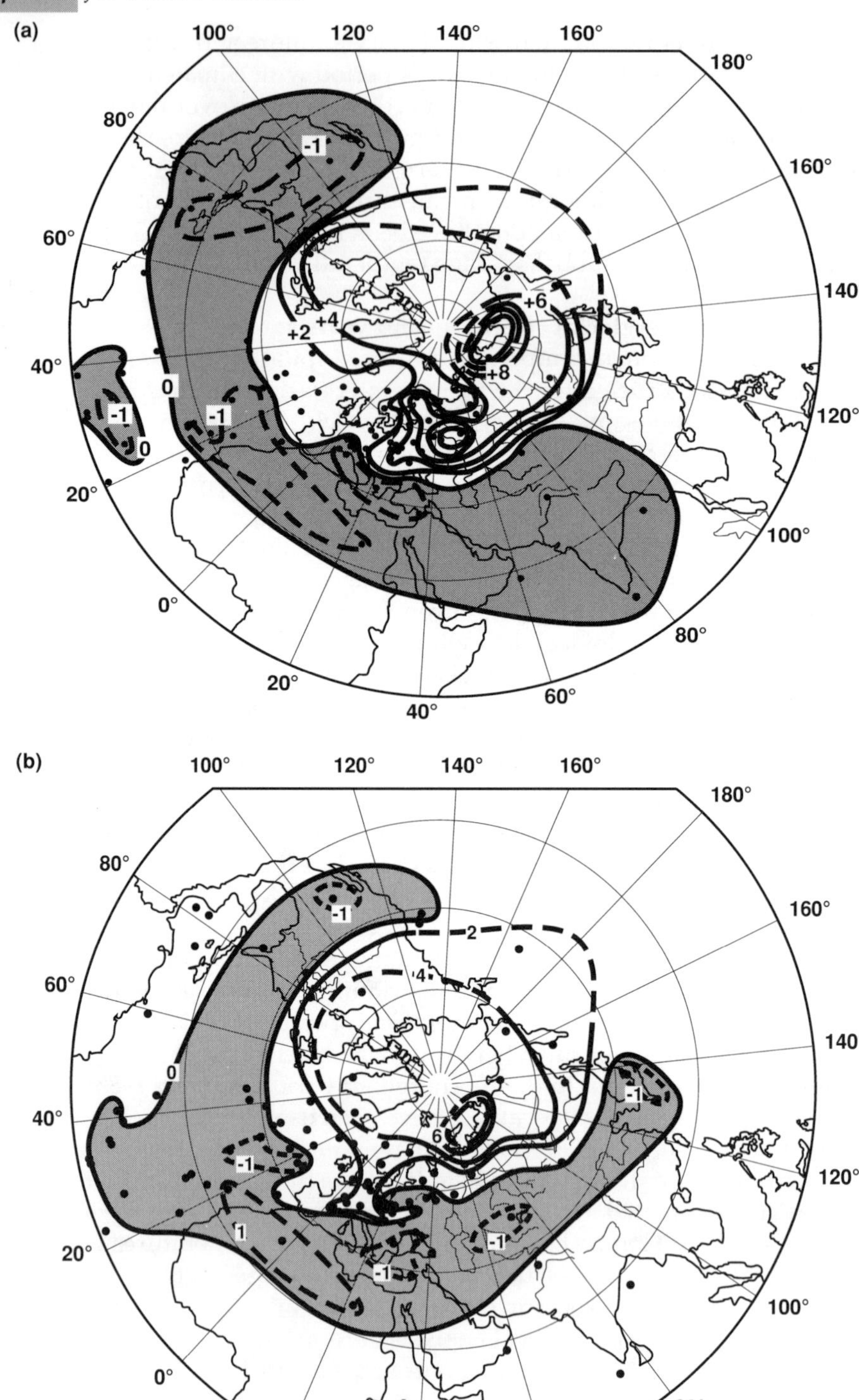
(a)
100°
120°
140°
160°
180°
80°
160°
60°
140°
40°
120°
20°
100°
0°
80°
20°
40°
60°
-1
0
+2
+4
+6
+8
(b)
100°
120°
140°
160°
180°
80°
160°
60°
140°
40°
120°
20°
100°
0°
80°
20°
40°
60°
-1
0
1
2
4
6

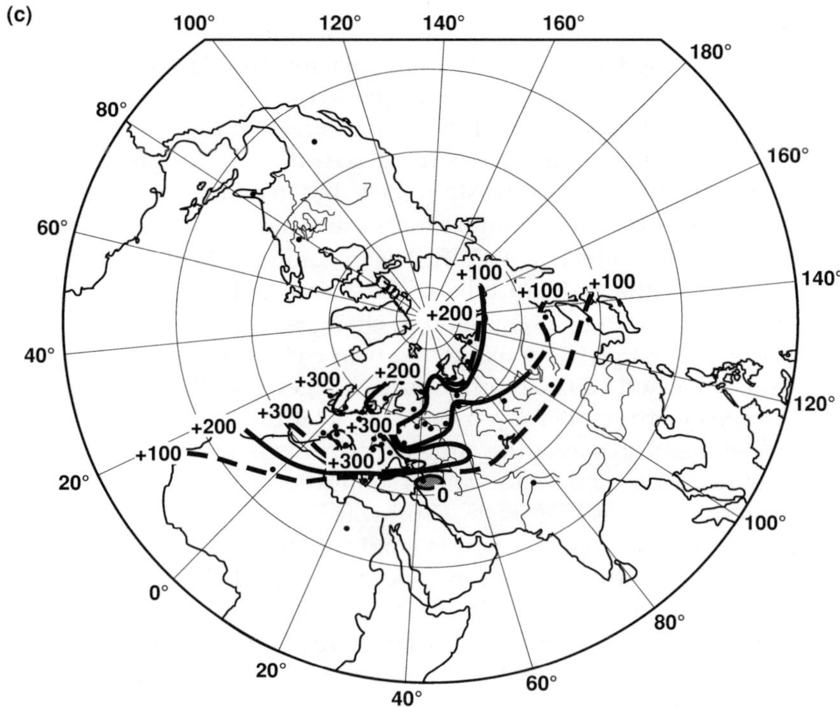


**Figure 2.6.** Paleoclimatic reconstructions for the Eemian interglacial optimum (about 125,000 years ago) showing departures from values characteristic of the second half of the nineteenth century for (a) surface air temperature (°C) for January–February, (b) surface air temperature (°C) for July–August, and (c) annual precipitation (mm/yr). Maps are based on data over the land from Velichko et al. (1982, 1983, 1984) and over the oceans from CLIMAP (1984) and Barash (1985). Data points are shown by the dots. Darkly shaded regions show negative departures.

Reconstruction of the Eemian interglacial climate shows two peculiarities: It was very warm and humid in the Northern Hemisphere, and the climate was characterized by more uniform latitudinal distributions of temperature and rainfall than occur at present (see Fig. 2.6). Based on the estimates by Velichko et al. (1984), high latitudes of the Northern Hemisphere experienced the most pronounced warming, with northern Eurasian temperatures increasing more in the east compared with the west. July temperatures in the northern British Isles exceeded modern values by 1 to 1.5°C, on the coasts of the North Sea by 2°C, and in northern Scandinavia and the east European part of the Arctic by 3°C. The most significant temperature changes occurred along the east Arctic Ocean coast of the Soviet Union (Yamal and Taimyr peninsulas), where the reconstructions suggest that July average temperatures were 6 to 8°C higher than the reference values characteristic of the second half of the nineteenth century.

The Soviet reconstructions suggest that the temperature departures were even larger during January, with a tendency toward higher temperatures over eastern Eurasia than over the west. In the British Isles, winter temperatures are estimated to have exceeded the reference value by only 2°C, while in Scandinavia the departure values reached 4 to 6°C and even 7°C in the northwestern part. In the eastern European part of the Arctic, reconstructed temperatures are estimated to have exceeded reference values by 2 to 4°C. Just as in summer, the intensified warming occurred for the high latitudes of the Asian continent along the Arctic Ocean coast. The largest temperature changes occurred farther to the east (in the Yana-Indigirka lowland) and near 65° to 70°N, where the departure values reached 12°C. Temperature changes over the Atlantic and Pacific Oceans were smaller, even in the high latitudes. However, reconstructed temperatures over the large ocean areas northward from 35° to 40°N did exceed the reference values somewhat.

In the interior regions of Eurasia, southward from 70°N and especially toward the east, reconstructions indicate that temperatures were appreciably higher than during the reference period. July departure values, however, were rather low even in the eastern regions, being only about 1 to 2°C higher than in Europe and southwestern Siberia. In northeastern Asia, July temperatures were higher by 2 to 3°C. January temperature departures in these regions reached 2 to 4°C in central Europe, 6 to 8°C (and even 10 to 11°C in some regions) in eastern Europe, and 6 to 10°C in Siberia.

South of 50°N, the positive departures were smaller, especially in the interior regions of Eurasia. In these areas, considerable changes in the distribution of temperature departures occurred. Summer periods had temperatures slightly lower than during the reference period (by 1 to 2°C). Equatorward of about 55°N, this area included the southern part of eastern Europe, and the northern part of central Asia (Kazakhstan and northern middle Asia). The area situated westward, along the northern part of the Mediterranean, was also characterized by small negative temperature departures, which also extended over some parts of the Atlantic Ocean (based on sediment core data). The northern boundary between latitudinal zones of positive and negative temperature departures shifted seasonally, being situated further south in winter. On the whole, evidence from continental regions in the tropics suggests that the low latitudes experienced areas of both weak positive and weak negative anomalies, with positive temperature departures prevailing. Unfortunately, there are too few data from the Southern Hemisphere to confirm that the warming was global and contemporaneous, leaving the possibility that the actual global temperature increase may have been significantly less than the Northern Hemisphere average. Latitudinal changes in precipitation distribution during the Eemian interglacial have been studied only for northern Africa and Eurasia. Based on these Soviet reconstructions, annual precipitation exceeded the reference levels in the high latitudes, with departures reaching 200 to 250 mm/yr (+25 to 30%) in the coastal regions of Northern Europe, and 200 mm/yr (+50 to 70%) in the Asian part of the Arctic (Velichko et al., 1983, 1984). Southward, from 65° to 45°N, precipitation was also higher than today, although the distribution varied unevenly. The most significant increase occurred in west-central Europe (300 to 500 mm/yr, or +30 to 50%). In east-central Europe, increases in precipitation were rather small (50 mm/yr, or +7 to 8%). In the interior of northern Asia, annual precipitation exceeded the reference values by 25 to 30%.

In spite of their small number, all of the analyzed data points in the now arid regions of Asia southward from 45° to 50°N (especially in the eastern regions) indicate that precipitation exceeded modern levels. In southern Kazakhstan and north-central Asia, precipitation was higher by 50% (100 mm/yr); in some regions of Eastern Europe that are now steppes, precipitation rose by almost 100% (500 to 600 mm/yr compared with 300 mm/yr now). Departure values were even higher farther westward (in the European Mediterranean regions), reaching 100% and more (800 to 900 mm/yr) on the western coast of the Black Sea and in southern France. In the regions of Africa close to the Mediterranean (especially in the west), the reconstructions suggest that departure values were also rather high, which is not characteristic of the modern extra-tropical regions.

In these reconstructions, areas with the highest positive precipitation departures coincided with areas where temperatures were lower than during the reference period. One of the possible explanations for this effect is cooling as a result of increased evapotranspiration. In these regions, data on vegetation assemblages indicate spreading of savanna vegetation into the area now occupied by the Sahara desert. For most of Africa and southern Asia, the precipitation exceeded the reference level. Data from the Shati Lake in Libya indicate that its maximum level was reached about 130 thousand years ago, suggesting that the Eemian optimum in this region coincided with a high rainfall period (Gaven et al., 1981; Velichko et al., 1984). In equatorial regions, the similarities between modern forest assemblages and those of that period make it rather difficult to develop estimates of precipitation departures.

By combining the regional results and the evidence of major ecosystem changes, these reconstructions of the Eemian interglacial indicate that the Northern Hemisphere mean temperature may have been about 2°C above the reference mean. For example, during this period, tundra was absent in most northern Eurasian regions. The forest zone boundary in eastern Europe was shifted northward by 200 to 300 km compared to its modern location, thereby reaching the Arctic Ocean coast. In some regions, the taiga zone boundary in northern Asia was shifted northward by 500 to 600 km. In southwestern Siberia and Kazakhstan, the forest-steppe boundary was shifted by 200 to 300 km and in eastern Europe by 500 to 600 km, with forests covering regions that are now steppes (Grichuk, 1982; Velichko et al., 1983). Farther westward in the Mediterranean region, a noticeable afforestation occurred in semi-arid regions, while savanna spread to cover some super-arid regions. In the tropical monsoon lands of Northern Africa and southern and eastern Asia, there is evidence, both from off-shore marine sediment records and from paleoclimate model experiments, that monsoon circulations strengthened during the Eemian optimum (Prell and Kutzbach, 1987).

### 2.5.3. The Pliocene Optimum (4.3 to 3.3 million years ago)

Budyko et al. (1985) estimate that the atmospheric $CO_2$ concentration was about 600 ppmv during the warmest part of the Pliocene (4.3 to 3.3 million years ago). This concentration was about 150 ppmv greater than the average concentration of $CO_2$ for the Pliocene period (see Table 2.1).

A preliminary reconstruction of the landscape pattern for this maximum optimum of the Pliocene has been developed by Borzenkova and Zubakov (1985) and Zubakov (1990) based on palynological and paleontological data. Their reconstruction, which was based on data from 80 locations, provides a basis for paleoclimatic analysis of the continents. To reconstruct air temperatures over the continents, they used data on the changes in extent of vegetation areas most dependent on summer or winter temperatures (Sinitsyn, 1967; Yasamanov, 1985). For 12 additional locations situated in western Europe, USSR, Iceland, and Alaska, summer and winter air temperatures were calculated by applying statistical methods to the Mid-Pliocene spore-pollen spectra of these regions and were processed by the method introduced by Liberman (Liberman et al., 1985; Zubakov and Borzenkova, 1988).

For the oceans, the main sources of information were microfaunal changes and oxygen-isotope data from deep-sea cores. Bandy et al. (1976) and Barash (1985) have shown that planktonic zones representing tropical, midlatitude, and subarctic waters shifted latitudinally during the last 10 million years. During the warmest periods, planktonic zones shifted poleward by about 10° of latitude; during the coldest periods, they shifted equatorward by about 10° of latitude.

The estimates of continental and ocean paleotemperatures are probably within ±1.5 to 2.0°C of actual values. However, in many cases the reliability of the data has not been adequately verified to ensure the accuracy of anomaly estimates to within a factor of two or more (Barash, 1985).

Figure 2.7 presents reconstructions of temperature differences for winter (Fig. 2.7a) and summer (Fig. 2.7b) between the Pliocene temperature optimum and reference values characteristic of the second half of the nineteenth century. In the Northern Hemisphere, winter temperatures are estimated to have exceeded nineteenth century temperatures by more than 4°C, summer temperatures to have exceeded reference temperatures by about 3°C, and the mean annual surface warming to have been almost 4°C. The largest temperature differences during the Pliocene climatic optimum occurred in the high latitudes (poleward of 70°N), where winter temperatures are estimated to have been as much as about 20°C warmer than during the reference period (as sea ice melted back), and summer temperatures were 7 to 8°C compared to this period. For such changes, however, the accuracy of the techniques must be more thoroughly investigated. There were also some regions where increased rainfall led to temperature decreases (e.g., Kazakhstan and the Sahara and Gobi deserts).

Figure 2.7c presents reconstructions of annual precipitation developed by Yefimova (1987) for the Pliocene optimum (as departures from values characteristic of the second half of the nineteenth century). These estimates were developed on a 5° by 5° grid over Northern Hemisphere continents by summing conditions when the air temperature was above 10°C and accounting for annual potential evapotranspiration. Estimates of annual precipitation were calculated using a relationship between the ratios of annual potential evapotranspiration and annual precipitation at the boundaries and within different modern landscape zones (Zubenok, 1976; Yefimova, 1987). Then, using data on potential evapotranspiration for the Pliocene and the map of landscape zonality by Zubakov and Borzenkova (1990), annual precipitation sums were estimated for certain Pliocene landscape zones and their boundaries with other zones. For example, precipitation is usually larger than potential evapotranspiration in forest zones by 0 to 300 mm/yr. Near the boundary between the forest and forest-steppe zones (savanna-steppes in the Pliocene), the total annual precipitation has been found to approximate potential evapotranspiration (Zubakov, 1986). In the forest-steppe or savanna-steppe zones, potential evapotranspiration is estimated to have exceeded precipitation by 200 mm/yr, and in the steppe zone to have exceeded precipitation by 200 to 400 mm/yr. At the boundary between steppe and semi-desert zones, the amount of precipitation can be half of the potential evapotranspiration. This method assumes that the modern (empirical) relationship between temperature, precipitation, and potential evaporation remains fixed (Budyko, 1984).

As suggested by the reconstructed precipitation map (Fig. 2.7c), for temperatures of the order of 4°C warmer than during the reference period, the vegetation types indicate that the precipitation must have been higher over the continental areas of the Northern Hemisphere. High and subtropical latitudes experienced annual precipitation increases of about 300 mm/yr, and middle latitudes experienced annual precipitation increases of about 100 to 150 mm/yr. Precipitation and moisture conditions were higher in areas of central Asia, the Middle East, and the Sahara that are now deserts. Areas that are now arid and semi-desert were replaced by savannas of different types.

(a)

100°
120°
140°
160°
180°
80°
60°
40°
20°
0°
20°
40°
60°
80°
100°
120°
140°
160°
2
4
6
8
10
12
14
16
18
20
–2
–4
0

(b)

100°
120°
140°
160°
180°
80°
60°
40°
20°
0°
20°
40°
60°
80°
100°
120°
140°
160°
2
4
6
8
10
12
14
0
–2

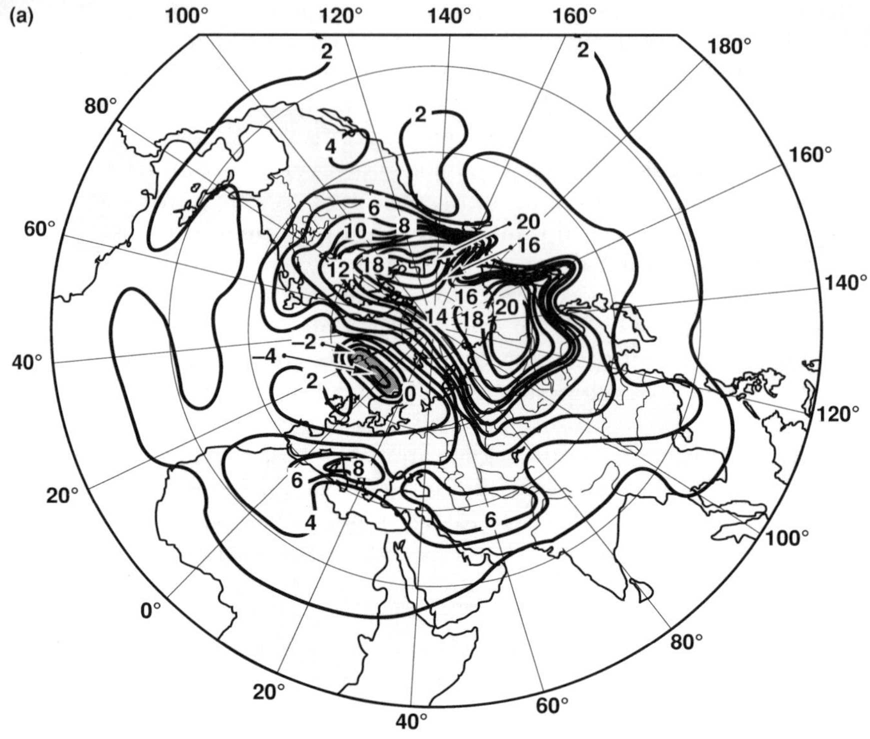

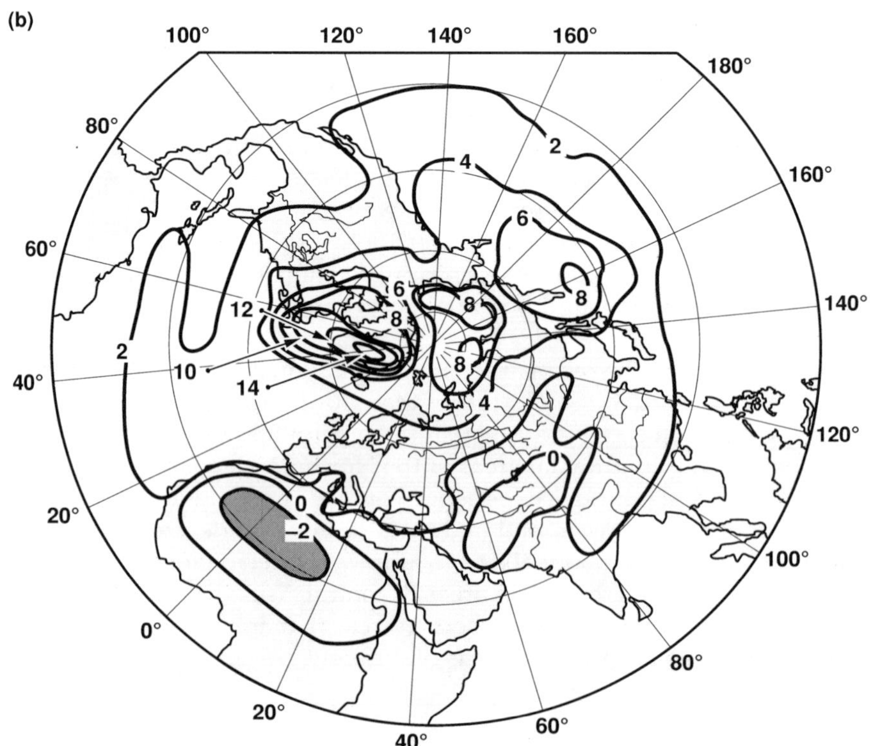

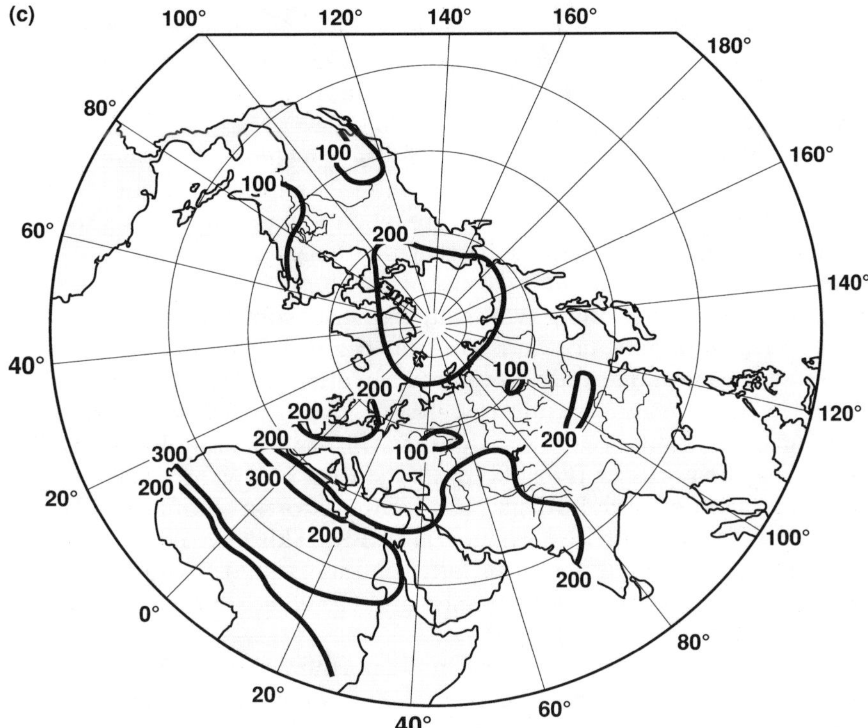


**Figure. 2.7.** Paleoclimatic reconstructions for the Pliocene optimum (about 4.3 to 3.3 million years ago) showing departures from values characteristic of the second half of the nineteenth century for (a) surface air temperature (°C) for January–February, (b) surface air temperature for July–August, and (c) annual precipitation (mm/yr). Maps for temperature are based on data from Borzenkova and Zubakov (1985) and Zubakov and Borzenkova (1988) and for precipitation on data from Yefimova (1987).

### 2.5.4. Similarities among Past Warm Periods

Table 2.2 summarizes the conditions of the three most recent warm intervals, including the Mid-Holocene optimum, the Eemian interglacial optimum, and the Pliocene optimum. Reconstructions of the Mid-Holocene optimum, which had the smallest increase in Northern Hemisphere temperature, indicate that July temperatures in some subpolar and midlatitude regions of North America and Eurasia may have been 1 to 2°C higher than during the second half of the nineteenth century. The hydrologic balance was markedly different in the northern monsoon areas of northern Africa and southern Asia, where the summer monsoons intensified. This occurrence of wetter conditions (an increased difference between precipitation and evaporation) is evident in greatly expanded tropical lakes and rivers, in higher lake levels, and in changes to more developed (mesic) vegetation in parts of northern Africa, the Middle East, and southern and eastern Asia. In the midlatitude continental interiors, the evidence is less clear. In some areas, it was drier (e.g., in central North America), whereas in Eurasia the changes in the hydrologic balance from the reference period were relatively smaller and both wetter and drier (Velichko, 1984, 1989; Kutzbach and Street-Parrott, 1985; Borzenkova, 1987; COHMAP, 1988; and others).

For the Eemian optimum, when the Northern Hemisphere average temperature was as much as 2°C warmer, summer temperatures in coastal regions near the Arctic Ocean were much higher than during the reference period and much higher than during the Mid-Holocene optimum. Along some parts of the coast of northern Asia, it was 6 to 8°C warmer; in winter the warming was even greater in some places, and tundra was absent in most northern Eurasian regions. The spatial and seasonal patterns of midlatitude temperature were more varied than at high latitude, but average conditions were generally 2 to 3°C warmer; in some regions, the changes were smaller than this, and in other locations it was cooler. The northern forest boundary shifted poleward by 200 to 500 km in parts of northern Eurasia. The southern forest boundary also shifted southward in southwestern Siberia and Kazakhstan to occupy territories that are now steppes. This latter boundary shift suggests increased precipitation both in this region and elsewhere in the middle latitudes (Velichko et al., 1984).

The hydrologic balance changed in the northern monsoon lands of North Africa and in southern and eastern Asia, where the Northern Hemisphere monsoon intensified. Lake levels rose, lakes formed where none exist today, savanna vegetation extended into present-day deserts,

**Table 2.2.** Warm intervals of the past 4 million years.

| Interval | Estimated increase of Northern Hemisphere air temperature from the modern value (°C) | Estimated $CO_2$ concentration |
|---|---|---|
| Holocene optimum | 1 | 270 to 280 (Lorius et al., 1985) |
| Eemian interglacial optimum | 2 | 280 to 300 (Lorius et al., 1985) |
| Pliocene climatic optimum | 3 to 4 | 550 to 600 (Budyko et al., 1985) |

and off-shore marine sediments show evidence of increased fresh-water runoff and increased monsoon-driven upwelling (Velichko et al., 1982, 1983; CLIMAP, 1984; Rossignol-Strick, 1985; Prell and Kutzbach, 1987).

Improved physical understanding of some of the causes of glacial-interglacial climate change has resulted from paleoclimatic simulations with climate models (Manabe and Broccoli, 1985; Manabe and Bryan, 1985; COHMAP, 1988; Mitchell et al., 1988). Model simulations of the Holocene optimum and Eemian interglacial show that changes in $CO_2$ concentration, orbital parameters, and albedo (including the effects of changes in ice sheet extent) all contributed to the past climatic changes evident in vegetation, lake level, and marine sediment data. Because $CO_2$ concentrations were comparable for both the Holocene optimum and Eemian interglacial, differences between these "warm" intervals must have resulted from changes in the other factors. The agreement between simulations of past climatic conditions and observed (reconstructed) climatic conditions is good in many, but not all, locations (COHMAP, 1988; Gallimore and Kutzbach, 1989). Prell and Kutzbach (1987) show climate changes (compared to present) that are similar in pattern but larger in magnitude for the Eemian interglacial than for the mid-Holocene optimum.

For the even greater increase in mean annual Northern Hemisphere temperatures of 3 to 4°C during the Mid-Pliocene optimum, high-latitude northern lands appear to have been as much as 5 to 15°C warmer than during the late nineteenth and early twentieth centuries, based upon fossil vegetation evidence. This degree of warming implies much less Arctic sea ice in winter and ice-free conditions in summer. There is also some sedimentary evidence to support this conclusion.

As exciting as the progress has been, many tasks remain to improve our understanding:

(1) Independent methods are needed to determine the accuracy and reliability of the climatic estimates.

(2) The poor areal coverage for the Southern Hemisphere must be significantly extended.

(3) The poor data coverage for the oceans, especially the Arctic Ocean, must be improved in order to better estimate the sensitivity of Arctic ice-cover to greenhouse warming.

(4) More paleoclimatic simulations for these periods must be made using improved climate models. The values of certain key climate variables must be estimated more accurately. For example, we know that the $CO_2$ and, possibly, $CH_4$ concentrations were relatively high during the Eemian interglacial, but the absolute $CO_2$ concentration must be verified. Moreover, stratigraphic analysis of the ice core data does not yet permit accurate determination either of the period of time between changes in $CO_2$ concentration and changes in temperature or of

which change occurred first. However, correlation coefficients of about 0.9 between these changes suggests that the changes were essentially contemporaneous.

(5) Physical understanding of the climates of the Pliocene is improving as models improve. A good example is the growing attention being given to modeling the carbon cycle (Berner et al., 1983; Budyko et al., 1985; Sundquist and Broecker, 1985; etc.). Likewise, further climate model studies are needed of the climatic effects of mountain uplift, changes in aerosol loading (Kutzbach et al., 1989), and the effects of changes in ocean gateways (Maier-Reimer et al., 1989). Changes in deep ocean circulation also need to be considered. Experiments with greatly increased $CO_2$ concentrations (up to 8 times present), made first by Manabe and Bryan (1985), must be extended and improved.

From this combination of studies, a better understanding of the possible range of factors that helped produce Holocene, Eemian, and the Pliocene warm periods should emerge.

## 2.6. LESSONS AND OPPORTUNITIES FOR THE FUTURE

This chapter provides a framework for studying the climates of the past. This section briefly summarizes some general conclusions in the four areas introduced in Section 2.1.

(1) *What have we learned to improve physical understanding of the causes of climate change?*

Paleoclimatic reconstructions and climate models indicate that the climate is very sensitive to changes in the $CO_2$ concentration and especially in more recent epochs to changes in the orbital parameters. For the periods discussed in this chapter, the climate has also been influenced by changes in geography and orography.

(2) *What have we learned about the patterns and about the magnitude and rate of natural climate change?*

Detailed maps of past climates now being developed can help address many types of practical questions. The magnitude of future climate change may be comparable to these past changes. During the warmest part of the Pliocene, when the $CO_2$ concentration was estimated to have been twice pre-industrial levels (a level that is also the projected level for the next century), the climate was very different from the present. Interpretation of these differences is complicated, however, because the future changes in climate may be extremely rapid compared to the apparently slow evolution of past climatic conditions.

(3) *What have we learned from the study of three prominent warm intervals (the Mid-Holocene, the Eemian, and the Pliocene optimum)?*

Significant high-latitude warming is characteristic of all three intervals, as are relatively small changes in low latitudes. For changes in precipitation, the somewhat different patterns in midlatitudes and tropical regions occur for reasons that are still only partly understood.

(4) *What have we learned about the accuracy of climate models from comparison of paleoclimatic data and paleoclimatic simulations?*

Many aspects of the different paleoclimatic intervals are simulated correctly by climate models. Simulations with increased $CO_2$ show high-latitude warming that is observed in the data. Model calculations with changed orbital parameters, which lead to amplification of the seasonal solar radiation cycle, suggest an intensification of seasonal monsoon circulations. Simulations with changed geography and orography show changes in climate patterns generally consistent with past climates, but also regions where models and reconstructions do not agree. More work is clearly needed both to improve models and to refine and reconfirm reconstructions.

Opportunities for future and increased collaboration between the US and USSR should be pursued to gain even more valuable and accurate information about past and future climates. Of most immediate importance:

(1) Joint studies should be devoted to developing methods of analysis and to comparing global paleoclimatic analyses for the Eemian interglacial.
(2) Joint studies and field trips should be devoted to questions of methodology and analysis of the Pliocene optimum.
(3) Joint workshops should be devoted to determining methods for verifying climate models based upon comparisons of paleoclimatic simulations and paleoclimatic data.

# *CHAPTER 3. RECENT GLOBAL CLIMATIC TRENDS*

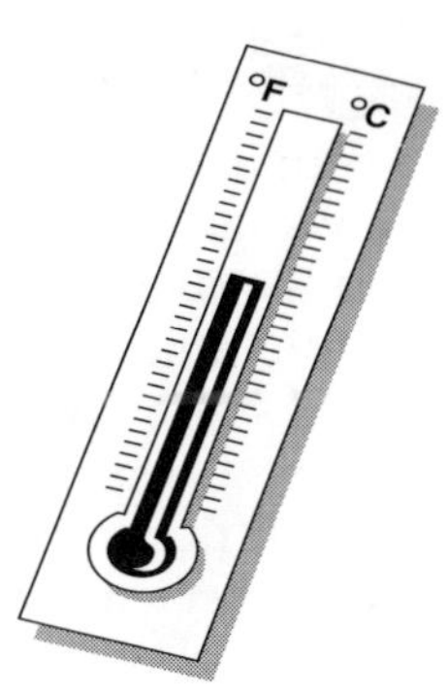


## 3.1. INTRODUCTION

This chapter describes climate changes that have occurred over about the last 100 years. This is the time interval for which sufficient instrumental measurements are available to allow estimation of global changes in the records of near-surface temperature and continental precipitation. Analyses of these records can provide important information about natural climate variability and about the initial response of the climate to recent changes in atmospheric composition.

Because the mean global surface air temperature is sensitive to many components of the Sun-Earth energy balance, it can be difficult to identify the specific causes of observed fluctuations of temperature. For larger and persistent changes in surface temperature, however, the most important factor on the time scale of centuries is atmospheric composition, especially the concentrations of greenhouse gases and, over few-year periods, injections of volcanic aerosols. The most important of the greenhouse gases is water vapor, which is controlled by natural climatic processes and without which our planet would be frozen. Changes in the amount of atmospheric water vapor affect the ability of the atmosphere to absorb solar energy and the radiative heat emitted from the Earth's surface, thereby affecting the vertical temperature profile and the surface temperature. Seasonal and long-term variations in the extent and characteristics of cloudiness and snow and ice cover can also change the net amount of solar energy that is absorbed. Variations in cloudiness and snow and ice cover can occur as a result of changes in temperature; thus, a feedback effect is produced whereby either more or less energy is reflected back to space.

Society's activities are leading to the emission of trace gases into the atmosphere and to the altering of atmospheric composition. The enhanced greenhouse effect that is occurring is of particular interest because it may significantly perturb the climate of the next century. The trace gases contributing to an enhancing of the greenhouse effect include carbon dioxide, ($CO_2$) methane, chlorofluorocarbons, ozone, and nitrous oxide (see Chapter 4). The presence of these trace gases causes changes in the atmospheric radiative fluxes, which then result in an increase in surface temperature. This change of temperature then affects the snow and ice cover, the amount of water vapor in the atmosphere, and the amount and characteristics of clouds. Such changes can then amplify or mitigate the initial response of the Earth-atmosphere system. This coupling of radiative fluxes to surface climate makes the globally averaged, mean annual surface air temperature a very sensitive measure of the climate in that it is closely linked to changes in the hydrologic cycle, atmospheric circulation (and cloudiness), ocean circulation, sea ice, and snow cover.

Twentieth-century variations in the surface air temperature are better documented and understood than are changes in any of the other climate elements. This chapter first discusses the natural factors influencing the natural variability, which serves as the background with which the enhanced greenhouse effect must compete. The chapter then analyzes changes in the thermometric record of surface temperature and, because of their great practical significance, evidence of modern-day changes in precipitation and sea level.

## 3.2. NATURAL CLIMATE VARIABILITY

Over the past 100 years, the global average surface temperature is believed to have been affected by a number of natural factors that can induce changes in the climate. These factors can be divided into two categories: (1) those that are best thought of as external to the climate system, and (2) those that are considered to be internal. External forcing factors are those that produce changes within the climate system, but originate outside the ocean/atmosphere system. Among the external factors that may have affected the modern temperature record are (1) volcanic activity and (2) changes in the solar irradiance. Major volcanic eruptions inject gaseous sulfur dioxide into the atmosphere that often reaches the stratosphere and is rather quickly converted to sulfuric acid aerosols. If present in sufficient quantities in the stratosphere where their removal rate is slow (the average lifetime of aerosol particles in the stratosphere is about 1 year), these aerosols can significantly affect the Earth's net radiation balance. Two internal factors that have led to climatic variations of the atmosphere/ocean system are the El Niño/Southern Oscillation (an ocean-atmosphere oscillation centered in the equatorial Pacific Ocean) and the chaotic behavior that can arise as a result of the highly complex, nonlinear behavior of and interactions between the atmosphere and ocean. It is quite possible that these external and internal factors have contributed to the natural variations in the modern temperature record in addition to any impact that man has had as a result of the enhanced greenhouse effect.

Volcanic eruptions produce aerosols that are in a constant state of evolution. Because the size distribution of these particles is very important in determining whether their influence will tend to warm or cool the surface (Toon and Pollack, 1980), estimates of their effect on the surface climate are strongly dependent on assumptions made about the number and size distribution of the aerosols. A number of empirical studies have been carried out to identify and estimate the effect of volcanic eruptions on surface temperatures over the last 100 years or more (Angell and Korshover, 1985; Sear et al., 1987; Bradley, 1988; Mass and Portman, 1989). Generally, these studies have concluded that major volcanic events, of which there were only about five during the past 100 years, may cause cooling. The evidence is not overly convincing, however, and the cooling caused by these major eruptions has probably been only a few tenths of a degree for a 1- to 2-year period after the event. This would imply that volcanic eruptions have not been a primary factor contributing to decadal and multi-decadal climate variability over the last 100 to 150 years. Unless a change in the character or frequency of explosive eruptions occurs in the future, it is unlikely that volcanic aerosols will be a factor that could substantially counterbalance the greenhouse effect. However, studies summarized by Budyko (1974), Vinnikov (1986) and Khmelevtsov (1986) indicate that atmospheric transparency fluctuations in the Northern Hemisphere derived from station measurements of direct solar radiation can induce not only drastic short-term changes, but also interdecadal trends of mean surface air temperature able to obscure trends caused by the anthropogenic increase in trace gases. Moreover, Robock (1979) and others suggest that increasing volcanic activity was the most probable cause of the temperature reduction during the Little Ice Age that made the seventeenth and eighteenth centuries anomalously cold.

Variations in the amount of solar energy received by the Earth also can affect the surface temperature. Direct measurements of solar irradiance began in 1967 using balloons and rockets. However, these measurements are inherently less reliable than those available since 1978, when satellite measurements of the solar irradiance were begun. These satellite measurements have demonstrated that solar luminosity variations of 0.1% are occurring in association with solar activity. Overall, data since 1967 suggest that there has been a regular variation of solar irradiance with an amplitude of about ±0.4% that is coincident with the magnetic variation of the Sun (Frohlich, 1988). Because of missing data between 1971 and 1976 and limitations in the quality of early balloon data, it is not possible to be certain whether this is a 22- or an 11-year variation.

The lack of any direct measurements of solar irradiance prior to 1967 has led to use of proxy measurements of solar output (e.g., sunspots, the size of the solar disk, $^{14}C$ measurements, etc.). Attempts have been made to relate these proxy measurements to the thermometric record, but their impact has been difficult to substantiate. This may be partly attributed to the large buffering of the short-term fluctuations, such as the 11- or 22-year solar periodicity, that occurs as a result of the large thermal inertia of the oceans (Wigley, 1988). Longer periodicities of approximately 80 years appear in the proxy record. The relationship of such periodicities to changes in solar irradiance and the climate record is, however, still subject to considerable uncertainty; reviews by Khromov (1973) and Pittock (1978, 1983) suggest that these periodicities may not be present or related.

Another mode of natural climate variability that has received considerable attention over the past several years is the El Niño/Southern Oscillation (ENSO), which occurs irregularly about every 3 to 7 years and persists for 1 to 2 years. Typically, an area of anomalously warm surface water expands westward to the dateline from the equatorial waters of the eastern Pacific off the coast of South America. Simultaneously, above-normal sea-level pressure occurring in the western Pacific weakens or reverses the easterlies in the eastern equatorial Pacific. The induced changes in atmospheric and oceanic circulations affect the circulation in many other portions of the globe. Conditions with opposite characteristics (anti El Niño/Southern Oscillation events) have also been observed between ENSO events. They also last for a period of a year or two. Empirical analyses have shown that these events can introduce considerable short-term climate variability into the global thermometric records. Jones and Kelly (1988) estimate that as much as 25 to 30% of the year-to-year variability of the climate record may be due to the ENSO phenomena.

The concept of climatic chaos (Lorenz, 1986) may be of importance with respect to the interpretation of interannual to decadal climate fluctuations. The term "chaos" refers to the random and erratic behavior of a deterministic or nearly deterministic system. Such variations can arise without any external changes to the system and make it more difficult to identify the initial stages of climatic perturbations.

## 3.3. CHANGES IN MEAN SURFACE AIR TEMPERATURE

### 3.3.1. Land-Based Analyses

The first attempt to estimate changes in the mean global surface air temperature was made at the end of the last century (Köppen, 1883). However, Köppen's work for the period 1731 to 1871, which was supplemented by Humphreys (1929) up to 1920, is now purely of historical interest. Several tens of studies made later have gradually improved the methodological basis and have further developed the information data base (Ellsaesser et al., 1986).

Modern ideas about the recent changes in global temperature based on land observations have developed from initial studies conducted by Mitchell (1961; 1963), Budyko (1969), Borzenkova et al. (1976), Vinnikov et al. (1980), Hansen et al. (1981), and Jones et al. (1982). The time series of global estimates of temperature obtained in these papers (and later updates) have been widely used and cited by other authors. The results of some of these investigations were summarized at the 1981 Leningrad meeting of US and USSR scientists on the effects of the increasing atmospheric concentration of carbon dioxide on climate (US-USSR, 1982). The participants at that meeting noted the good agreement between the estimates of changes in the Northern Hemisphere surface air temperature that were obtained by Soviet and American scientists using different methods and data sets.

Over the past few years, additional work has reduced the uncertainties in the data record and improved the analysis techniques used to compile estimates of the changes in the mean global surface air temperature. Independent research teams located at the University of East Anglia (Jones et al., 1986a, 1986b; Jones, 1988), the National Aeronautics and Space Administration (NASA) Goddard Institute for Space Studies (Hansen and Lebedeff, 1987, 1988), and the State Hydrological Institute (Vinnikov et al., 1987, 1990) have recently produced new estimates of land-based temperature change. The primary sources of data for all of these studies were the magnetic-tape archives of climatic information based on publications from the World Meteorological Organization (WMO) and other national climate publications, including the *World Weather Records* and *Monthly Climatic Data for the World* prepared by the US National Climatic Data Center.

Although each of the research teams used the same basic data sources, each used quite different methods. Except for large random errors (which they exclude), Hansen and Lebedeff (1987, 1988) implicitly assume that if there are any inhomogeneities in the data, they are not systematic and will sum to near-zero over large areas. They test for (but do not remove) the impact of locally generated, urban heat islands. Area-averages of temperature are derived by Hansen and Lebedeff (1987) using a procedure in which each station is assumed to be representative of changes over a large area (i.e., out to an influence radius of 1000 km).

Jones et al. (1986a, 1986b) attempt to adjust for all forms of biases in the data, including urban heat island effects and random errors. They use a procedure to area-weight temperatures so that stations can contribute to only one grid box of 5° latitude by 10° longitude. Jones et al. (1985) provide an estimate of the quality of the station data they use, as well as recommendations for their corrections. Both Jones et al. (1986a, 1986b) and Hansen and Lebedeff (1987, 1988) base their results on observations from about 2000 stations.

Compilations by Vinnikov et al. (1987, 1990) differ in that they are based on fewer stations (301 in the Northern Hemisphere and 265 in the Southern Hemisphere) and apply a statistically optimum method for spatial averaging of meteorological fields. The fundamentals of this technique have been described by Gandin and Kagan (1976) and Kagan (1979). These methods enable the Soviet researchers to use a smaller number of stations than the other groups when the spatial correlation function is isotropic and time is invariant. Vinnikov et al. (1987, 1990) have corrected their data for obvious inhomogeneities in the temperature records as well as for the presence of random errors. They considered the effects of urban heat islands on the climate record to be insignificant if the population of the city in which a station is located was less than 1 million (Groisman and Koknayeva, 1989).

Although the approaches used by each of the research teams vary, the hemispheric and global estimates of the change in mean surface air temperature over the past 100 years are all in close agreement. Figures 3.1 through 3.3 show changes in mean annual surface air temperature derived by the three independent research teams. The estimates are presented for the 1881 to 1987 (or 1988) period. Before the second half of this century, there were no observational data from the Antarctic regions; therefore, the estimates for the Southern Hemisphere cover only the zone 0° to 60°S and for the globe the zone 90°N to 60°S. The data presented in these figures represent the land-based thermometric record. The figures demonstrate the considerable agreement among the three independent studies. It should be noted that the reference periods (i.e., the interval of years over which the average departure is assumed to be zero) are different for the three studies; as a result, only the changes from year to year are directly comparable in that the zero departure lines have different offsets.

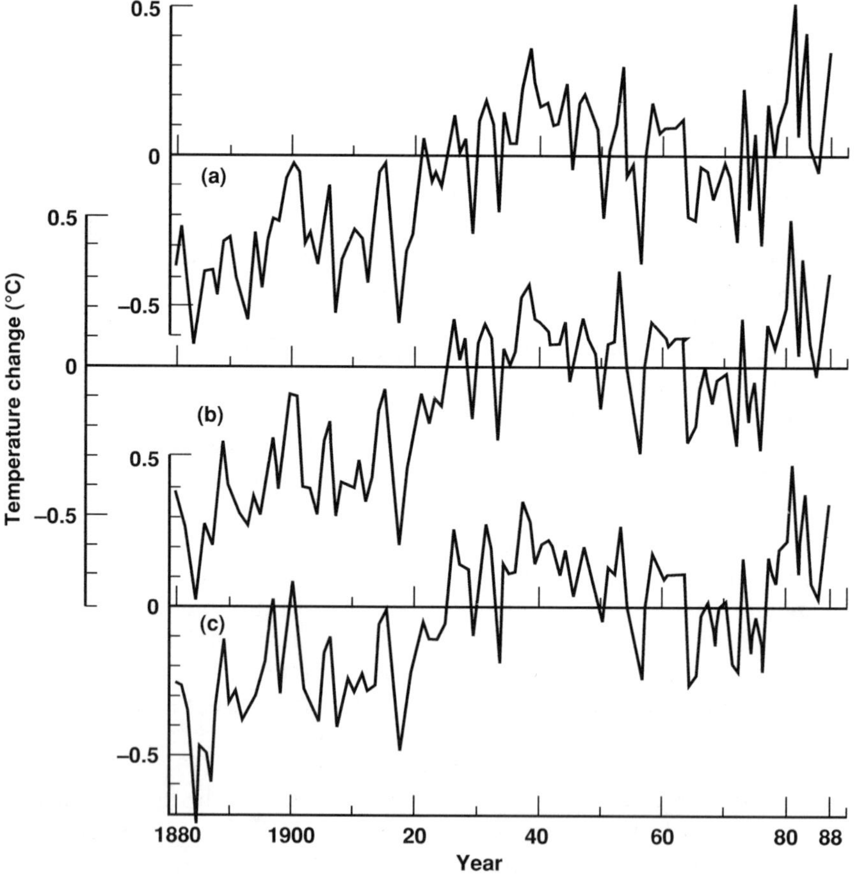


**Figure 3.1.** Estimates of present-day changes in mean annual surface air temperature in the Northern Hemisphere. (Note that reference periods are different.) Data are from (a) Jones et al. (1986a) and later updates, (b) Hansen and Lebedeff (1987), and (c) Vinnikov et al. (1990).

Correlation coefficients between the estimates of mean annual surface air temperature for the Northern Hemisphere range from 0.95 to 0.97; for the Southern Hemisphere from 0.93 to 0.95; and for the globe (90°N to 60°S) from 0.97 to 0.98.

Table 3.1 gives standard deviations and linear trends of the mean air temperature over the period 1881 to 1987 for the records of various research groups. Differences appear to be insignificant. A linear trend line fit through these observations indicates an increase in globally averaged surface air temperature of about 0.4 to 0.5°C over this 100-year time period for each of the three series. The rate of warming has not been continuous in time or space. For example, Fig. 3.4a indicates that the warming trend decreases and nearly vanishes for the period from the 1930s to 1987. Likewise, the rate of temperature increase is significantly lower over the full period of record compared to the years 1881 to the 1940s (Fig. 3.4b). These results serve to illustrate the difficulty in interpreting the observed increase of temperature over the past 100 years or so as a simple monotonic increase.

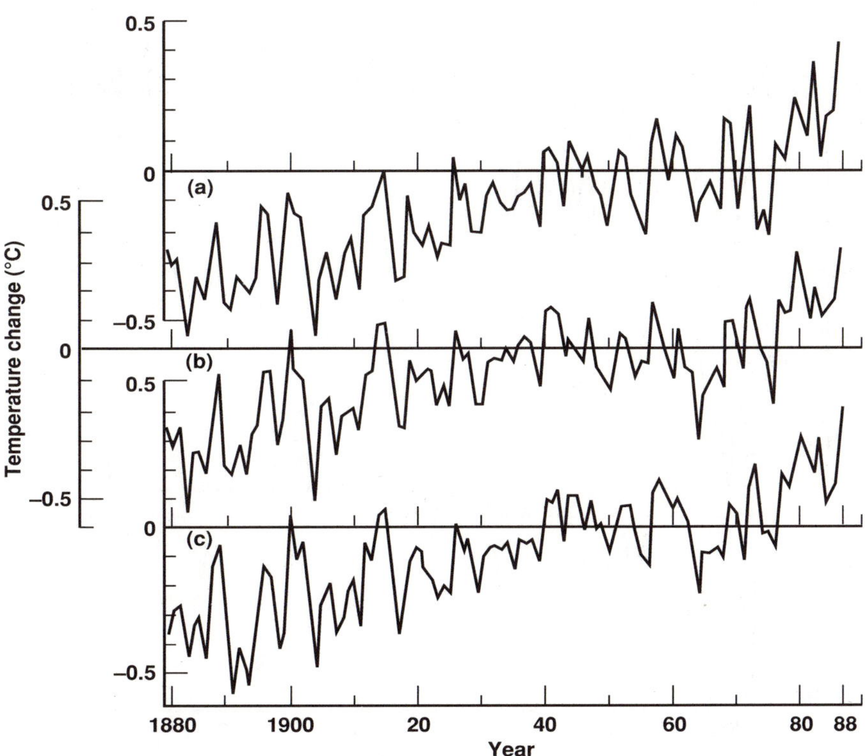


**Figure 3.2.** Estimates of present-day changes in mean annual surface air temperature of the Southern Hemisphere (0° to 60°S). (Note that reference periods are different.) Data are from (a) Jones et al. (1986b) and later updates, (b) Hansen and Lebedeff (1987), and (c) Vinnikov et al. (1990).

Both the variability in the trend and the sensitivity of the calculation to the specified starting and ending years cannot be over-emphasized. For example, if instead of calculating the 100-year Northern Hemisphere trend over the years 1881 to 1987, the trend were calculated over the period 1901 to 1987 (87 years), the trend would be reduced to a rate of 0.4°C/100 yr. Obviously, great caution must be taken in interpreting linear trend estimates. This is especially true with respect to trends of temperature calculated starting in the 1880s, because this was probably one of the coldest decades of the second half of the nineteenth century (Jones et al., 1986a). On hemispheric space scales, warming in the Northern Hemisphere was interrupted between the early 1940s and the mid-1960s by weak cooling. Concurrently, a small decrease in the positive trend of temperature occurred in the Southern Hemisphere. Since the early 1970s, however, the warming has resumed in both hemispheres such

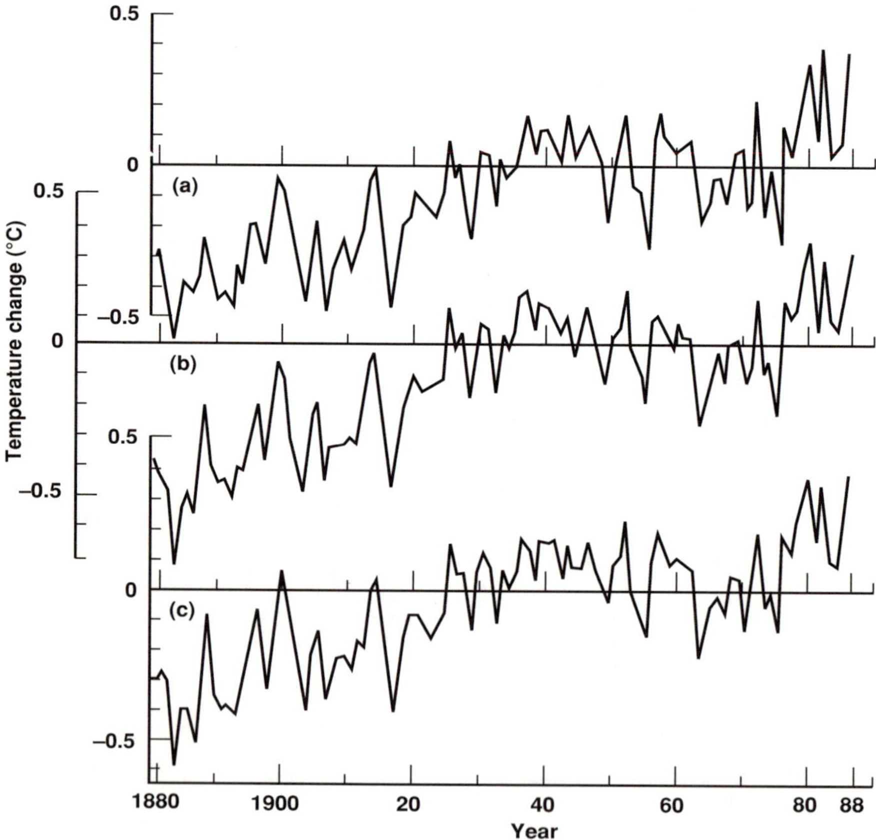


**Figure 3.3.** Estimates of present-day changes in mean global annual surface air temperature (90°N to 60°S). (Note that reference periods are different.) Data are from (a) Jones et al. (1986c) and later updates, (b) Hansen and Lebedeff (1987), and (c) Vinnikov et al. (1990).

**Table 3.1.** Standard deviations and linear trends of the mean annual surface air temperature for the period 1881 to 1987 for Northern Hemisphere (NH, 0° to 90°N), Southern Hemisphere (SH, 0° to 60°S), and the global (90°N to 60°S) domains based on data from the three compilations cited.

| | Standard deviation (°C) | | |
|---|---|---|---|
| Compilers of data set | NH | SH | Global |
| Jones et al. (1986a, 1986b, 1988) | 0.24 | 0.20 | 0.21 |
| Hansen and Lebedeff (1987,1988) | 0.26 | 0.18 | 0.22 |
| Vinnikov et al. (1987, 1990) | 0.23 | 0.19 | 0.20 |
| | Linear trend (°C/100 yr) | | |
| | NH | SH | Global |
| Jones et al. (1986a, 1986b, 1988) | 0.50 | 0.50 | 0.50 |
| Hansen and Lebedeff (1987,1988) | 0.61 | 0.41 | 0.51 |
| Vinnikov et al. (1987, 1990) | 0.48 | 0.49 | 0.48 |

that the global average temperature for the 1980s was about 0.25°C higher than that in the 1970s.

Figure 3.5 displays the changes in the mean annual surface air temperature for the six latitudinal zones 60° to 90°N, 30°to 60°N, 0° to 30°N, 0° to 30°S, 30° to 60°S and 60° to 90°S. The data are from Vinnikov et al. (1990). These graphs show that the trends in the mean zonal surface air temperature increase with increasing latitude. In the zone 60° to 90°N, the mean warming rate has been approximately 1°C/100 yr (1881 to 1987), which is about twice as large as the global mean. Because the year-to-year variability of temperature in the high latitudes is several times larger than that observed in other areas, however, the statistical significance of the warming is actually greater in the lower latitudes.

As for the global domain, the rates of change in the different latitudinal zones do not show a simple monotonic or constant increase in temperature. Not only are there latitudinal variations in the trends, but there are seasonal variations as well. Figure 3.6 shows the estimated trends for the mean monthly air temperature for each hemisphere based on the results from Jones et al. (1986a, 1986b), Jones (1988), and Vinnikov et al. (1987, 1990). Although there are small differences in the estimates due to random errors in data and sampling variability, the general features are quite consistent. The warming takes place in all months of the year in both hemispheres, but it is most pronounced in winter months. Part of the reason for the smaller amplitude of seasonal

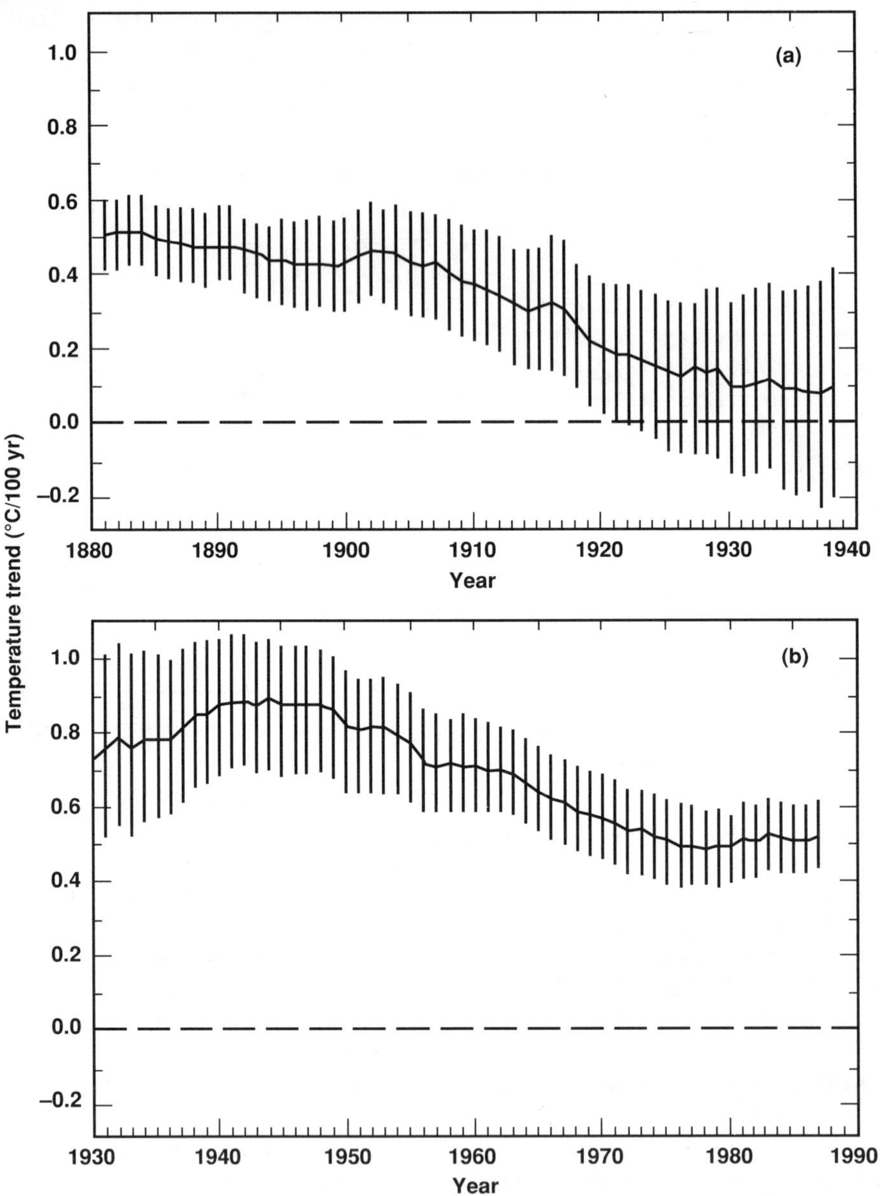


**Figure 3.4.** Linear trends of Northern and Southern Hemisphere land temperatures (from Jones, 1988) and their associated 95% confidence intervals expressed as rates of change over 100 years: (a) ending year for all trends is 1987, and the beginning year is given on the x-axis; and (b) ending year for all trends is given along the x-axis, and the beginning year is 1881. Trends reflect changes due to changes in both atmospheric composition and to other factors.

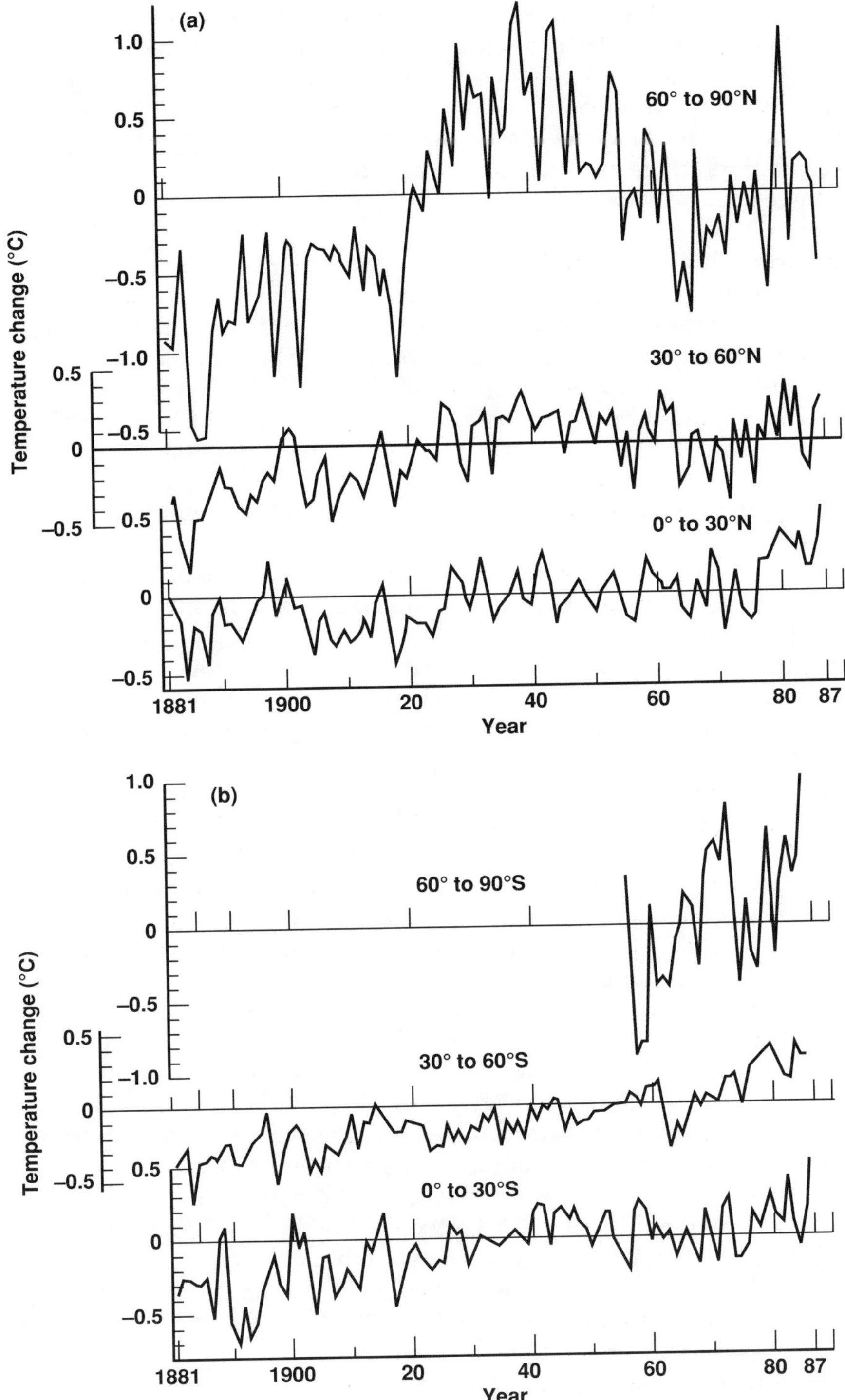


**Figure 3.5.** Mean annual surface air temperature of three latitudinal belts in (a) the Northern Hemisphere and (b) the Southern Hemisphere (Vinnikov et al., 1990).

variations in trends in the Southern Hemisphere as compared to the Northern Hemisphere can be attributed to a lack of data for the zone 60° to 90°S. Figure 3.7 shows the changes of seasonal surface air temperature over land areas from the data of Jones et al. (1986a, 1986b). Since 1880, when the station-data coverage first became minimally acceptable for global calculations, there has been an upward trend of surface air temperatures in all seasons.

### 3.3.2. Uncertainties in the Land-Based Observations

Although there is a general consistency among the results of the three analyses, questions about the certainty of the estimates of global, hemispheric, and regional trends of temperature have been raised by a number of authors (e.g., Callendar, 1961; Mitchell, 1963; Bradley et al., 1985; Ellsaesser et al., 1986; Karl and Quayle, 1988; Karl et al., 1989; Karl and Jones, 1989; Parker, 1989; Wood, 1990). Questions of data reliability focus on three aspects:

(1) Insufficient and changing spatial coverage of the station network.

(2) Rapid increases in urbanization around many of the observing stations, which might produce a spurious warming in the climate record.

(3) Inhomogeneity of the time series due to varying types of instrument shelters, to station relocation, and to changes in the methods of observation, measurement, and processing of thermometric data.

The changing density of observations over the land has been addressed by sensitivity studies using so-called "frozen" networks (Jones et al., 1986a, 1986b). The results of these studies indicate that changes in station density since about 1900 seem to have had little effect on the estimates of hemispheric temperature trends. However, between 1880 and 1900, the decadal uncertainty of global average temperature anomalies could be as large as 0.1 to 0.2°C.

A considerable amount of work has been focused on estimating the effects of increased urbanization on the thermometric record. Recent work by Jones et al. (1989) estimates that perhaps as much as 0.1°C of the 0.5°C warming over the past 100 years may be due to urbanization. Hansen and Lebedeff (1987) also addressed this problem by recalculating their estimates of global average temperatures after omitting all stations in cities with populations in excess of 100,000. This reduced their estimate of the global trend by about 0.1°C/100 yr, down to 0.4°C/100 yr. They hypothesized that perhaps as much as another 0.1°C/100 yr of urban-induced warming may remain in their data due to urban warming effects at stations with populations below 100,000. This would produce a global warming trend of about 0.3°C/100 yr.

In another recent study, Karl et al. (1988) showed that urban biases of annual average temperatures as large as 0.1°C can be detected at stations in towns with population as small as 10,000. This result was

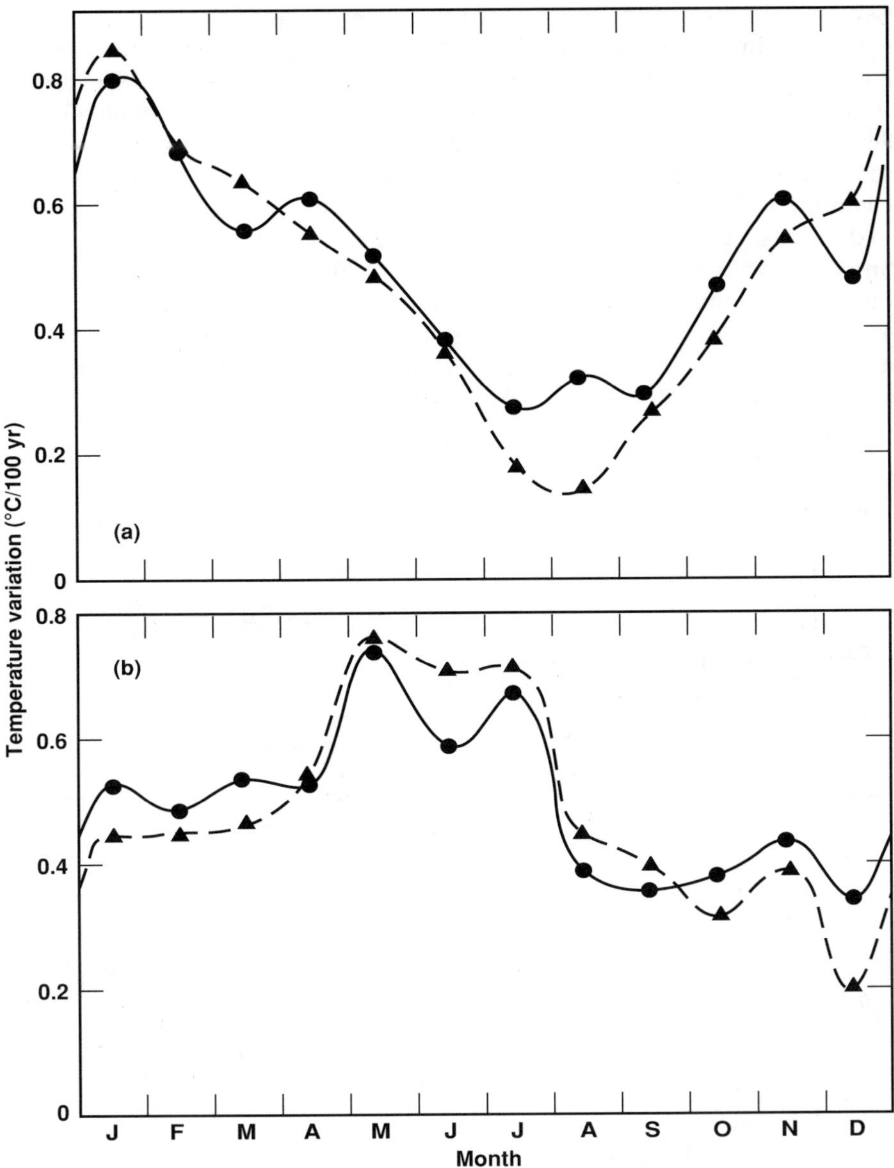


**Figure 3.6.** Annual variations in estimates of linear trends of mean monthly surface air temperatures for the (a) Northern (0° to 90°N) and (b) Southern (0° to 60°S) Hemispheres from 1881 to 1987. The solid curve is based on Jones et al. (1986a, 1986b), and the dashed curve is based on Vinnikov et al. (1990).

based on a comprehensive network of mostly rural stations in the US Historical Climate Network (HCN), data which have been corrected to the extent possible for other station inhomogeneities (Karl et al., 1986; Karl and Williams, 1987; Quinlan et al., 1987). Groisman and Koknayeva (1989) compared the United States area-average temperature trends derived from this network to estimates from Vinnikov et al. (1990) for the continental United States. The differences in the trends between the two data sets were quite small (about 0.05°C/100 yr), despite the fact that many of the stations used by Vinnikov et al. (1990) were located in and around major metropolitan areas that have experienced rapid growth over the past century. The other two global land-based thermometric data sets were also compared for the contiguous United States with the HCN data. Both the Jones et al. (1986a, 1986b) and the Hansen and

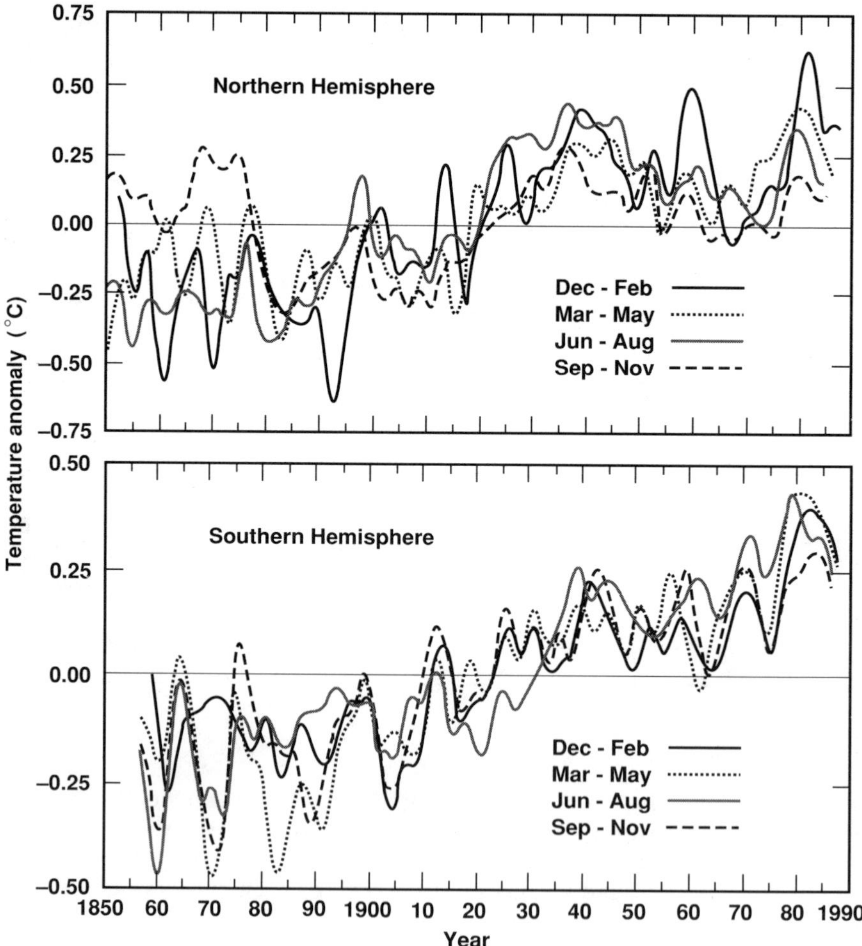


**Figure 3.7.** Seasonal land-surface air temperature anomalies (based on data from Jones et al., 1986a, 1986b) relative to means for 1881 to 1987. A nine-point binominal filter has been applied to the data (beginning and end segments are derived by using double weights of the last 5 years of data).

Lebedeff (1987) data were found to warm at a faster rate than the HCN by about 0.1 to 0.2°C/100 yr (Hansen et al., 1988; Karl and Jones, 1989; Karl and Jones, 1990). Comparison of mean air temperature trend estimates from a network of rural stations in the USSR (Vinnikov et al., 1987, 1990) show no noticeable urban bias (Groisman and Koknayeva, 1989).

A crucial question is why the results over the United States using the rural HCN data set are so similar to the results of Vinnikov et al. (1990). A likely reason is that other inhomogeneities in the thermometric record have compensated for the urban warming bias found in the HCN analysis. Examples of such opposing systematic biases that could compensate for each other include station relocation to airports versus development of urban heat islands. Other forms of inhomogeneities in the thermometric record stem from changes in the observation times used to record the daily temperatures and from changes in instrument shelters used to house thermometers. However, changes in observation times and in instrument shelters are likely to have had a substantial effect on global or hemispheric temperature only if systematic changes occurred in large countries. Based on current information, we do not expect that these effects have introduced significant biases in the globally and hemispherically averaged changes of temperature, but caution must be exercised.

Thus, based on the close agreement of all three research groups, the global average surface temperature is estimated to have risen by between 0.4°C and 0.5°C from the late nineteenth century to the present.

### 3.3.3. Ocean Surface Temperatures

Data sets of ocean surface temperature are also valuable sources of information about climate changes. Systematic measurements of sea surface and marine air temperatures began in the mid-nineteenth century, and over 100 million of these observations have been transferred to computer-readable magnetic tapes. Although many observations date back to the mid-nineteenth century, over three-fourths of them have been taken since World War II. Most observations were taken by ships of opportunity (i.e., ships that just happened to be sailing in a particular location) following routes that favored preferred navigational transects across the global oceans; these routes have changed over the past century. As a result, large geographic areas of the oceans have not been adequately sampled, and even today, one-third of the global oceans are not represented in the National Oceanic and Atmospheric Administration (NOAA) Comprehensive Ocean-Atmosphere Data Set (COADS) (Slutz et al., 1985).

There are few observations from the tropical areas of the Pacific Ocean until the 1950s, and few from oceanic regions poleward of 40°S until even more recently. Even when observations are available for a given area, there may be only one or two values for a given month, and adjustments may be required to convert the resulting estimate to a mid-month or mid-area equivalent. On the other hand, the rather poor coverage of marine observations across the global oceans is compensated for to some extent by the fact that the trends of ocean temperature anomalies tend to have a substantial amount of spatial coherence; that is, warming and cooling trends tend to occur on large spatial scales in the oceans (much more so than for surface temperature anomalies over land).

For marine air temperatures, biases exist for a variety of reasons. Ships have become larger over the years, and the height of the deck above sea level has increased. As ships have become larger, their absorption of insolation has increased and caused a bias in the observations of daytime marine air temperature. For sea surface temperature measurements, biases are known to have been introduced by changes in the methods by which the temperature is measured. Only in recent decades have observations of sea surface temperatures included records of the measurement procedure used; however, the depth of the measurement, shallow for small or empty ships, deeper for large, fully laden ships, is still not documented. Oceanographic measurements of sea temperature (i.e., data from research cruises) are well-documented and precise, but far fewer in number than voluntary ship observations and very poorly distributed for long-term global studies.

In spite of these difficulties, observations of both sea surface temperature and marine air temperature have been used to construct time series (see Fig. 3.8) that show a warming trend of approximately the same magnitude as that of the land-based data. However, with so few long-term observations of temperature over the oceans, particularly in the Southern Hemisphere, and because of the differing measurement practices, global and hemispheric time series extending back into the nineteenth century should be used circumspectly. Nevertheless, the

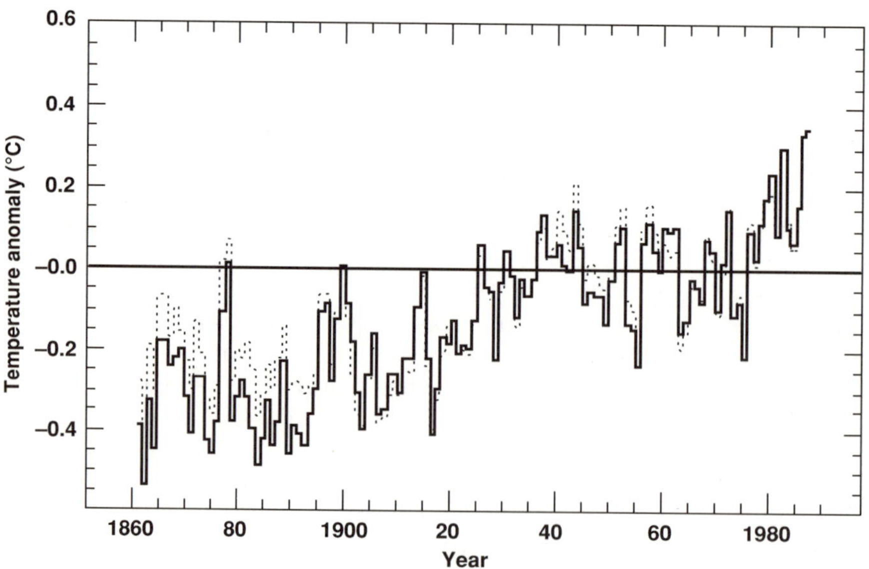


**Figure 3.8.** Estimates of global temperature anomalies (land and ocean) from Folland and Parker (1989) (dotted) versus Jones et al. (1986c) (solid). Updates to 1988 were provided by Folland (1989).

observations from both the land and marine data provide us with unequivocal evidence that the near-surface global average of temperature has increased between 0.4° and 0.5°C over the last 100 years.

## 3.4. CHANGES IN PRECIPITATION OVER THE CONTINENTS

Along with air temperature, precipitation is the most important meteorological element characterizing the state of the natural environment and influencing the conditions of human economic activity. The world precipitation network was quite well-developed by the beginning of the twentieth century, encompassing more than 20,000 gauging stations. In some countries, the network was very dense. For example, in Germany at the start of this century, precipitation was observed at over 3000 gauging stations, in Australia at 2250, in Great Britain at 4000, in India at 2700, etc. (Vannari, 1911). At present, precipitation is measured at about 40,000 fully-equipped meteorological stations and 140,000 gauging stations. However, the regular international exchange of precipitation information comes from only about 1500 stations. Observations from only approximately 1250 stations are published at regular intervals in the international publication *Monthly Climatic Data for the World*. Thus, it is not possible at this time to use the full set of meteorological data in climatological studies even though there are a great number of operating meteorological stations and gauging sites.

### 3.4.1. Observed Changes in Precipitation

One of the first attempts to estimate recent trends in mean precipitation over the extratropical part of the Northern Hemisphere was by Apasova and Gruza (1982). Their work was based on maps of relative anomalies of monthly precipitation sums. They concluded that total precipitation amounts for January, for July, and for the year as a whole have increased over the last 100 years. Because the technique for constructing maps did not consider the homogeneity of the precipitation time series, however, this must be viewed as a preliminary result.

The recent work by Bradley et al. (1987) and Diaz et al. (1989) has been the most comprehensive, in terms of the quantity of data used and the size of the territory analyzed. In these studies, measurements were used from about 1500 stations. However, it was not the absolute amounts of precipitation or their anomalies that were averaged, but a precipitation index based on percentiles of the statistical distribution. These indices vary in the range of 0 to 1 and represent the probabilities of receiving a specified amount of precipitation. It is important to understand that this time series of mean indices does not give a measure of the total integrated water accumulating over a region. However, the index used by Bradley et al. (1987) is a useful measure of the areal extent that receives either more (≥0.5) or less (≤0.5) than normal precipitation. If total precipitation is of interest, then it is most meaningful to compute the precipitation indices for areas of relatively uniform precipitation characteristics. For such areas, the indices are approximately proportional to changes in mean precipitation amount (Bradley et al., 1987).

Examples of the differences between area-averages of total precipitation and an index of precipitation are shown in Figs. 3.9 and 3.10. This particular index of precipitation has been derived by standardizing the precipitation anomalies using the gamma distribution over various portions of the United States that have similar rainfall characteristics. Although there are important year-to-year differences, the overall trend, or lack thereof, is similar for both treatments of the data. The overall trend is positive, but not statistically significant. The data in Fig. 3.9 are derived from a fine-scale national network of 5000 stations in the United States (Karl, 1988), and the data in Fig. 3.10 are extracted from Bradley et al. (1987) and a denser set of Soviet stations (Bradley and Groisman, 1989). Further studies of precipitation indices and actual precipitation anomalies for different regions of the world confirm that both approaches provide similar estimates of year-to-year changes in precipitation and in long-term trends (Bradley and Groisman, 1989).

Spatial averages of the percentile values from the work of Bradley et al. (1987) are shown in Fig. 3.11 for annual precipitation indices in three latitudinal zones: 35° to 70°N, 5° to 35°N, and 0° to 5°N. As can be seen, the precipitation index at continental stations in the extratropical part of the Northern Hemisphere has increased over the last 100 years, particularly over the last 30 to 40 years. In lower latitudes (5° to 35°N), there is no systematic trend in the precipitation index before the early 1950s, but thereafter the precipitation index begins to decrease. In the subequatorial belt from 0° to 5°N, there is no obvious trend in the annual index.

Similar changes have been observed by other studies using mean total precipitation data (not indices) for the continents of the Northern Hemisphere in the band 35° to 70°N (Groisman, 1990; Vinnikov, et al., 1990). In their study, the anomalies of mean annual precipitation and mean precipitation for the warm period of each year (May to September) have been calculated. The records of mean annual precipitation for the Soviet Union, for western Europe, for North America in the 35° to 55°N zone, and for all continents in the 35° to 70°N belt (excluding Canada north of 55° and China) are shown in Fig. 3.12. Sloping lines correspond to estimates of linear trends. The estimates are based on data from about 300 stations in North America and western Europe, and from more than 600 stations in the USSR.

Table 3.2 presents statistical parameters and estimates of the linear trend for these regions (Groisman, 1990; Vinnikov et al., 1990). An upward tendency is apparent in all regions except western Europe. For the USSR and North America (35° to 55°N), the mean annual precipitation rates have increased at about 9%/100 yr and 7%/100 yr, respectively. However, the increase of precipitation in North America (35° to 55°N) is not well represented by a linear trend because precipitation was high at

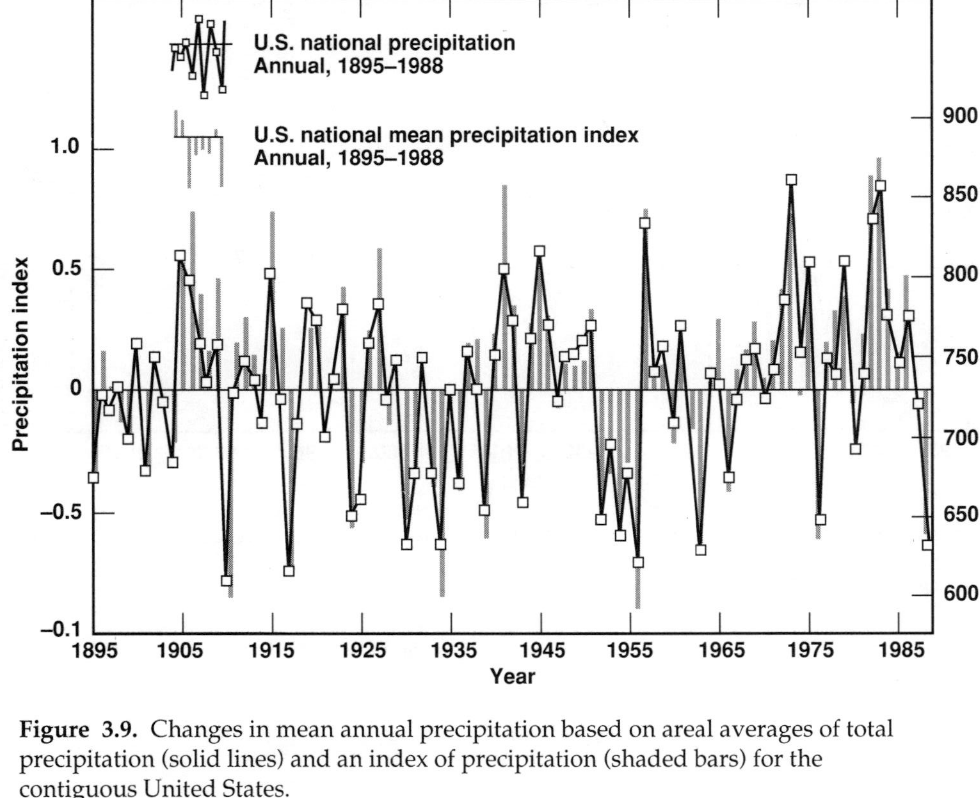


**Figure 3.9.** Changes in mean annual precipitation based on areal averages of total precipitation (solid lines) and an index of precipitation (shaded bars) for the contiguous United States.

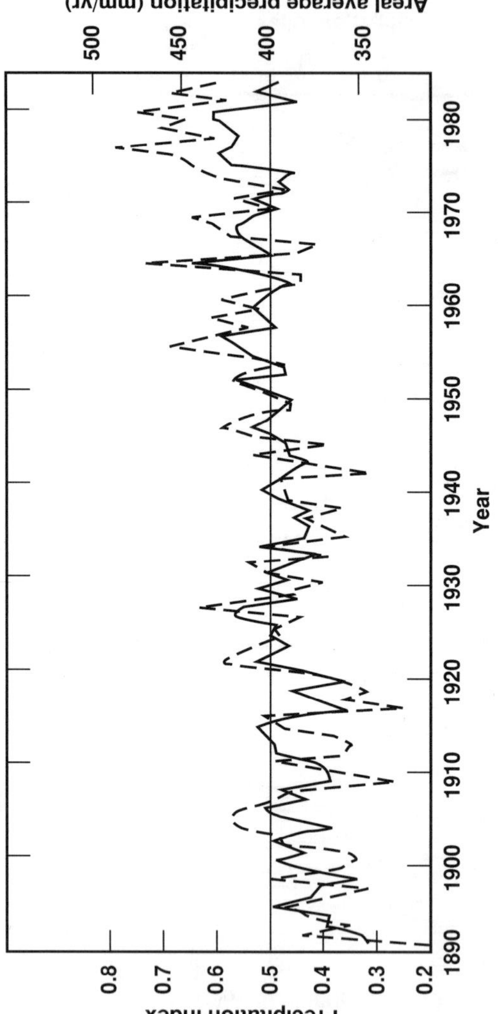


**Figure 3.10.** Changes in annual precipitation based on areal averages of total precipitation (solid lines) and an index of precipitation (dashed lines) for the USSR.

the beginning of the twentieth century, and the trend is heavily influenced by the multi-year wet period during the early 1980s. Karl (1988) points out the importance of multi-year climate fluctuations and the difficulty of ascribing physical significance to general trends in the data when these multi-year fluctuations are embedded in the climate record.

Based on the work of Groisman (1990) and Vinnikov et al. (1990), estimates for precipitation changes for the warm part of the year, when snow cover is absent almost everywhere, are similar to the estimated trends for annual precipitation, except for the area of the contiguous United States where little or no trend is apparent (2% increase over 100 years). On the average, for the entire continental region in the 35° to 70°N zone, the annual precipitation increase over the past 100 years was approximately 6% and 4% for the warm part of the year.

### 3.4.2. Uncertainties in the Precipitation Record

Although the abundance of measurements would seem to make estimates of precipitation trends a relatively easy task, a great deal of

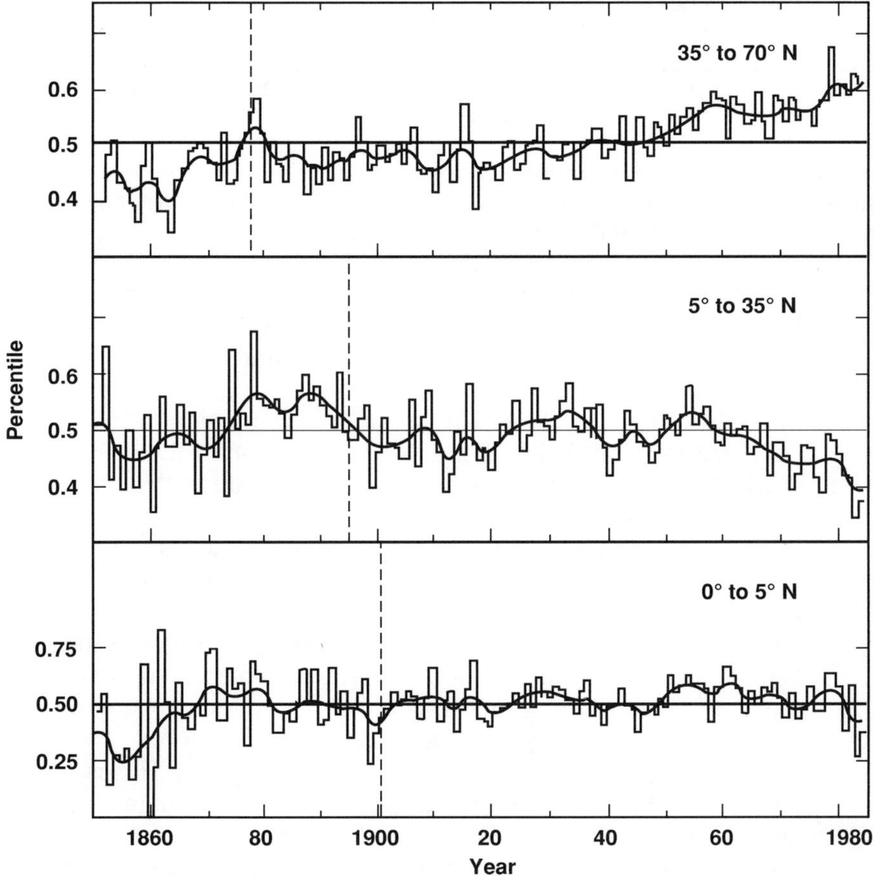


**Figure 3.11.** Precipitation indices for zones 35° to 70°N, 5° to 35°N, and 0° to 5°N from Bradley et al. (1987).

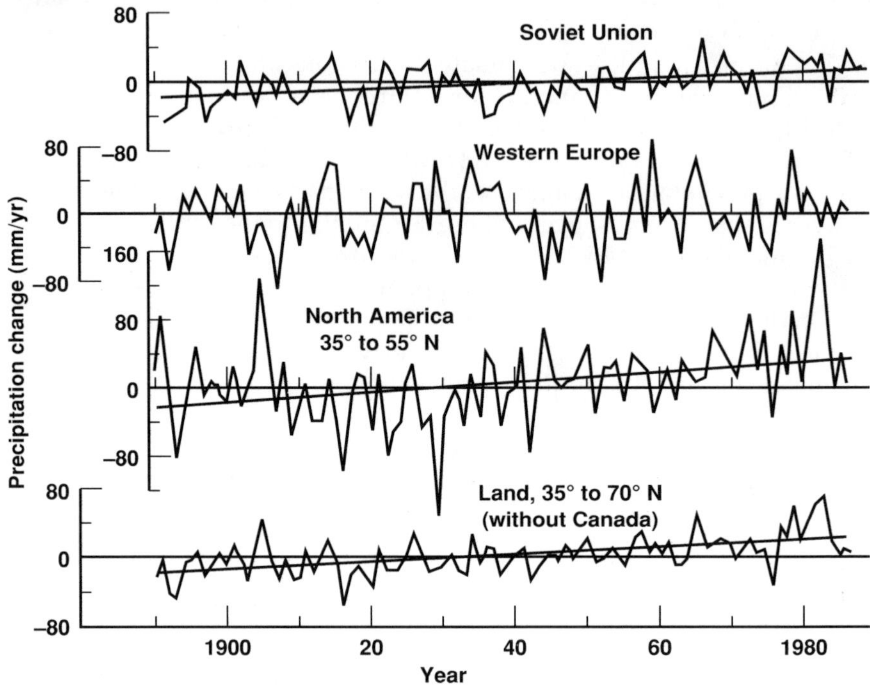


**Figure 3.12.** Changes in mean annual precipitation on the continents of the extratropical part of the Northern Hemisphere. The inclined lines correspond to estimated linear trends (Vinnikov et al., 1990).

information obtained in the past remains unavailable for computer-based analysis. In addition, there are other difficulties in detecting global precipitation trends related to their specific temporal and spatial statistical structure:

(1) Monthly and annual precipitation amounts vary significantly over small spatial scales. The spatial correlation declines markedly as the distance increases up to several hundred kilometers.

(2) The statistical distribution of mean monthly and annual precipitation for many regions of the Earth is significantly different from normal.

(3) Precipitation-gauge measurements are biased estimates of the actual amounts of precipitation falling over a particular area (Sevruk, 1989). Throughout the time of record, new techniques have been developed that allow collection of an ever-increasing portion of the actual precipitation falling. This is particularly important in recording solid precipitation (e.g., snow, hail). As a result, the series of measurements at individual stations probably contains instrumental trends not corresponding to real changes in precipitation.

**Table 3.2.** Estimates of trends and other statistical parameters of precipitation on the continents of the Northern Hemisphere (Groisman, 1990; Vinnikov et al., 1990).

| Territory | Normal (mm/yr) | Standard deviation (mm/yr) | Trend (mm/100 yr) | Trend (% of normal /100 yr) | Fraction of variance accounted for by the trend |
|---|---|---|---|---|---|
| Annual precipitation: | | | | | |
| USSR | 400 | 19 | 35 | 9 | 26 |
| Western Europe | 760 | 36 | 6 | 1 | 0 |
| North America (35° to 55°N) | 720 | 45 | 53 | 7 | 9 |
| Continents in the zone 35° to 70°N (excluding Canada north of 55°N, and China) | 570 | 20 | 34 | 6 | 22 |
| May through September precipitation: | | | | | |
| USSR | 240 | 11 | 9 | 4 | 4 |
| Western Europe | 300 | 20 | 0 | 0 | 0 |
| North America (35° to 55°N) | 320 | 28 | 23 | 7 | 4 |
| Continents in the zone 35° to 70°N (excluding Canada north of 55°N, and China) | 280 | 12 | 12 | 4 | 7 |

To decrease the influence of noise and random errors in measurements, the values measured are averaged over space and time, using as many station records as can be assembled for a given region. In this way, it is assumed that nonsystematic errors will tend to cancel out. Where systematic errors are known to be present in the data, adjustments can be made. For example, Vinnikov et al. (1990) removed inhomogeneities from the USSR precipitation record. Because methods for correcting Soviet data (based on information mostly unavailable to other scientists) were not used by Bradley et al. (1987), this could have affected their results to some extent. However, the estimated trend is similar in sign in both studies. For the US national data (Fig. 3.9), only small differences were found between the national network of 5000 stations and a fixed network of 900 HCN stations that had been adjusted for station inhomogeneities (Karl and Williams, 1987).

### 3.4.3. Sensitivity of Precipitation to Changes of Temperature

The sensitivity of mean precipitation to global warming is very important. In the zone 0° to 35°N, there is a nonsignificant, inverse relationship between temperature anomalies and precipitation indices, temperatures having increased overall while precipitation has decreased (see Figs 3.5 and 3.11). For the Northern Hemisphere as a whole and for the zone 35° to 90°N, temperatures and precipitation have both increased over the past 100 years; however, their increase has not been consistent in time. Moreover, the relationship between these two elements is weak (Table 3.3). Note that the 1920s were warm and dry, while the 1960s were cool and wet. Nonetheless, there appears to have been a 0.4°C increase in the mean temperature of the Northern Hemisphere and an increase in precipitation of about 6% over the continents in the 35° to 70°N zone from 1891 to 1986.

Model results described in Chapter 5 also indicate that an increase in global precipitation will be associated with $CO_2$-induced global warming, although the regional pattern of the precipitation changes are complex. It is interesting that precipitation increases have been observed in instrumental data over the last century when global temperatures have also increased. However, it would be unwise to extrapolate this (or, for that matter, any) empirically estimated climate trend for purposes of predicting future precipitation. Further research on the relationship between temperature and precipitation for various regions is warranted in view of the importance of these parameters for many aspects of human activity.

**Table 3.3.** Correlation coefficients of annual anomalies of temperature and precipitation indices for continental land areas. Italicized values are statistically significant at the $\alpha = 0.05$ level.

| Interval | 0° to 35°N | 35° to 90°N | 0° to 90°N |
|---|---|---|---|
| 1891 to 1920 | –0.21 | 0.18 | 0.13 |
| 1921 to 1950 | –0.31 | 0.01 | –0.15 |
| 1951 to 1980 | –0.05 | 0.05 | *0.32* |
| 1981 to 1986 | –0.09 | *0.26* | *0.26* |

## 3.5. CHANGES IN SEA LEVEL

Global mean sea level has long been considered an indicator of climate change. For example, sea level has varied over a range of about 100 m during past glacial-interglacial cycles, being depressed as ice built up on the continental areas during the very cold glacial periods. Climate warming, either natural or anthropogenic, can logically be expected to cause a rise in global sea level as a consequence of thermal expansion of ocean waters and of the melting of continental ice sheets in polar regions and mountain glaciers in temperate regions. However, the problem of detecting climate-induced sea level change is complicated by the large number of factors to which sea level responds and by the limitations of the technology that has been used to measure it.

Tide-gauge records are the traditional source of data used to analyze long-term changes in sea level. These records show variations on various time scales due to tides, storm surges, ocean-atmosphere effects such as the ENSO, and vertical land movement (including such effects as mountain uplift, volcanism, and glacial rebound). Other effects, such as either changes in the prevailing winds, tides, and ocean currents or the subsidence of land due to the withdrawal of water, oil, or natural gas, are less easily identified. Thus, changes in sea level recorded by tide gauges may reflect either real changes in the level of the oceans, or simply changes in the elevation of the land on local or regional scales.

The importance of sea level rise as an indicator of global climate change was noted by Etkins and Epstein (1982), among others. Prospective increases in sea level may have serious consequences for coastal communities all over the world, yet all quantitative estimates of future sea level rise as a consequence of the projected global warming have very important limitations.

Over the last half century, various researchers have estimated the rate of change of global mean sea level. Their estimates generally fall within the range of 10 to 30 cm/100 yr, as shown in Table 3.4. The results of studies by Gornitz and Lebedeff (1987) and by Barnett (1988) are shown in Fig. 3.13. These increases appear to be in excess of trends over the past 1000 years, which are believed to have been quite small. A recent study by Peltier and Tushingham (1989), utilizing both linear regression and empirical orthogonal function analyses of tide gauge records corrected for glaciostatic land motion, indicates a global rate of rise of 24 ± 9 cm/100 yr for the period 1920 to 1970.

Accelerated warming and consequent melting of the existent mountain glaciers may lead to a further increase of 5 to 15 cm to sea level over the next 60 years, and deterioration of the ice shelves emerging from the polar ice caps in Greenland and Antarctica may either add or subtract up to about 20 to 30 cm to sea level over this period, depending on whether the rate of increase of snow accumulation is greater than the rate of increase of melting (Polar Research Board, 1985; Meier, 1990). There is however, the potential for modest reduction in sea level if the rate of

**Table 3.4.** Estimates of sea level rise from various analyses.

| Rate (cm/100 yr) | | Comments | References |
|---|---|---|---|
| Uncorrected | Corrected[a] | | |
| 11 ± 8 | | Many stations, 1907 to 1939 | Gutenbert (1941)[b] |
| 12 to 14 | | Combined methods | Kuenen (1950)[b] |
| 11 ± 4 | | Six stations, 1807 to 1943 | Lisitzin (1958)[b] |
| 12 | | Cryologic estimate | Wexler (1961)[b] |
| 12 | | Selected stations, 1900 to 1950 | Fairbridge and Krebs (1962) |
| 30 | | Many stations, 1935 to 1975 | Emery (1980) |
| 12 | | Many stations, grouped into regions, 1880 to 1980 | Gornitz et al. (1982) |
| | 10 | Many stations, grouped into regions | Gornitz et al. (1982) |
| 15 ± 1.5 | | Selected stations, 1903 to 1969 | Barnett (1983) |
| 14 ± 1.4 | | Many stations, grouped into regions, 1881 to 1980 | Barnett (1984) |
| 23 ± 2.3 | | Many stations, grouped into regions, 1930 to 1980 | Barnett (1984) |
| | 10 ± 1 | Mean of regional means, 1880 to 1982 | Gornitz and Lebedeff (1987) |
| | 12 ± 3 | Arithmetic mean, 1880 to 1982 | Gornitz and Lebedeff (1987) |
| | 11 | East coast only | Peltier (1986) |
| | 24 ± 9 | 40 stations, 1920 to 1970 | Peltier and Tushingham (1989) |

[a] Correction attempted for crustal and/or glacial isostatic motion.
[b] In Lisitzin (1974), after Barnett (1983).

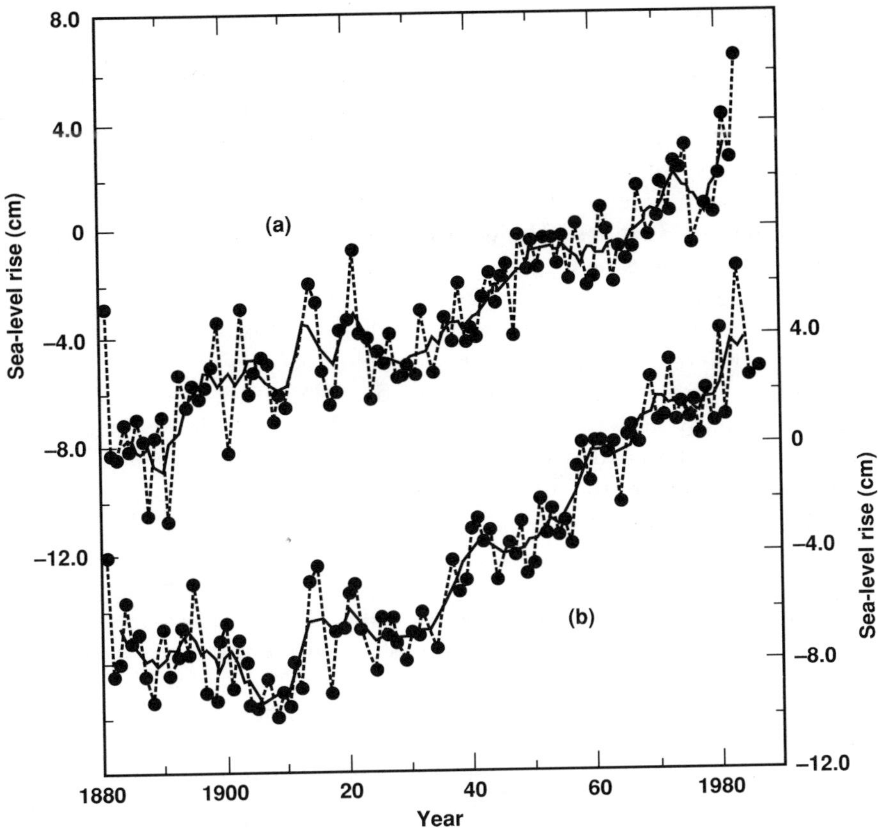


**Figure 3.13.** Global mean sea level rise over the last century. The baseline is obtained by setting the average for the period 1951 to 1970 equal to zero. The dashed line represents the annual mean; the solid line represents the 5-year running mean. Data are from (a) Gornitz and Lebedeff (1987) and (b) Barnett (1988).

snow accumulation in polar regions increases markedly. Warming of the ocean waters could add as much as 25 cm to sea level by the year 2050, depend-ing on the intensity of vertical mixing and changes in the ocean circulation (see Chapter 5). Taken together, such changes, (which are quite uncertain) could raise sea level to heights not experienced in historical times and start to cause inundation in low-lying areas and along coastlines subject to storm surges.

## 3.6. LESSONS AND OPPORTUNITIES FOR THE FUTURE

Climates of the past 100 years provide context for considering climate change over the next 100 years.

(1) *What have we learned about how the climate has varied over the past 100 years?*

Measurements of surface air temperature, which is the best available, long-term indicator of the global climate, indicate that warming has been occurring in both the Northern and Southern Hemispheres, but that it has been more uniform in time and space in the Southern Hemisphere. The increase in the global mean surface air temperature is estimated to have been about 0.4 to 0.5°C over the past 100 years.

(2) *What have we learned about the seasonal and latitudinal pattern of the changing climate?*

The warming has generally been largest in the winter and concentrated in high latitudes. The warming has not, however, been spatially uniform, with some regions experiencing little change while other areas have experienced warming or even cooling.

(3) *What have we learned about changes in precipitation?*

Although the record is much more variable, analyses of rain-gauge measurements suggest a tendency for a considerable increase in precipitation in many midlatitude regions on the Northern Hemisphere's continents. However, because multi-year fluctuations have been large, it is still difficult to interpret the meaning of these trends.

(4) *What have we learned about other recent changes in the climate system?*

The global mean sea level responds to changes in the amount of glacial ice and the temperature of the ocean. Over the past 100 years, sea level has risen by 10 to 20 cm, which is consistent with indications that warming is occurring.

There remain a number of critical gaps in our understanding of the climate variations of the past 100 years. To accelerate the rate at which these gaps are closed, the following actions are recommended:

(1) Quantify the impact of urban warming in the temperature record and develop a strategy to prevent its impact on future climate monitoring.
(2) Make better use of the extensive operational networks of precipitation measurements, placing emphasis on ensuring data homogeneity and on reducing the uncertainties introduced into spatial averages of precipitation by temporal changes in the spatial density of stations.
(3) Expand research on the relationship between change in global and regional temperature and precipitation using both modeling and empirical studies.
(4) Continue monitoring the solar irradiance to better understand the fluctuations in climate caused by variations in solar energy; also, continue analyses of data from dendroclimatology and other proxy measures of past solar behavior.
(5) Conduct much better statistical analyses of the changes of temperature and precipitation over the past 100 years to analyze the causes and mechanisms of global and regional climate change, including the role of chaotic climatic behavior and other natural fluctuations.

(6) Expand the collection and monitoring of data to include additional variables (e.g., soil moisture) that are expected to change or that are of importance in impact analyses.

# *CHAPTER 4. CHANGES IN ATMOSPHERIC COMPOSITION*

## 4.1. INTRODUCTION

Atmospheric concentrations of greenhouse gases are changing as a result of human activities. Concentrations of carbon dioxide ($CO_2$), methane ($CH_4$), nitrous oxide ($N_2O$), carbon monoxide (CO), and several chlorinated and brominated halocarbons are known to be increasing. Stratospheric concentrations of ozone are decreasing, whereas tropospheric concentrations appear to be increasing. In general, the changing concentrations of these gases are thought to be primarily attributable to human-related causes. For example, the concentration of $CO_2$ is increasing largely as a result of fossil-fuel burning and deforestation. Table 4.1, adapted from Wuebbles and Edmonds (1988) and Wuebbles et al. (1989), summarizes the characteristics and anthropogenic sources of many of the greenhouse gases.

Many of these gases have direct effects on climate through their absorption of infrared radiation (see Table 4.2). Although $CO_2$ has received the most attention, climate models indicate that the sum of radiative effects from the growing atmospheric concentrations of the other gases, along with induced effects on the atmospheric distribution of ozone, could be comparable with the radiative effects projected for $CO_2$ alone.

In addition to direct radiative effects, many of these gases have indirect radiative effects on climate through their interactions with atmospheric chemical processes (see Table 4.2). Both measurements and theoretical models indicate that the distribution of ozone in the troposphere and stratosphere is changing as a result of such interactions. Ozone not only is a greenhouse gas, but also is the primary absorber of ultraviolet (UV) radiation at wavelengths from about 200 to 300 μm. The concern about changes in the amount of UV radiation reaching the Earth has led to much of the recent and past scientific and public interest in the condition of the ozone layer.

As indicated in Table 4.2, there are other concerns regarding chemical interactions in the atmosphere. Increasing levels of $CH_4$ are projected to cause an increase in the stratospheric concentration of water vapor (each 1-ppmv increase in the $CH_4$ concentration[1] is expected to

[1] ppmv = parts per million by volume.

**Table 4.1.** Important atmospheric greenhouse gases having significant anthropogenic sources and affecting global atmospheric composition and climate (based on Wuebbles and Edmonds, 1988; Wuebbles et al., 1989).

| Gas | Common name | Current surface concentration (ppmv)[a] | Current trend in atmospheric concentration (% per year)[a] | Atmospheric lifetime (years) | Primary anthropogenic sources |
|---|---|---|---|---|---|
| $CO_2$ | Carbon dioxide | 351 (1988) | ~0.4 | ~250[b] | Fossil-fuel burning land-use conversion |
| $CH_4$ | Methane | 1.7 | ~1 | ~10 | Ruminant animals; rice paddies; biomass burning; gas and mining leaks |
| CO | Carbon monoxide | 0.12 (N.H.)<br>0.06 (S.H.) | ~1 (N.H.)<br>~0 (S.H.) | ~0.3 | Energy use; agriculture; forest clearing |
| $N_2O$ | Nitrous oxide | 0.31 | ~0.3 | 150 | Cultivation and fertilization of soils |
| $NO_x$ (= NO + $NO_2$) | Reactive odd nitrogen | 1 to $20 \times 10^{-5}$ | Unknown | ≤0.02 | Fossil-fuel burning; biomass burning |
| $CFCl_3$ | CFC-11 | $2.6 \times 10^{-4}$ | ~4 | ~0 | Chemical industry |
| $CF_2Cl_2$ | CFC-12 | $4.4 \times 10^{-4}$ | ~4 | ~120 | Chemical industry |
| $C_2Cl_3F_3$ | CFC-113 | $3.2 \times 10^{-5}$ | ~10 | ~90 | Chemical industry |
| $CHF_2Cl$ | CFC-22 | $9 \times 10^{-5}$ | ~7 | ~15 | Chemical industry |

**Table 4.1.** (Continued)

| Gas | Common name | Current surface concentration (ppmv)[a] | Current trend in atmospheric concentration (% per year)[a] | Atmospheric lifetime (years) | Primary anthropogenic sources |
|---|---|---|---|---|---|
| $CH_3CCl_3$ | Methyl chloroform | $1.4 \times 10^{-4}$ | ~4.5 | ~6 | Chemical industry |
| $CF_2ClBr$ | Ha-1211 | $1 \times 10^{-6}$ | ~12 | ~12 | Fire extinguishers |
| $CF_3Br$ | Ha-1301 | $1 \times 10^{-6}$ | ~15 | ~110 | Fire extinguishers |
| $SO_2$ | Sulfur dioxide | $1$ to $20 \times 10^{-5}$ | Unknown | ~0.02 | Coal and petroleum burning |
| COS | Carbonyl sulfide | $5 \times 10^{-4}$ | <3 | 2 to 2.5 | Biomass burning; fossil-fuel burning |

[a] Corresponds to mid-1980s values unless otherwise noted. Concentrations given are estimated global-average values.

[b] The $CO_2$ lifetime is computed as the ratio of the current atmospheric burden to net annual removal. Net annual removal is taken to be estimated emissions less atmospheric accumulation. This value is only a rough estimate of the $CO_2$ lifetime because its removal rate varies in time after initial emission.

**Table 4.2.** Direct radiative effects and indirect chemical interactions between trace gases and climate.

| Gas | Greenhouse gas? | Is its tropospheric concentration affected by chemistry? | Effects[a] on tropospheric chemistry? | Effects[a] on stratospheric chemistry? |
|---|---|---|---|---|
| $CO_2$ | Yes | No | No | Yes, affects $O_3$ |
| $CH_4$ | Yes | Yes, reacts with OH | Yes, affects OH and $O_3$ | Yes, affects $O_3$ and $H_2O$ |
| CO | Yes, but weak | Yes, reacts with OH | Yes, affects OH and $O_3$ | Not significant |
| $N_2O$ | Yes | No | No | Yes, affects $O_3$ |
| $NO_x$ | Yes | Yes, reacts with OH | Yes, affects OH and $O_3$ | Yes, affects $O_3$ |
| $CFCl_3$ | Yes | No | No | Yes, affects $O_3$ |
| $CF_2Cl_2$ | Yes | No | No | Yes, affects $O_3$ |
| $C_2Cl_3F_3$ | Yes | No | No | Yes, affects $O_3$ |
| $CHF_2Cl$ | Yes | Yes | No | Yes, affects $O_3$ |
| $CH_3CCl_3$ | Yes | Yes, reacts with OH | No | Yes, affects $O_3$ |
| $CF_2ClBr$ | Yes | Yes, photolysis | No | Yes, affects $O_3$ |
| $CF_3Br$ | Yes | No | No | Yes, affects $O_3$ |
| $SO_2$ | Yes, but weak | Yes, reacts with OH | Not significant | Yes, increases aerosols |
| COS | Yes, but weak | Yes, reacts with OH | No | Yes, increases aerosols |
| $O_3$ | Yes | Yes | Yes | Yes |

[a] Effects on atmospheric chemistry are limited to effects on constituents having a significant influence on climate, and may occur as a consequence of either direct chemical effects or indirectly through influence on radiative fluxes and temperature (and thence on temperature-dependent reaction rates).

increase the stratospheric $H_2O$ concentration by about 2 ppmv). Because water vapor is also a greenhouse gas, this increase will result in additional climate change. Many of the species of interest react with hydroxyl (OH) in the troposphere; increasing emissions of $CH_4$ and CO can reduce OH concentrations, resulting in longer lifetimes for these and other greenhouse gases. Even $CO_2$, which has no known chemical interactions of significance within the troposphere or stratosphere, can affect the concentrations of stratospheric ozone through its radiative cooling of the stratosphere. This cooling results in a slowing down of the catalytic destruction of ozone and a net increase in the ozone concentration.

To provide the basis for determining future climate, it is therefore essential to document and understand past changes in atmospheric composition and to develop the capabilities for projecting future concentrations. This chapter will discuss our understanding first of the global carbon cycle (Section 4.2) and then of the other important chemically and radiatively active constituents of the atmosphere (Section 4.3). Section 4.4 then discusses future concentrations, and Section 4.5 describes the perturbations to the radiative fluxes that these changes will cause.

## 4.2. CARBON DIOXIDE AND THE GLOBAL CARBON CYCLE

### 4.2.1. Observed Changes in Atmospheric Concentration

Continuous measurements of the $CO_2$ concentration at the Mauna Loa Observatory in Hawaii were begun in 1958 by Keeling and coworkers (Keeling et al., 1976) and have been continued as part of the US National Oceanic and Atmospheric Administration monitoring program (Komhyr et al., 1985). These measurements provide the best record of the increase in the atmospheric concentration of this important greenhouse gas. Monthly and annual average $CO_2$ concentrations at Mauna Loa since 1958 are plotted in Fig. 4.1. Continuous measurements of the $CO_2$ concentration are now made routinely by the United States at a network of remote stations at Mauna Loa, Hawaii, at Barrow, Alaska, at American Samoa, and at the South Pole. Flask samples are collected at 20 additional sites. Data collected by this sampling program reveal spatial patterns in $CO_2$ variations that show the influences of exchanges between the atmosphere and other reservoirs. Interestingly, the Mauna Loa measurements, the longest record available, appear to represent a reasonable approximation of the global $CO_2$ concentration and its seasonal variation in the Northern Hemisphere. Between 1959 and 1988, the annual average $CO_2$ concentration at Mauna Loa increased from about 316 ppmv to 351 ppmv. This 11% increase in concentration corresponds to an increase in the carbon content of the atmosphere of about 75 PgC.[2] (Measurements being analyzed to determine the 1989 concentration of $CO_2$ at Mauna Loa are ezpected to show a 1.5- to 2.0-ppmv increase over 1988.)

[2] PgC = $10^{15}$g of carbon = 1 billion tonne.

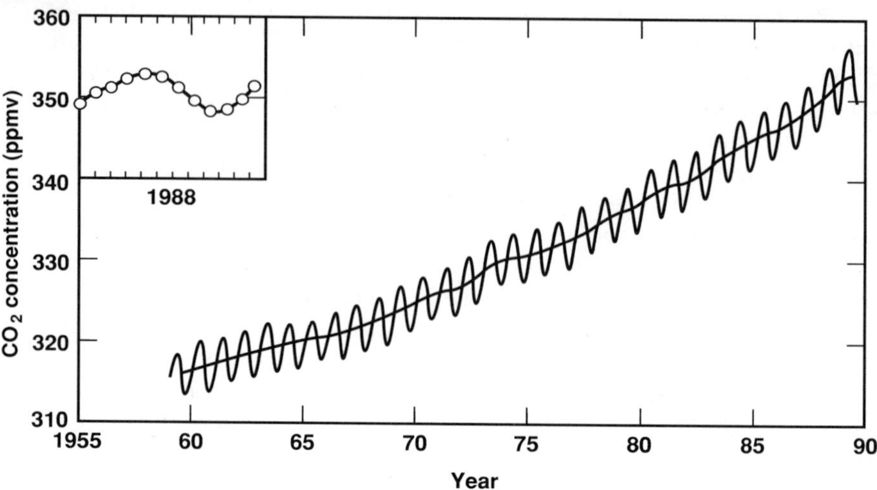


**Figure 4.1.** Measurements of the $CO_2$ concentration at Mauna Loa, Hawaii, since 1958. The seasonal variations are associated with photosynthesis-respiration cycles of the land-biosphere superimposed on long-term trends caused by fossil-fuel emissions and deforestation. The year 1988 is the latest for which data showing the full annual cycle are available, as shown in detail in the inset.

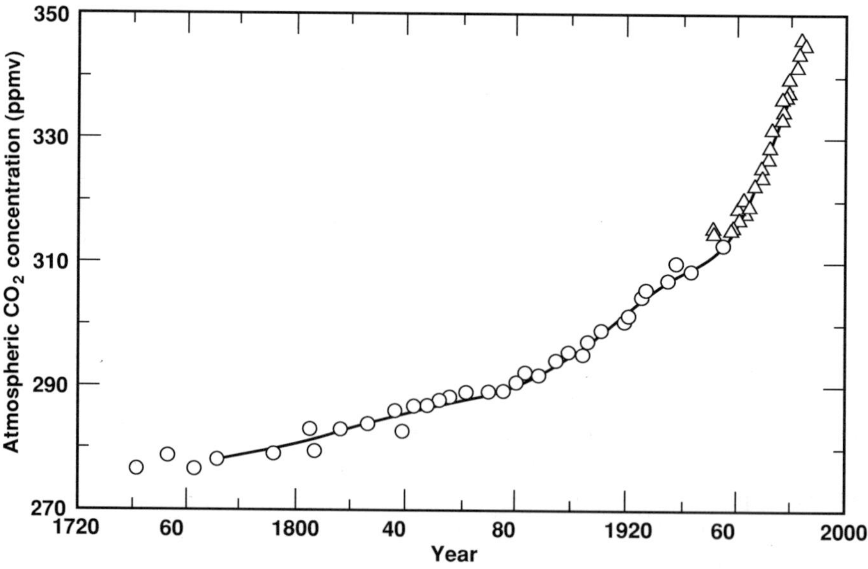


**Figure 4.2.** Carbon dioxide concentration over the last 200 years based on the $CO_2$ concentration in ice bubbles in a core from the Antarctic as determined by Neftel et al., (1985), and extended to the present using the Mauna Loa record.

The $CO_2$ concentration in air bubbles trapped in ice records provides a record of atmospheric concentrations before direct measurements began in 1958. Data obtained from an ice core extracted at Siple Station, Antarctica (Neftel et al., 1985; Friedli et al., 1986), record the history of $CO_2$ from the middle of the eighteenth century, prior to the onset of significant fossil-fuel use, to the time of the Mauna Loa record. This ice-core record, which meshes well with modern measurements, indicates a mid-eighteenth century $CO_2$ concentration of about 278 ppmv (Fig. 4.2); the 1988 concentration of 351 ppmv represents a 26% increase.

For even longer periods, the most remarkable record is drawn from the ice core extracted near the Soviet East Antarctic Station at Vostok, Antarctica. This record spans 160,000 years (Lorius et al., 1985) and provides simultaneous records of temperature (derived from changes in the deuterium isotope concentration in the ice layers) and $CO_2$ concentration (measured in gas bubbles trapped in the ice layers) (Barnola et al., 1987). A comparison of the Vostok $CO_2$ and deuterium time series shows that the atmospheric $CO_2$ concentration, which ranged from about 200 to 280 ppmv, and the temperature varied in near-lockstep over the last 160,000 years (see Fig. 2.3). Both the $CO_2$ and temperature fluctuations appear to have been in phase with the insolation periodicities of the high-latitude Northern Hemisphere over the entire period of record (see Chapter 2). The 10°C range in air temperature at Vostok (which is 3.5 km above sea level) is consistent with the 2 to 3°C change in the global mean ocean temperature evident from sediment data. The in-phase $CO_2$ and temperature variations suggest that much of the interglacial warming and glacial cooling during the period represented by the ice core can be attributed to the $CO_2$ greenhouse feedback.

Glacial-interglacial variations in the concentration of $CH_4$, another greenhouse gas, are also found in gas bubbles trapped at Vostok (see Fig. 2.4 and Section 4.3.1). The concentration of $CH_4$, which has risen from 0.7 to about 1.7 ppmv in the last 300 years, is also highly correlated with the Vostok temperature record. Raynaud et al. (1988) estimate that, although the major glacial-interglacial greenhouse feedback is from $CO_2$, as much as 25% of the total might be attributed to changes in the concentration of $CH_4$.

### 4.2.2. The Natural Carbon Cycle

The response of the Earth System to the relatively rapid $CO_2$ forcing by anthropogenic fossil-fuel burning—the "large scale geophysical experiment" of Revelle and Suess (1957)—is embedded in the long-term flow of natural variations. For example, the current increase in atmospheric $CO_2$ (Fig. 4.1) from burning of fossil fuels and land use changes must be considered from the perspectives of a much more slowly varying $CO_2$ background concentration, of global warming, and of emergence from the last glacial maximum about 18,000 years ago (Barnola et al., 1987). These

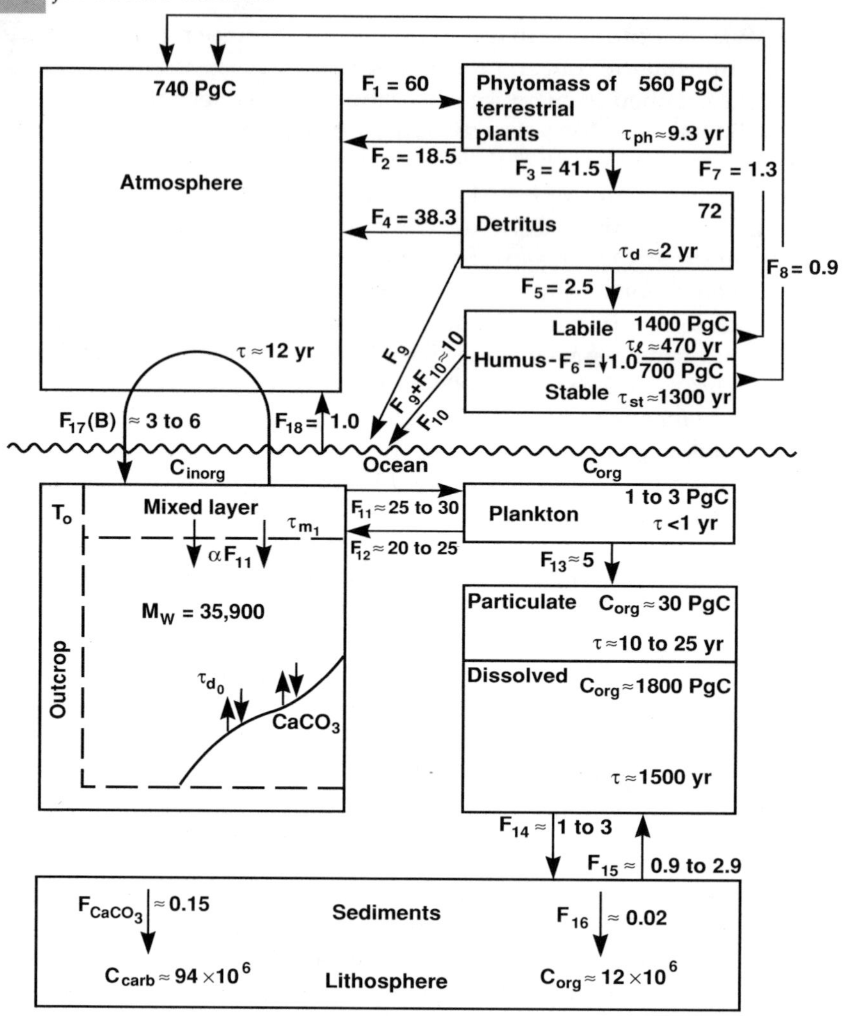


**Figure 4.3.** Diagram depicting the reservoirs and fluxes composing the global carbon cycle (adapted from Kobak, 1988). The estimated values of carbon storage in the major reservoirs (in PgC) and the estimated fluxes of carbon (F in PgC/yr) are indicated as well as the average lifetime of carbon in these reservoirs ($\tau$, in years). The lifetimes given here apply to individual molecules rather than to changes in the net amount in the reservoir (refer to text and Tables 4.1 and 4.4). The various fluxes are: $F_1$ – net assimila-tion of $CO_2$ in photosynthesis by continental plants; $F_2$ – root respiration; $F_3$ – carbon input from litter fall onto the soil surface; $F_4$ – mineralization of litter; $F_5$ – new formation of humus; $F_6$ – stable humus formation; $F_7$, $F_8$ – oxidation of humus; $F_9$ – terragenic transfer of organic matter from land to the ocean by suspension and windblown dust; $F_{10}$ – transfer of organic carbon ($C_{org}$) by river runoff; $F_{11}$ – photosynthesis by phytoplankton; $F_{12}$ – mineralization of $C_{org}$; $F_{13}$ – transfer to reservoir of particulate and dissolved $C_{org}$; $F_{14}$ – deposition of $C_{org}$ on the ocean floor; $F_{15}$ – mineralization of $C_{org}$ on the ocean floor; $F_{16}$ – transfer of $C_{org}$ to the deep lithosphere; $F_{17}$(B) – net exchange of $CO_2$ between the atmosphere and ocean; $F_{18}$ – mineralization of $C_{org}$ on the coastal shelf and elsewhere; $M_w$ – total amount of inorganic carbon in the ocean waters; $\tau_{ml}$ – mixed layer temperature; $\tau_o$ – outcrop region temperature; $\tau_{do}$ – deep ocean temperature; $\alpha F_{11}$ – flux of new soft organic tissues from the ocean mixed layer to the deep ocean.

glacial-interglacial variations of the atmospheric $CO_2$ concentration are themselves embedded in longer-term concentration variations and in changes in the carbon fluxes between carbonaceous rocks and the atmosphere. A final and philosophically compelling reason for considering the "geological" perspective is that some fraction of the carbon emissions from fossil fuels will persist in the atmosphere for time scales overlapping those of longer-term biogeochemical processes and could alter these processes well beyond the fossil-fuel era of human history (Hoffert, 1974; Sundquist, 1985, 1986; Kasting and Walker, 1989).

The natural carbon cycle involves exchanges of carbon between reservoirs in the atmosphere, oceans, lithosphere, and biosphere (Fig. 4.3). These carbon transfers occur over a wide range of time scales and with widely differing inter-reservoir fluxes; they have been occurring throughout the Earth's history. During this time, atmospheric $CO_2$ levels are believed to have fluctuated markedly in response to natural factors (see also Section2.2). Changes in solar insolation, continental drift, glaciations, ocean circulation, and biological interactions, as well as the greenhouse effect of atmospheric $CO_2$ itself, are all believed to have influenced the atmospheric $CO_2$ concentration and the climatic evolution of the Earth long before *homo sapiens* entered the scene.

#### *4.2.2.1. The Geological Cycling of Carbon*

It is believed that Venus, Earth, and Mars all formed with comparable endowments of elemental carbon, although these planets presently have atmospheres with quite different amounts of $CO_2$, owing to differences in their climatic histories and in the evolution of their carbon cycles. Venus has a 90-bar, largely $CO_2$ atmosphere; Earth has a 1-bar, $N_2$-$O_2$ atmosphere with about 0.0003 bar of $CO_2$; and Mars has a 0.006-bar $CO_2$ atmosphere (Kasting and Toon, 1989). In contrast to Venus, the overwhelming fraction of the Earth's carbon, and possibly that on Mars, is locked up in the lithosphere rather than being present in the atmosphere.

The estimated distribution of carbon on the present-day Earth is given in Table 4.3. These values are fairly representative, although estimates of certain reservoirs may vary by ±25%. Figure 4.3 gives a less aggregated picture of the reservoirs in the biosphere, including a breakdown of the terrestrial carbon biomass into phytomass, detritus, and humus, and of the marine biomass into planktonic, particulate, and dissolved organic carbon. Estimates of the fluxes between reservoirs are also indicated (Kobak, 1988).

The small size of the atmospheric reservoir compared with the oceanic and carbonaceous rock reservoirs means that relatively small carbon perturbations in the latter can exert a large influence on the atmospheric $CO_2$ concentration. On geological time scales, the biospheric components run as a relatively fast, nearly closed cycle in which >99.9% of photosynthetically fixed organic carbon respires back to $CO_2$ within a few decades (see Section 4.2.2.3.). Organic carbon in rocks ($C_{org}$) is unoxidized (i.e., "reduced") carbon leaked from the terrestrial and marine biospheres. This carbon is present in much greater amounts than in the living and dead organic carbon pools because the burial of small fractions of organic material in anoxic environments results in an enormous accumulation over geologic time. Sedimentary carbonate rocks ($C_{carb}$) are oxidized carbon compounds accumulated in even greater amounts. Virtually all the carbon that ever cycled through the atmosphere is in these rock reservoirs.

The normal form of $C_{org}$ is kerogen—an insoluble organic molecular material with no regular structure. A small fraction of kerogen (about 6000 PgC) has been transformed underground over hundreds of millions of years under very specific temperature, pressure, and geologic reservoir conditions into economically recoverable fossil fuels—coal, oil, and natural gas. This miniscule fraction of $C_{org}$ forming the fossil-fuel reserve is still eight times the amount of carbon in the atmosphere.

The carbon cycle on time scales of 1 to 10 million years and greater, including on the time scales of continental drift (100 Myr) and planetary evolution (1000 Myr), is thought to be controlled by the balance between atmospheric $CO_2$ recycled from sedimentary rocks and the geochemical weathering of silicate rocks by $CO_2$ in the presence of liquid water to form carbonate carbon ($C_{carb}$) (Walker et al., 1981; Berner et al. 1983; Lasaga et al., 1985). Table 4.3 indicates that 90% of the carbon in rocks is $C_{carb}$. The rate-controlling step of the weathering process is the reaction between the $CO_2$ dissolved in rainwater and the silicate minerals in soil, which can be accelerated by respiration of soil bacteria in the roots of vascular land plants and by the soil-forming bacteria themselves. The weathering process releases calcium and magnesium ions and converts $CO_2$ into

**Table 4.3.** Approximate content of the carbon reservoirs of the Earth System, summarized from Fig. 4.3.

| Reservoir | Carbon amount (in PgC) |
|---|---|
| Atmosphere (present value) | 740 |
| Terrestrial biosphere | 560 |
| Detritus and labile humus | 1472 |
| Stable humus and peat | 700 |
| Marine biosphere | 2 |
| Dead marine organic matter[a] | 1,830 |
| Oceans | 35,000 |
| Organic carbon rocks | 12,000,000 |
| Carbonate rocks | 94,000,000 |

[a] This marine carbon pool is poorly understood, with recent observations by Druffel et al. (1989) suggesting that it is larger and has more variable lifetimes than previously thought (Toggweiler, 1990).

bicarbonate ions. Dissolved species are transported by river runoff to the oceans, where the calcium recombines with bicarbonate to form calcium carbonate ($CaCO_3$). The result is a net sink for atmospheric $CO_2$.

Calcium carbonate in the present ocean is formed by calcareous plankton (e.g., foraminifera) whose shells precipitate down through the ocean column and accumulate on the sea floor as sediments. Prior to their evolution, this task may have been accomplished by mat-forming stromatolites, or even abiotically (Kasting and Toon, 1989).

In the absence of a mechanism for returning $CO_2$ to the atmosphere, carbonate deposition would deplete the atmospheric plus oceanic reservoirs in only 400 thousand years (Berner et al., 1983). In reality, a return loop of the $C_{carb}$ cycle exists that is driven by plate tectonics: The sea floor, bearing its load of carbonate sediment, spreads from midocean ridges to continental margins, where it is subducted into the mantle. In the process, $C_{carb}$ is subjected to high pressures and temperatures, and $CO_2$ is driven off to the atmosphere through volcanic emissions. At the same time, silicate minerals re-form by "metamorphic decarbonation."[3] The net effect is the reverse of the $CO_2$ weathering described above.

The balance between the rate at which silicate rocks are chemically weathered to carbonate by atmospheric $CO_2$ and the rate at which $CO_2$ degasses to the atmosphere by metamorphic decarbonation of carbonate rocks controls the atmospheric $CO_2$ level. (On these million-year time scales, the oceans play the passive role of "instantly" adjusting $C_{carb}$ sedimentation rates to weathering rates.) The flux due to weathering is the product of two factors: the weathering rate per unit area (which is dependent on the atmospheric $CO_2$ concentration and the temperature) and the continental area. The degassing rate is in turn the product of the rate of sea floor spreading and the susceptibility of dolomite and calcite reservoirs to decarbonation. Atmospheric $CO_2$ levels are therefore high when sea-floor spreading rates are large and continental area is small. This occurred in the Mid-Cretaceous, 100 million years ago, when geochemical models indicate that the $CO_2$ concentration was about 10 times the pre-industrial level (Lasaga et al., 1985; Volk, 1987). The Mid-Cretaceous can be considered a "worst-case" analog to massive injections of $CO_2$ from the burning of fossil fuel. Available geologic evidence indicates such elevated $CO_2$ levels and warmer climates did indeed characterize Mid-Cretaceous climates (Budyko and Izrael, 1987). Geologic data and models indicate that global-mean temperature then was about 8°C or more warmer than today, due primarily to the enhanced $CO_2$ greenhouse effect (Barron and Washington, 1985; Budyko et al. 1985).

[3] Metamorphic rocks require heat and pressure to form, but do not require remelting (as igneous rocks do) or precipitation in water (as sedimentary rocks do).

#### 4.2.2.2. *Interglacial Transitions: The Milankovitch-Vostok-Ocean Connection*

Since the Huronian glaciation 2,500 million years ago, the Earth has experienced several ice ages (see also Chapter 2). The Antarctic and Greenland ice caps influence global climate even today, but as recently as 18 thousand years ago (the last glacial maximum), the Northern Hemisphere was ice-covered to a much greater extent, and global-mean temperatures were several degrees colder (global-mean air temperatures were about 5°C colder, ocean temperatures were about 2 to 3°C colder). Variations in the relative concentration of $^{18}$O in sea-floor cores indicate that the ice volume and climate have varied markedly on time scales of less than a million years, with seven major climatic cycles and many minor ones being evident in the record of the last 700 thousand years (Broecker, 1982).

The orbital insolation explanation for glacial and interglacial cycling was first proposed by Milutin Milankovitch (Milankovitch, 1920; Hays et al., 1976). He held that cyclic changes in seasonal solar radiation caused by periodic variations in the Earth's orbit drive climate fluctuations on 10-thousand to 100-thousand-year glacial-interglacial time scales. The seasonal insolation cycle arises mainly from the tilt of the Earth's spin axis relative to a normal to the orbital plane (ecliptic). The tilt, or obliquity of the ecliptic, is presently $\varepsilon = 23.45°$. Seasonal cycles are also affected by orbital eccentricity; presently, $e = 0.017$. The seasonal irradiance cycle at any latitude depends in a small but nontrivial way on the timing of Earth-Sun distance variations relative to the equinoxes, as measured by the precession parameter, $e \cdot \sin \omega$, where $\omega$ is the orbit angle between vernal equinox and perihelion. It is known from orbital mechanics that $\varepsilon$, $e$, and $e \cdot \sin \omega$ vary quasi-harmonically with periods of about 100 thousand, 41 thousand, and 21 thousand years, respectively. The Milankovitch theory appeared to be dramatically vindicated in the early 1970s, when spectral analysis of core data exhibited distinct $^{18}$O peaks at these periods (Imbrie and Imbrie, 1979). This was strong evidence that astronomical forcing was the "pacemaker" of ice ages.

On closer examination, however, problems have developed with the Milankovitch theory. Orbitally driven insolation variations occur at specific latitudes and with specific seasonality. Because the global-mean, annually averaged solar forcing remains virtually constant, some feedback mechanism is needed to convert seasonal- and latitude-specific perturbations to a global climate response. The original mechanism assumed that the interannual buildup (or melting) of high-latitude Northern Hemisphere ice sheets was induced by local temperature changes. A net ice-albedo change would presumably persist from one seasonal cycle to the next. The change in the latitude of the ice line would thereby change the global-mean albedo, which in turn would feed back on the global-mean radiation balance until a new equilibrium climate was attained.

This land ice albedo-temperature feedback, by itself, was found to be too weak to explain the amplification of the orbital effect (see North and Coakley, 1979; Broccoli and Manabe, 1987). The lowered obliquity of 25 thousand years ago (i.e., $\varepsilon = 22.25°$ versus the current $23.45°$) is thought to have triggered the last glacial advance; however, ice albedo-temperature feedback would only have led to an equatorward shift of the ice line of about 3°. Albedo-temperature feedback from insolation changes produced by precession of the equinoxes would also have been too small to explain the glacial buildup and the 15° poleward shift of the Northern Hemisphere ice line since the last glacial maximum (North and Coakley, 1979). Moreover, the hemispheres are in phase with regard to the tilt cycle, but out of phase with regard to the precession cycle that is believed to have triggered the sharp transition 11 thousand years ago from the cold conditions evidenced by Northern Hemisphere glacial ice to the much warmer global climate of 6 thousand years ago.

Broecker (1982) took a critical step toward a modified orbital theory of climate change by observing that ice-age terminations occur nearly synchronously in both hemispheres. Because ice ages are planet-wide, he argued that the feedback chain amplifying local insolation changes must be capable of initiating a glacial advance in the Northern Hemisphere, of propagating the cooling rapidly to the Southern Hemisphere, and of being sufficiently strong to account for observed worldwide climatic variations.

A possible explanation to this puzzle is the periodic absorption and release of $CO_2$ by the oceans that is triggered by localized insolation changes. Atmospheric $CO_2$ mixes rapidly between the hemispheres (on a time scale of about 1 year) and provides a climate feedback amplifier through variations in the intensity of the greenhouse effect. Ice cores from both hemispheres show simultaneous increases in the atmospheric $CO_2$ concentration from about 200 ppmv to pre-industrial concentrations, increases that are coincident with global warming and emergence from the last glacial maximum (Broecker, 1982). For example, the Vostok $CO_2$ variation range described earlier (i.e., from 200 to 280 ppm) implies the transfer of about 170 PgC between the atmosphere and other carbon-cycle reservoirs during the glacial-interglacial cycling.

Boyle (1988) presented evidence indicating that a contributing factor to the change in carbon fluxes is the redistribution of nutrients to deep ocean waters during pre-glacial times. Higher concentrations of $CO_2$ in the deep ocean increase the *in situ* carbonate ion content of the deep ocean, resulting in higher carbonate dissolution rates. This creates an imbalance between carbonate input from continental weathering and output to deep-sea sediments. In the Boyle scenario, the total alkalinity (i.e., the total concentration of strong ions in the ocean) rises in both the deep and surface ocean, and the $CO_2$ concentration drops on time scales of 2.5 to 6 thousand years in order to balance carbonate inputs and outputs. Broecker and Peng (1986, 1989) argue that this process is too slow to account for some of the ice-core data, and that in any event it would only depress the $CO_2$ concentration by about 40 ppmv. They claim that the remaining 40-ppmv drop needed to explain atmospheric $CO_2$ depressions of from 280 to 200 ppmv during transitions to glacial climates results from a shutoff in the flow of North Atlantic Deep Water (NADW), which has a low total alkalinity, to the Southern Ocean. This shutoff raises the total alkalinity of the surface ocean not only at high latitudes but also worldwide, thereby depressing the $CO_2$ concentration and cooling the planet. The reverse process occurs during transitions to interglacials when NADW "turns on," the $CO_2$ concentration rises, and the Earth warms. A temporary shutoff of NADW may have occurred during the so-called "Younger Dryas" from 10.8 thousand to 10 thousand years ago when fresh water from the Mississippi drainage basin may have been diverted to the North Atlantic, punctuating the gradual warming since the last glacial maximum by a sharp, but short-lived, global cooling. In general, however, these processes require that orbital insolation variations in the Northern Hemisphere be the main triggers of changes in the oceanic circulation and biology, perhaps involving high-latitude evaporation, surface salinity, and sea-ice extent.

Although the details are still unclear, research over the past few years thus suggests oceans are not simply passive reservoirs in the natural carbon cycle; rather, they can take an active role in changing the atmo-spheric $CO_2$ concentration when there are changes in ocean biology or circulation. These effects have yet to be included in representations of the anthropogenically driven global carbon cycle.

### 4.2.2.3. *Natural Cycling of Carbon Between the Atmosphere, Biosphere, and Oceans*

The main factors influencing the atmospheric $CO_2$ concentration on time scales of 10 to 1000 years are changes in the terrestrial biomass, in productivity of the marine biomass, in the circulation patterns of the oceans, and in the total amount of carbon in the atmosphere and oceans (e.g., see Sundquist and Broecker, 1985).

The oceans (primarily the deep ocean) currently contain about 50 times more carbon than the atmosphere and serve as the most important sink for carbon added to the atmosphere. The net rate of $CO_2$ uptake by the oceans is determined by three processes: (1) $CO_2$ exchange across the air-sea boundary, which is dependent on water temperature and wind speed; (2) the incorporation of surface-water carbon into compounds other than $CO_2$ (primarily carbonate and bicarbonate; only about 1% is present as dissolved $CO_2$); and (3) ocean mixing and circulation, which transport carbon from surface waters into deeper ocean layers where it can be sequestered from atmospheric exchange (Baes, 1982; Volk and Hoffert, 1985; Boyle, 1988).

Carbon is also sequestered in the biosphere. Estimates of the carbon content of various marine and terrestrial ecosystems and their net primary production (NPP) are given in Table 4.4. Unfortunately, such

**Table 4.4.** Estimated biomass (including soil carbon having relatively rapid turnover times) and net primary productivity (NPP) of global ecosystems compiled from various sources by Ajtay et al. (1979). NPP is the photosynthesis less respiration rate of autotrophic plants.

| Ecosystem type | Biomass (PgC) | NPP[a] (PgC/yr) | Turnover time[b] (yr) |
|---|---|---|---|
| Tropical rain forest | 765 | 16.8 | 46 |
| Tropical seasonal forest | 260 | 5.4 | 48 |
| Temperate evergreen forest | 175 | 2.9 | 60 |
| Temperate deciduous forest | 210 | 3.8 | 55 |
| Boreal forest | 240 | 4.3 | 56 |
| Woodland and shrubland | 50 | 2.7 | 19 |
| Savanna | 60 | 6.1 | 10 |
| Temperate grassland | 14 | 2.4 | 6 |
| Tundra and alpine | 5 | 0.5 | 10 |
| Desert and semidesert scrub | 13 | 0.7 | 18 |
| Extreme desert (rock, sand, ice) | 0.5 | 0.03 | 16 |
| Cultivated land | 14 | 4.1 | 3 |
| Swamp and marsh | 30 | 2.7 | 11 |
| Lake and stream | 0.05 | 0.4 | 0.1 |
| Total terrestrial | 1837 | 53 | 35 |
| Open ocean | 1.0 | 41.5 | 0.02 |
| Upwelling zones | 0.008 | 0.2 | 0.04 |
| Continental shelf | 0.27 | 9.6 | 0.03 |
| Algal beds and reefs | 1.2 | 1.6 | 0.75 |
| Estuaries (excluding marsh) | 1.4 | 2.1 | 0.67 |
| Total marine | 3.9 | 55 | 0.07 |
| Total terrestrial plus marine | 1841 | 108 | 17 |

[a] Net primary productivity is uniformly taken to be 45% of dry-matter production.

[b] Turnover time is assumed to equal biomass divided by NPP.

estimates are typically known only to within a factor of 2, owing to large regional variations and generally limited sampling. For example, these estimates, based on Ajtay et al. (1979), differ somewhat from those in Table 4.3, which were derived from different sources. Terrestrial organic carbon is regenerated to $CO_2$ by bacterial decomposers on turnover times estimated by dividing the biomass by its NPP. An NPP of 108 PgC/yr for the planet as a whole (Table 4.4) implies that the entire 740 PgC (as $CO_2$) in the atmosphere passes through the biosphere in less than 7 years. This short time scale contrasts sharply with the much longer time of 100 to 500 years for the atmosphere-ocean-biota system to reach a new equilibrium if sources or sinks of $CO_2$ are changed.

The terrestrial biosphere contains virtually all of the highly active biomass. The organic carbon of the land ecosystems is distributed between living plants (about 25 to 30%) and organic compounds in the soil. In general, the extent of the terrestrial biosphere is limited mainly by the availability of water, but also is affected by temperature, by light levels, and possibly by the $CO_2$ concentration. The estimates of terrestrial biomass are also complicated by changing land use (deforestation, regrowth, etc.). Table 4.4 shows that the turnover times of terrestrial carbon generally average a few decades, but are quite variable, being longer in forests than in grasslands.

Most soil carbon is contained in humus and peat; detritus contains only a few percent of total soil carbon. The largest amount of soil carbon is concentrated in the boreal belt (about 35% of the total storage); the tropical soil pool is in second place with 27.5%. Newly formed humus contributes about 2.5 PgC/yr, or 0.1% of the total organic carbon content in soils. The largest amount of humus is formed in soils in the tropical belt and the smallest in the polar zone (Fig. 4.4). However, of the total amount of newly formed humus, only about 40% (1 PgC/yr) is strongly bound with mineral particles in the soil (stable humus), whereas about 60% (1.5 PgC/yr) is labile (biologically active) humus that is subject to further microbiological reprocessing and oxidation.

Just as for ecosystems, the turnover times of organic carbon in different parts of the soil pool are unequal (see Fig. 4.3). Detritus has the shortest lifetime (about 2 years), whereas labile humus has a much longer lifetime (about 480 years). The stable part of the soil pool is characterized by the least mobility; there, the residence time of carbon is about 1350 years. The turnover time for the total soil humus is, on average, about 850 years. Comparisons between turnover times obtained for different soils and data based on radiocarbon dating of these soils show good agreement (Chichagova, 1985).

The marine biosphere is mainly nutrient-limited; thus, it depends critically on the ability of ocean currents and mixing to recycle nitrate and phosphate to the surface. In the surface photic zones, where light can penetrate, ocean plankton collectively photosynthesize carbon at rates comparable to the land biota. Indeed, current estimates of NPP by

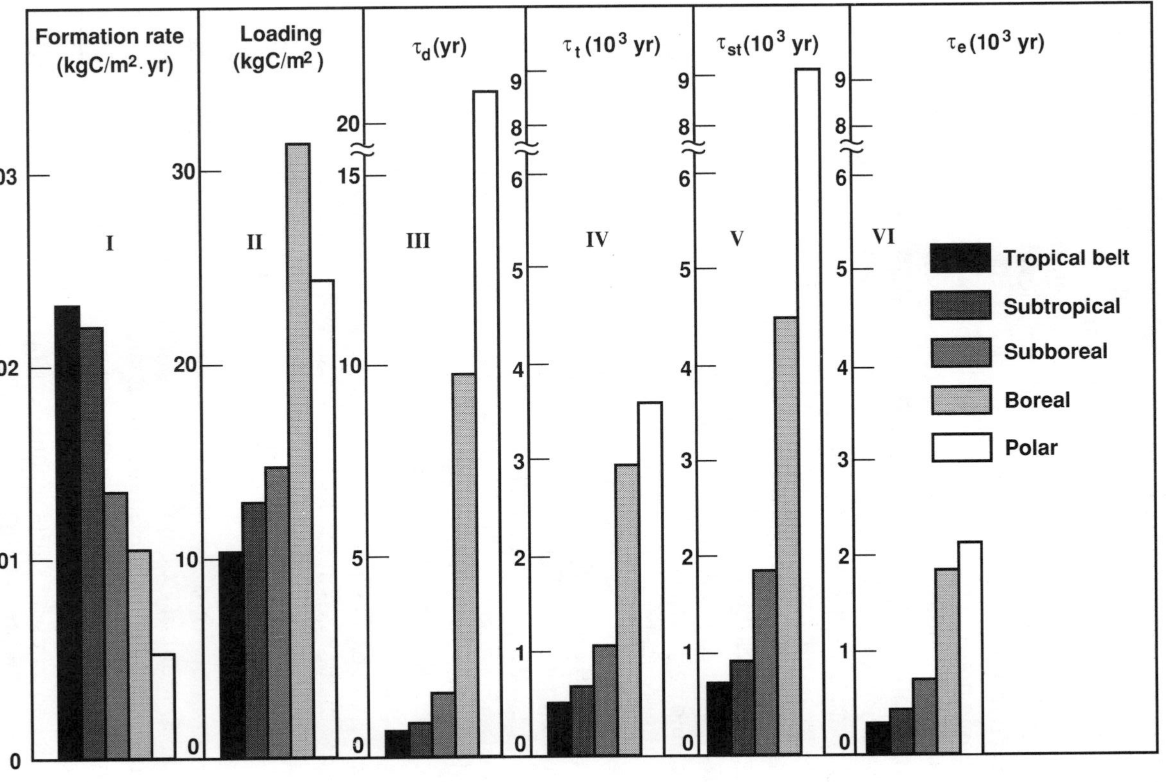


**Figure 4.4.** Estimated organic carbon content and the intensity of its transformation in soils of different climatic belts. (From Kobak, 1988.)

I. Formation rate of new organic carbon (humus and peat) in soils.

II. Organic carbon loading in soils.

III. Residences time of carbon in the detritus making up the upper layer of soils ($\tau_d$).

IV. Residence time of carbon in humus and peat ( $\tau_t$).

V. Residence time of carbon in the stable part of the soil humus ($\tau_{st}$) where humus is strongly bound with soil mineral particles.

VI. Residence time of carbon in the labile part of the humus ($\tau_\ell$).

phytoplankton are closer to 100 PgC/yr than to the 55 PgC/yr cited in Table 4.4 (Sundquist, 1985). These plants are eaten, and about 20% of their $C_{org}$ and nutrients is incorporated into organic detritus and fecal pellets that fall to the deeper ocean layers (Fig. 4.3). As the pellets fall through the water column, their organic carbon is oxidized (resulting in a minimum in dissolved oxygen at a depth of about 1 km), and their nutrients are reintroduced at depth in dissolved form. Because these nutrients are recycled by upwelling and ocean mixing, the biological productivity of the ocean can be affected by a change in ocean circulation.

The biological productivity of the surface ocean can also affect $CO_2$ levels in a somewhat surprising way. The present level of marine productivity creates an inorganic carbon sink in the surface layer, and dissolution of falling organic matter creates a source in deep water. This results in a profile of dissolved inorganic carbon that increases with depth. The concentration of atmospheric $CO_2$ depends on the dissolved organic carbon in the surface layer because the relative amounts of $CO_2$, carbonate, and bicarbonate are in near thermodynamic equilibrium as governed by the chemistry of the carbonate ion. If, for example, the productivity of plankton were to decrease, the result would be a more uniform profile of dissolved organic carbon and a higher $CO_2$ concentration in the atmosphere. This would result from the change in the profile of the inorganic carbon. Such indirect changes in atmospheric $CO_2$ through a change in ocean circulation (which affects nutrient supply) and meridional temperature gradient could produce atmospheric $CO_2$ changes mediated by the marine biosphere (Baes, 1982; Volk and Hoffert, 1985; Byutner, 1986; Verbitsky and Lapenis, 1987; Boyle, 1988).

There is also the prospect of changes in atmospheric $CO_2$ arising from geochemical changes at the depth at which $CaCO_3$ sediments are dissolved (the lysocline). On time scales of the deep ocean circulation (about 1000 years), the introduction of additional carbon into the ocean-atmosphere system will shift the equilibrium and cause $CaCO_3$ in existing sediments to dissolve. The resulting change in oceanic composition, in combination with changes in the levels of dissolved organic carbon, would affect the partitioning of carbon between the atmosphere and oceans in the long-term equilibrium (Plass, 1972; Byutner, 1986; Sundquist, 1986).

The implications of such changes could be profound. A significant increase in the amount of $CO_2$ introduced from fossil-fuel burning relative to what is currently in the atmosphere could shift the equilibrium of the system such that elevated $CO_2$ levels could persist for periods as long as 100 thousand years (Kasting and Walker, 1989). Indeed, the long-term effects of such geochemical shifts would depend primarily on the amount of $CO_2$ input from fossil-fuel emissions, and not on the rate of injection. In addition, just as Broecker and Peng (1986, 1989) argue that the $CO_2$ concentration changes when the NADW carbon pump switches on and off during ice-age transitions, it is also possible that fossil-fuel-driven changes in climate and ocean circulation could lead to unexpected $CO_2$ changes, even to "unpleasant surprises in the greenhouse" (Broecker, 1987). Thus, any assessments of the climatic impact of scenarios of fossil-fuel emissions should also include consideration of the long-term geochemical perspective, because the environment of our remote descendants could be impacted by the global energy policies followed over the next 50 years.

### 4.2.3. The Effects of Human Activities on the Carbon Cycle

Ice core measurements indicate that the atmospheric $CO_2$ concentration has risen from about 278 ppmv to about 350 ppmv since the early 1700s. Several analytical methods have been used to indicate that this change has been due to human activities, the primary contributing factors being the burning of fossil fuel and changes in land use (e.g., deforestation, agriculture). Historical records indicate that fossil-fuel burning has led to emission of about 170 PgC from 1700 to 1985 (Marland et al., 1989). Current fossil-fuel emission rates are about 5.6 ±0.5 PgC/yr, implying emissions of almost 200 PgC through 1989. Figure 4.5 shows the history of fossil-fuel emissions as compiled by Keeling (1973) and extended by Marland et al. (1989). Note that the near exponential increase in emissions has been interrupted only by the two World Wars and by the sharp oil-price fluctuations of the 1970s and early 1980s.

The contribution of land-use changes is much less certain. Isotopic and some inventory-based estimates of emissions from the massive deforestation of Northern Hemisphere continents since 1700 indicated emission of as much as 150 PgC or more during this period due to the

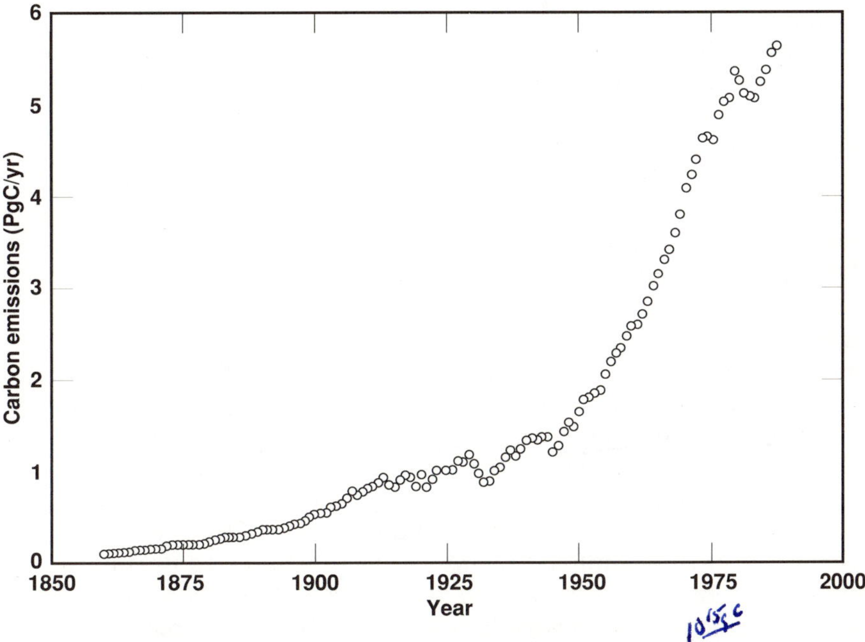


**Figure 4.5.** Reconstruction of fossil-fuel emissions since 1850 in PgC/yr. Estimates for 1850 to 1949 are from Keeling (1973) and for 1950 to 1986 from Marland et al. (1989)

"pioneer effect" (Houghton et al., 1983; Houghton, 1986). Later studies indicate that deforestation and land use changes may have contributed 90 to 120 PgC (Houghton, 1989). Model studies are most consistent with biospheric releases this low (Siegenthaler and Oeschger, 1987) or even lower; studies described in Budyko and Izrael (1987) suggest only 60 to 90 PgC from 1860 to 1963, of which deforestation and land use change contribute about 75% and soil cultivation about 25%. These results thus indicate a carbon transfer to the atmosphere equivalent to about 10 to 20% of the land phytomass, but only about 1% of the soil pool.

Since the pre-industrial period, the atmospheric amount of $CO_2$ has risen from about 590 PgC (278 ppmv) to about 740 PgC (350 ppmv), an increase of about 153 PgC. With fossil-fuel and biospheric contributions being about 200 PgC and 90 PgC, respectively, anthropogenic emissions have been roughly half of the pre-industrial atmospheric burden. The increase in the atmospheric burden from pre-industrial levels, however, has been only about 150 PgC (i.e., 740 to 590 PgC), or about 25% of the pre-industrial burden. That is, the fraction of anthropogenic emissions remaining in the atmosphere (the airborne fraction) has been only 150 PgC out of 290 PgC, or about half. The balance of the anthropogenic emissions has, presumably, been absorbed by the oceans or added to the land biomass or soil pool in undisturbed regions. In general, the instantaneous airborne fraction depends on the injection rate versus the removal rate to other reservoirs and will decrease slowly after emissions have ceased; thus, an airborne fraction of about 50% cannot be used to make long-term projections, especially if there are significant changes in future emission rates.

Because of the potentially important role of the biosphere in estimating the impact of human activities on the $CO_2$ concentration, it is essential to gain a better understanding of current exchanges among the atmosphere, ocean, and biosphere. Three different approaches have been invoked thus far to estimate whether the land biomass is now changing; these results are needed to determine whether changes in land plants are resulting in a net source or sink of carbon to the atmosphere:

(1) Global deforestation less regrowth.

Analyses by Kobak (1988) of land-use data in many countries (changes in forest area and arable land, transformation of forest and arable land areas, changes in the carbon content of soils in cultivated areas, etc.) suggest that the ecosystems of the temperate and boreal zones have been a net sink of atmospheric carbon to the extent of 0.6 to 0.7 PgC/yr over the last 30 years. Analyses based on land-clearing data typically emphasize deforestation of rain forests in the tropics of Latin America, Asia, and Africa (WRI, 1990). The present destruction of tropical rainforests at a rate of 2 to 3 million hectares per year would lead not only to a biospheric carbon source, but also to disruption and extinction of irreplaceable biological species as the standing carbon pools that are their natural habitats (the tropical forests) are depleted. Brazil typifies a major region in which rain forests are subjected to the land clearing associated with "slash and burn agriculture" and urban development. According to Houghton et al. (1987), just one of Brazil's nine Amazon provinces accounts for more carbon exchange than all temperate and boreal regions combined. There are acknowledged difficulties with using this method for sampling and accounting for regrowth, however. Trabalka (1985) summarizes various data sets and estimates that changing patterns of land use lead to $CO_2$ emissions of 0.6 to 2.6 PgC/yr. Another recent analysis of global deforestation data indicates that the terrestrial biosphere emits 1.0 to 2.0 PgC/yr (Houghton, 1986, 1989; Houghton et al., 1987).

(2) Variation of the $^{13}C/^{12}C$ ratio in atmospheric $CO_2$.

Owing to preferential fractionation of the heavy carbon isotope by land plants, $^{13}C$ in atmospheric $CO_2$ is sensitive to $CO_2$ exchanges with the land biosphere, although only slightly affected by oceanic exchange. Keeling et al. (1989) find that carbon isotope data indicate a terrestrial biospheric source of order 1.0 PgC/yr and hypothesize that this comes from a combination of accelerating tropical deforestation and warming of boreal soils.

(3) Latitudinal gradient of atmospheric $CO_2$.

The partial pressure of $CO_2$ in the oceans is much more nonuniform with latitude than the $CO_2$ in the atmosphere because of equatorial upwelling, temperature, and marine biological effects. This leads to air/sea gas exchanges such that oceans are a source of $CO_2$ to the atmosphere in the tropics and a sink at higher latitudes, with the small residual representing net $CO_2$ uptake by the sea. Apart from ocean sources and sinks, a pole-to-pole variation in atmospheric $CO_2$ concentration is generated by the fossil-fuel combustion sources that predominate in the industrialized Northern Hemisphere (Hoffert, 1974). In a recent study, Tans et al. (1990) argue that the observed interhemispheric $CO_2$ gradient, some 2.5 ppm, is too small to be explained by an interhemispheric mixing time of one year and the present Northern Hemisphere fossil-fuel source of more than 5 PgC/yr. Tans et al. infer therefrom that the temperate forests of the Northern Hemisphere constitute a carbon sink as large as 2.0 to 3.4 PgC/yr that must be partially balancing the fossil-fuel source. This conclusion is opposite to that of the prior analyses, which find that the land biosphere is a carbon source.

Note that the net global biospheric sources or sinks of ±1 to 3 PgC/yr proposed by these researchers are only a few percent of the global NPP, and much less than the uncertainty in the global primary productivity itself. Although several ingenious methods have been devised to estimate the role of the land biosphere, even the sign of the net biospheric carbon exchange with the atmosphere remains controversial. Deforestation is certainly a carbon source; however, at least one analysis based on matching observed latitudinal concentration gradients to an interhemispheric atmospheric $CO_2$ transport model suggests that the terrestrial biosphere is a carbon sink. This issue must be resolved for confidence to be maintained in the predictions of carbon-cycle models, particularly those used in complex scenarios of future energy use.

## 4.3. TRACE GREENHOUSE CONSTITUENTS

This section describes the budgets, distributions, and trends of the non-$CO_2$ greenhouse gases, with emphasis on those whose concentration trends suggest a potentially significant effect on the future climate. In addition, aerosols and other atmospheric constituents that can also influence climate are discussed.

### 4.3.1. Methane

Although its atmospheric abundance is less than 0.5% that of $CO_2$, $CH_4$ is an important greenhouse gas. The globally averaged atmospheric concentration of $CH_4$ is now about 1700 ppbv[4] (see Fig. 4.6), and it is increasing by about 16 ppbv/yr (or about 1%/yr) according to measurements from Khalil and Rasmussen (1987), Blake and Rowland (1988), Dianov-Klokov and Yurganov (1989), and Dianov-Klokov et al. (1989). Measurements indicate that the $CH_4$ concentration is highest at latitudes polewards of 30°N. Major uncertainties exist regarding the cause (or causes) of the increase in the $CH_4$ concentration and its likely future growth rate.

Estimates of the sources and sinks of atmospheric $CH_4$ are shown in Table 4.5. Methane is produced naturally via anaerobic decomposition in biological systems. Approximately 60% of the emissions of $CH_4$ into the atmosphere, however, are related to human actions (Andronova and Karol, 1986; Cicerone and Oremland, 1988; Wuebbles and Edmonds, 1988; Khalil and Rasmussen, 1990). Major anthropogenic sources of $CH_4$

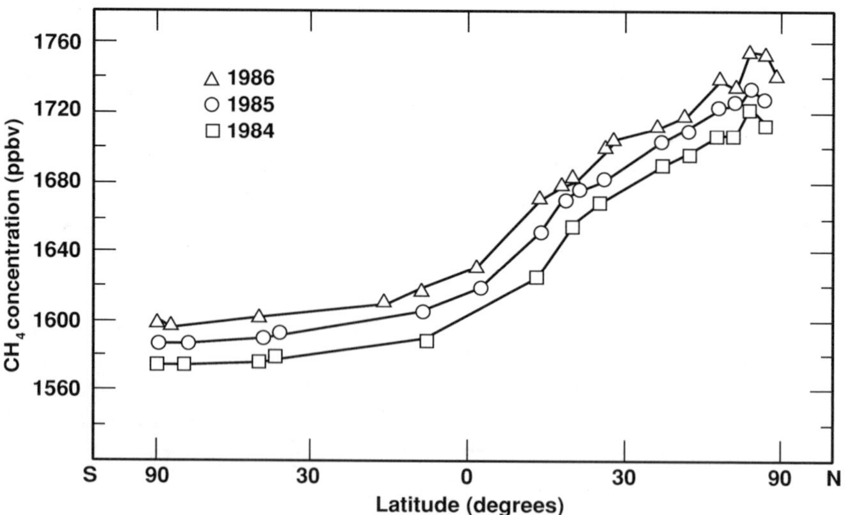


**Figure 4.6.** Observed $CH_4$ concentrations as a function of latitude from 1984 to 1986 based on data from the NOAA Global Monitoring for Climate Change (GMCC) program (Peterson, 1988).

[4] ppbv = parts per billion by volume.

**Table 4.5.** Sources and sinks of atmospheric methane in teragrams of methane per year ($TgCH_4/yr$) based on recently updated versions by Cicerone and Oremland (1988) and Wuebbles and Edmonds (1988).

| Sources and sinks of atmospheric methane | Methane flux[a] ($TgCH_4/yr$) | |
|---|---|---|
| | Best estimate | Range |
| Sources | | |
| *Natural:* | | |
| Enteric fermentation (wild animals) | 4 | 1 to 7 |
| Wetland (swamps, etc.) | 110 | 60 to 160 |
| Lakes | 4 | 2 to 6 |
| Tundra | 3 | 2 to 5 |
| Oceans | 10 | 0 to 20 |
| Termites (and other insects) | 25 | 5 to 45 |
| Methane hydrates | 5 | ? |
| Other | 40 | 0 to 80 |
| *Manmade:* | | |
| Enteric fermentation (domesticated animals) | 77 | 40 to 110 |
| Rice paddies | 70 | 40 to 100 |
| Biomass burning | 70 | 30 to 110 |
| Natural gas and mining losses | 50 | 25 to 75 |
| Solid waste | 30 | 0 to 60 |
| Sinks | | |
| Reaction with tropospheric OH | 350 | 250 to 450 |
| Transport to and reaction with OH, Cl or O in stratosphere | 50 | 30 to 70 |
| Microorganism uptake by soils | 32 | 16 to 48 |
| Accumulation | | |
| Atmospheric increase | 45 | 40 to 50 |

[a] Best estimate and range.

include domesticated ruminant animals (cows, etc.), rice paddies, biomass burning, natural gas and mining losses, and solid-waste emissions. Approximately 20% of the total emissions are estimated to be related to energy production (Wuebbles and Edmonds, 1988). Analysis of $^{14}C$ in $CH_4$ indicates that 21 ± 3% of atmospheric $CH_4$ had been derived from fossil carbon at the end of 1987 (Wahlen et al., 1989).

Destruction of atmospheric $CH_4$ occurs primarily through reaction with hydroxyl (OH). Reduction in the amount of global atmospheric OH during recent decades may provide an explanation for as much as 20 to 50% of the $CH_4$ increase, but this is highly uncertain given the paucity of measurements of tropospheric OH concentrations. Methane chemistry can also affect ozone concentrations in the troposphere and stratosphere. Further, destruction of additional $CH_4$ in the stratosphere increases the stratospheric water vapor concentration. Atmospheric measurements indicate that the global atmospheric lifetime of $CH_4$ is about 10 years.

Ice-core data going back 160 thousand years indicate that the concentration of $CH_4$ in the pre-industrial atmosphere was less than half the present concentration (Rasmussen and Khalil, 1984; Pearman et al., 1986; Khalil and Rasmussen, 1987; Raynaud et al., 1988). The $CH_4$ concentration was about 700 ppbv until the beginning of the nineteenth century (see Fig. 4.7). As indicated in Fig. 2.4, concentrations during glacial periods were even smaller, as low as 350 ppbv (Raynaud et al., 1988).

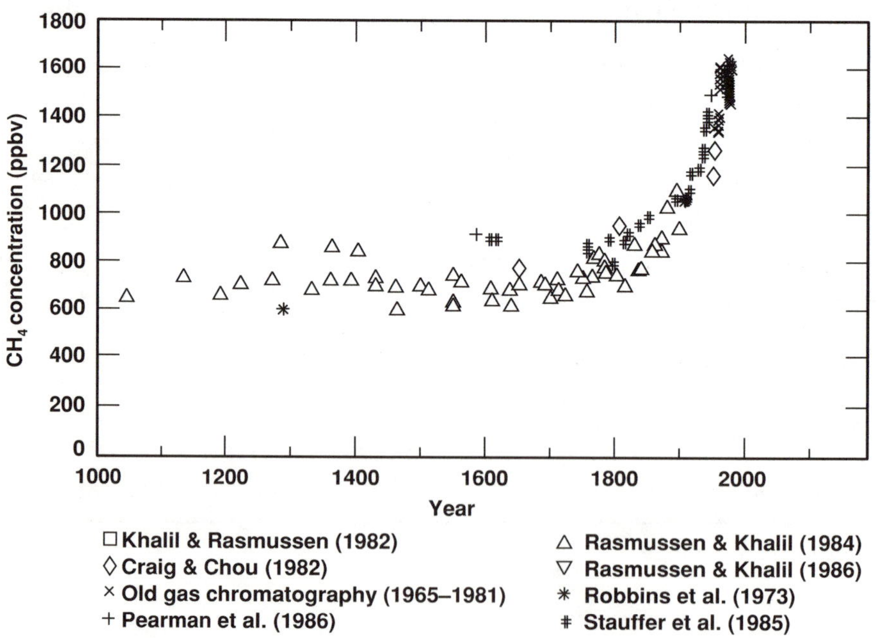


**Figure 4.7.** Ice core and atmospheric measurements of $CH_4$ concentration for the last 1000 years. (Based on Khalil and Rasmussen, 1987.)

#### 4.3.2. Halocarbons

Halogenated hydrocarbons, or halocarbons, have received recent attention because of their increasing concentrations and potentially significant effects on stratospheric ozone. The most important are the chlorofluorocarbons (CFCs), in particular $CFCl_3$ (referred to as CFC-11) and $CF_2Cl_2$ (CFC-12).

Chlorofluorocarbons CFC-11 and CFC-12 have the highest concentrations of the man-made chlorocarbons, 0.26 and 0.44 ppbv in 1989, respectively. The surface air concentrations of these two gases are currently increasing at a rate of over 4%/yr (see Fig. 4.8). CFC-11 is used primarily as a blowing agent for plastic foams and (other than in the US) as a propellant in aerosol cans; CFC-12 is used primarily as a refrigerant and as a propellant in aerosol cans. Although natural sources are not thought to be important in the global budgets of these compounds, natural sources of CFCs from volcanic emissions have been suggested (Isidorov and Ioffe, 1986), although abyssal ocean values and oceanic general circulation models (OGCMs) suggest that such sources must represent much less than 1% of current man-made emissions.

Other important chlorocarbons include CFC-113 ($CF_2ClCFCl_2$), CFC-22 ($CHF_2Cl$), and methyl chloroform ($CH_3CCl_3$). Concentrations of CFC-113 are increasing by about 10%/yr; present surface concentrations are about 0.03 ppbv. Abundances of CFC-22 and $CH_3CCl_3$ are 0.09 and 0.14 ppbv; concentrations are increasing at rates of about 7%/yr and

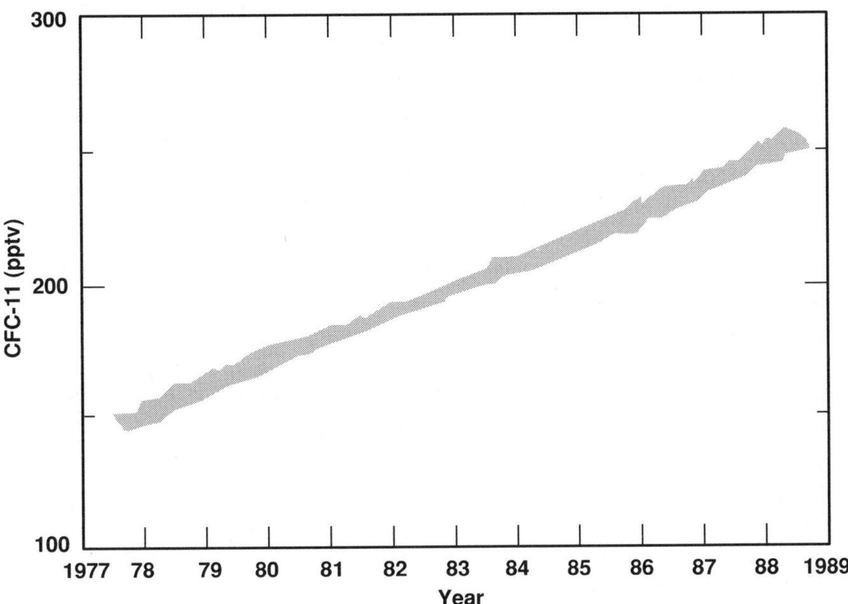


**Figure 4.8.** Measurements of the CFC-11 concentration at Mauna Loa Observatory since mid-1977, based on NOAA GMCC data. Shading shows range of measurements.

4.5%/yr, respectively. Chlorofluorocarbons CFC-113 and $CH_3CC1_3$ are used primarily as solvents; CFC-22 is used in air conditioning and refrigeration.

All of these man-made chlorocarbons have a relatively long atmospheric lifetime (i.e., a relatively slow removal rate). Methyl chloroform has the shortest chemical lifetime, about 6 to 7 years, while CFC-12 has the longest lifetime, about 120 years. The long lifetimes contribute to the rapidly increasing concentrations of these gases. The fully halogenated chlorocarbons, particularly CFC-11 and CFC-12, are of primary concern in ozone depletion studies because these CFCs are destroyed when all of their chlorine is released in the stratosphere, primarily by photolysis.

With the signing of the Montreal Protocol in 1987 by most CFC-producing nations,[5] there should be a significant reduction in CFC-11 and CFC-12 production in coming years. The Protocol currently calls for a 50% reduction in production of these chemicals by the major producers by 1998. At the treaty review meeting in June 1990, the participants agreed to the total phaseout of the most destructive chlorofluorocarbons and halocarbons because of increased recognition of the potential for significant depletion of stratospheric ozone.

Because many of the applications of these chemicals provide important societal needs, there has been an intense effort to find replacements, particularly for refrigeration, air conditioning, and foam-blowing. The suggested replacement compounds, such as HCFC-123, HCFC-141b, and HFC-134a, are generally halogenated hydrocarbons, some containing chlorine (the HCFCs) and others not (the HFCs). All of the suggested replacements have shorter lifetimes, generally less than 20 years, and their concentrations will, as a result, increase more slowly in the atmosphere than the compounds they are replacing. Nonetheless, most of these compounds are greenhouse gases, and could affect climate if concentrations grow to be large enough (WMO, 1989).

Recent research studies (AFEAS, 1989; WMO, 1989) have defined halocarbon Greenhouse Warming Potentials (GWPs) for many of the CFC replacement compounds as the relative effect on climate (i.e., radiative forcing) of a unit mass emission of a halocarbon relative to a unit mass emission of CFC-11. All of the HCFCs and HFCs have smaller GWPs than the CFC compounds they would replace. For example, HCFC-123, HCFC-141b, and HFC-134a have GWPs of 0.02, 0.09, and 0.25, respectively. The GWPs for these halocarbons are a strong function of their atmospheric lifetimes and their infrared absorption features.

### 4.3.3. Nitrous Oxide

The atmospheric $N_2O$ concentration is increasing at a rate of about 0.3%/yr (Khalil and Rasmussen, 1988b; Watson et al., 1988). The current tropospheric concentration is about 307 ppbv. Ice core data indicate that

---

[5] Both the US and USSR are signatories to the Montreal Protocol.

the concentration of $N_2O$ from 3000 to 150 years before present was about 285 ppbv (Pearman et al., 1986; Khalil and Rasmussen, 1988b).

The two most likely causes of the observed increase are: (1) nitrification and denitrification of nitrogen from industrially produced fertilizers, and (2) high-temperature combustion, such as occurs in coal-burning power plants. Recent evidence is contradictory as to whether combustion is the major cause or totally unimportant (Hao et al., 1987; Muzio and Kramlich, 1988; Linak et al., 1989), although the most recent papers suggest that combustion sources are much smaller than previously estimated. Khalil and Rasmussen (1990) have used ice core data to constrain the human-related contribution to 10 to 30% of total emissions; Wuebbles and Edmonds (1988) estimate that approximately 24% of the $N_2O$ sources are human-related if the combustion source is zero. Table 4.6

**Table 4.6.** Sources and sinks of atmospheric $N_2O$ in teragrams of nitrogen per year (TgN/yr) based on a recently updated version by Wuebbles and Edmonds (1988).

| Sources and sinks of atmospheric $N_2O$ | Nitrogen flux (TgN/yr) | |
|---|---|---|
| | Best estimate | Range |
| Sources | | |
| *Natural:* | | |
| Oceans and estuaries | 2.0 | 1 to 3 |
| Natural soils | 6.5 | 3 to 10 |
| Aquifers | 0.8 | 0.5 to 1.1 |
| *Manmade:* | | |
| Fossil-fuel combustion | 0.0 | 0 to 3.2 |
| Biomass burning | 0.7 | 0.5 to 0.9 |
| Fertilized soils | 0.8 | 0.6 to 1.0 |
| Cultivated natural soils | 1.5 | 1 to 2 |
| Sinks | | |
| Photolysis and reaction with $O(^1D)$[a] in stratosphere | 10.5 | 7.5 to 13.5 |
| Accumulation | | |
| Atmospheric increase | 3.5 | 3 to 4 |

[a] Oxygen atom in an excited state.

provides a budget indicating estimated sources and sinks of atmospheric $N_2O$. If the combustion sources of $N_2O$ are as small as in current estimates, additional sources may need to be found to explain the present increasing trend.

The primary sink for $N_2O$ is photolysis in the stratosphere; it is not chemically reactive in the troposphere. Recent estimates of the $N_2O$ lifetime range from 120 to 150 years (WMO, 1989). The reaction of $N_2O$ with excited oxygen atoms is the primary source of stratospheric nitrogen oxides, which catalytically react with stratospheric ozone. Thus, an increasing $N_2O$ concentration could lead to a net destruction of stratospheric ozone.

### 4.3.4. Ozone

Ozone is a natural greenhouse gas. About 90% of atmospheric ozone is present in the stratosphere (which extends from approximately 10 to 50 km above the Earth's surface), with most of the rest in the troposphere (approximately the lowest 10 km above the surface). Unlike the other trace constituents considered, the vertical distribution of ozone is of major importance in determining its impact on climate.

Ozone is produced naturally in the stratosphere as a consequence of chemical reactions initiated by the photolysis of oxygen molecules. Destruction results primarily from catalytic reactions with chlorine oxides, nitrogen oxides, and hydrogen oxides, especially in the stratosphere. Although concentrations of these compounds are much smaller for ozone, the catalytic reactions allow a single molecule to destroy a thousand or more molecules of ozone. Tropospheric ozone comes from downward transport of stratospheric ozone and from local photochemical production, and is destroyed both by surface reaction and chemical reaction.

Ozone in the global troposphere and stratosphere is measured using satellite and ground-based instruments. Both the US and USSR have ground-based stations participating in the global ozone measurement network. Data from these measurements are analyzed using varying approaches to determine trends (e.g., Karol et al., 1987, 1990b; Reinsel et al., 1987, 1988; Watson et al., 1988).

Measured changes in ozone from ground-based measurements and satellite data indicate that ozone concentrations in the upper stratosphere are decreasing, and that overall total ozone amounts at northern midlatitudes have decreased since 1969 (Watson et al., 1988; DeLuisi et al., 1989; Fioletov, 1989; WMO, 1989). The observed decrease in the ozone concentration in the upper stratosphere is comparable to that projected over this time period as a consequence of the effects of CFCs, other trace gases, and solar flux variability (i.e., the 11-year solar cycle) (Wuebbles and Kinnison, 1988; DeLuisi et al., 1989). From 1969 to 1986, the observed decrease in the annually averaged total-column ozone amount was 1.7 to 3.0% at latitudes between 30° and 64°N, with the largest decrease occurring in the winter months at the higher latitudes.

Detailed analyses of mean monthly total ozone trends for 1975 to 1986 (Karol et al., 1990a, 1990b) show positive and negative trends, respectively, in the 1975 to 1980 and 1981 to 1986 subperiods of the increase and decrease in the flux of ultraviolet radiation through the 11-year solar cycle. Maximum changes of up to 1%/yr occurred during the summer months for midlatitude and tropical air masses. Tropospheric concentrations of ozone appear to have increased consistently with human-related emissions of nitrogen oxides and hydrocarbons (Logan, 1985; Tiao et al., 1986; WMO, 1989), although uncertainties and local variations are too large to establish a global trend.

In the Southern Hemisphere, a large, sudden decrease in the abundance of spring-time Antarctic ozone has been noted to occur over the

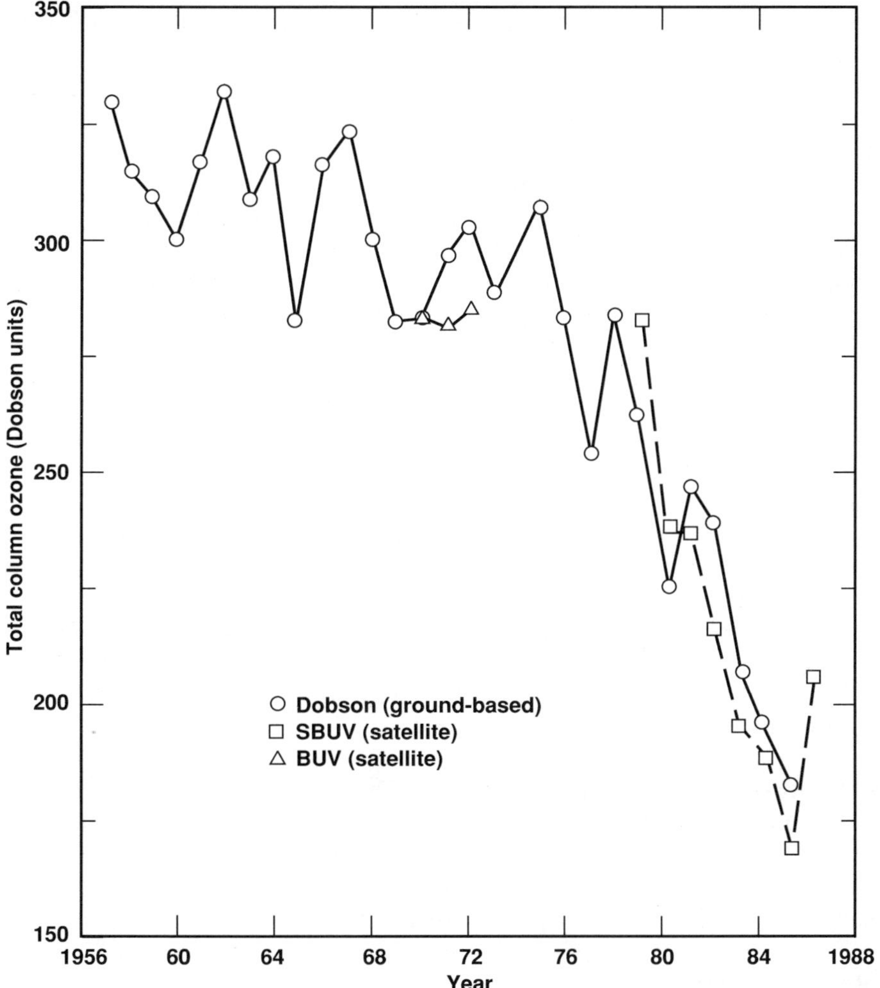


**Figure 4.9.** Observed decrease in spring-time total column ozone amount (in Dobson units) over the Halley Bay Station in Antarctica (Solomon, 1988; Watson et al., 1988) based on results from a ground-based Dobson spectrophotometer and satellite-based solar backscattered ultraviolet (SBUV) and backscattered ultraviolet (BUV) radiation.

last decade (see Fig. 4.9). This sudden decrease has been termed the Antarctic "ozone hole." Decreases of more than 50% in the total column ozone amount as compared to historical values have been observed. Measurements made in 1987 showed that more than 95% of the ozone in the region from 15 to 20 km had disappeared during the September–October period (WMO, 1989). Observations also indicate that the unique meteorology during the winter and spring over Antarctica sets up special conditions that force an isolated air mass (the polar vortex). Polar stratospheric clouds form if the temperatures are cold enough within this vortex region, as is often the case near the Antarctic tropopause. Once the Sun appears in early spring, chemical reactions on the cloud particles allow chlorine to be in a very reactive state with respect to ozone. The weight of the evidence strongly indicates that man-made chlorine species (Cl produced from the dissociation of CFC-11, CFC-12, etc.) are primarily responsible for the observed ozone decrease within the vortex.

The magnitude of the Antarctic ozone decrease was much smaller in 1988 than in 1987. Garcia and Solomon (1987) suggest such variations are expected as a result of the dynamical effects of the quasi-biennial oscillation. Other authors indicate that interannual variability can result from changes in planetary-wave activity (Jadin and Lysenko, 1988). Satellite data from September to November 1989 show that the Antarctic ozone hole was deeper than in 1988 and similar in magnitude to that in 1987.

Because temperatures are not generally as low, and, because the vortex in the Northern Hemisphere tends to break down earlier, a springtime ozone decrease in the Arctic has not yet been measured. Measurements made in the winter of 1988/89 by US and Soviet researchers suggest such a decrease could be possible in the future; high ClO concentrations, necessary for major $O_3$ destruction, were found in the vortex (WMO, 1989). Although highly speculative, it is possible that the greenhouse effect from $CO_2$ and other gases, which causes a cooling of stratospheric temperatures, could further intensify the Antarctic "ozone hole" and lead to a hole occurring in the northern polar region. Such cooling could also lead to more favorable conditions for the occurrence of polar stratospheric clouds.

### 4.3.5. Stratospheric Water Vapor

The increase of water vapor concentrations in the troposphere is expected to be an important positive feedback mechanism in determining future climate change. Tropospheric water vapor concentrations will increase with increasing tropospheric temperatures. The effect of an increased amount of tropospheric water on stratospheric concentrations is unknown because of uncertainties in the mechanism determining the transport of tropospheric $H_2O$ to the stratosphere. However, changes in the concentration of stratospheric water vapor can also affect climate. Such changes could occur as a result of chemical interactions in the stratosphere. Because of a poorly understood freeze-drying process at the tropical tropopause, very little tropospheric water vapor normally penetrates into the stratosphere. The lower stratosphere is very dry (3 to 4 ppmv) compared to the lower troposphere (where the concentrations are about a thousand times higher). Concentrations of water vapor increase to about 6 ppmv in the upper stratosphere because $H_2O$ is produced by the local oxidation of $CH_4$. Thus, stratospheric water vapor concentrations should increase as $CH_4$ increases. Because water vapor is one of the primary absorbers of infrared radiation, along with $CO_2$ and $O_3$, an increase in water vapor would further add to the greenhouse effect of the increasing $CH_4$ (Wuebbles et al., 1989). Changes in the mechanisms determining the concentration of water vapor in the lower stratosphere could also further affect climate.

### 4.3.6. Other Greenhouse Gases

The hydroxyl radical (OH) is not itself a greenhouse gas, but it is extremely important as a chemical scavenger of many trace gases in the troposphere. Hydroxyl is the primary tropospheric scavenger of $CH_4$, CO, $CH_3CCl_3$, $CH_3Cl$, $CH_3Br$, $H_2S$, $SO_2$, DMS (dimethylsulfide), and other hydrocarbons and hydrogen-containing halocarbons. The concentration of OH affects the atmospheric lifetime of these gases, and therefore contributes to their impacts on ozone and climate. Reactions with OH in the troposphere limit the amount of $CH_4$ and halocarbons reaching the stratosphere, where these species can cause changes in the ozone distribution. In the troposphere, OH and other related hydrogen oxides play a central role in the production of ozone. Hydroxyl is formed in the troposphere through the photolysis of ozone and subsequent reaction of the excited oxygen atom with water vapor. Its primary removal is through reactions with $CH_4$ and CO. The global distribution of OH has not been well measured (Prinn et al., 1987; WMO, 1989).

Although carbon monoxide (CO) is not a greenhouse gas, it is important because of its reactivity with OH. Peak CO concentrations of about 200 ppbv occur at high northern latitudes, with minimum concentrations of about 50 ppbv found throughout the Southern Hemisphere. Concentrations are increasing by about 1%/yr globally, although there is little evidence of an increase occurring in the Southern Hemisphere (Cicerone, 1988; Khalil and Rasmussen, 1988a; Dianov-Klokov et al., 1989; Dianov-Klokov and Yurganov, 1989). Carbon monoxide has a relatively short lifetime on the order of a few months. Nearly half its production comes from human-related sources (40 ±20%). As such, its concentrations over land are often affected by local sources, leading to difficulty in establishing long-term trends. Nonetheless, if the CO concentration continues to grow, the resulting decrease in OH concentrations could increase the lifetime (and produce larger concentrations) of important greenhouse gases such as $CH_4$. An estimate of the budget for the sources and sinks of CO is given in Table 4.7; the imbalance in sources and sinks with the observed accumulation rate is an indication of substantial uncertainty in these terms.

Emissions of nitrogen oxides from combustion sources have probably increased over recent decades, resulting in increased tropospheric concentrations over continents, especially in the boundary layer and in the flight corridors used by commercial aircraft. Nitrogen oxide emissions in the troposphere can influence local ozone concentrations and may play an important role in the apparent increase in the concentration of ozone in the troposphere. Nitrogen dioxide is an important absorber of visible solar radiation, and could affect climate directly if concentrations become large enough.

Several non-methane hydrocarbons (NMHC) are greenhouse gases, but it is unlikely, because of their short atmospheric lifetimes, that their concentrations will become large enough to have a major direct effect on

**Table 4.7.** Sources and sinks of atmospheric CO in teragrams of carbon per year (TgC/yr) based on a recently updated version by Wuebbles and Edmonds (1988).

| Sources and sinks of atmospheric CO | CO fluxes (TgC/yr) | |
|---|---|---|
| | Best estimate | Range |
| Sources | | |
| *Natural:* | | |
| Plant emissions | 55 | 20 to 90 |
| Oxidation of natural hydrocarbons | 250 | 50 to 500 |
| Forest wildfires | 10 | 5 to 20 |
| Oceans | 20 | 10 to 200 |
| Methane oxidation | 260 | 75 to 450 |
| *Manmade:* | | |
| Energy use | 200 | 150 to 250 |
| Agriculture | 110 | 40 to 170 |
| Forest clearing | 160 | 80 to 350 |
| Oxidation of anthropogenic hydrocarbons | 40 | 0 to 80 |
| Sinks | | |
| Soil update | 110 | 80 to 140 |
| Reaction with OH | 820 | 500 to 1100 |
| Accumulation | | |
| Atmospheric increase | ~2 | |

climate, at least within the next several decades. However, NMHC emissions do affect tropospheric chemistry, and could influence future amounts of $O_3$ and OH.

### 4.3.7. Aerosols

An increase in atmospheric aerosols (particulate matter) could change climate, although in a manner different from $CO_2$ and the other greenhouse gases. High-altitude aerosols, such as those injected into the stratosphere by volcanic eruptions, tend to reduce solar radiation reaching the lower atmosphere, causing a net climatic cooling. An increase in stratospheric sulfuric aerosols from increased carbonyl sulfide (COS) emissions would have a similar effect, but with longer-term implications (Wang et al., 1986).

The climatic effects of aerosols injected into the troposphere are more difficult to quantify because of their strong regional variation in concentration and the wide range of their radiative effects. Such regionally dependent aerosol types as soot, sulfate, desert soil, Arctic haze, and maritime aerosols may contribute to regional effects on climate. However, trends in the distributions of such aerosols are poorly known.

Aerosols in the troposphere result from a variety of sources. Natural sources over continents include combustion (e.g., forest fires), chemical reactions of natural trace gases, soil and dust particles raised from the Earth's surface, and pollen plus volatile hydrocarbons emitted by plants (Prospero et al., 1983). Anthropogenic sources include $SO_2$ and elemental carbon emissions. Over the oceans, aerosol sources include sea salt and oceanic emission of DMS (Bates et al., 1987). Conversion of DMS to form aerosols is thought to provide the major production of cloud condensation nuclei (CCN) over the ocean; changes in DMS emissions could provide an important feedback on climate through effects on cloudiness and albedo (Charlson et al., 1987; Penner, 1989). There are many existing uncertainties associated with the magnitude and even the sign of potential aerosol effects on future climate; however, many of the studies suggest that the overall effects of anthropogenic aerosols injected into the troposphere could produce a net cooling of the global climate. This may be particularly important over land areas, where the cooling effect of aerosols produced by sulfur emissions may already be affecting the observed rate of temperature change.

## 4.4. PROJECTIONS OF FUTURE CONCENTRATIONS

The global chemical system is so complex that simple extrapolation of trends is not always reliable and, as a consequence, numerical models of this system must be constructed. Such models must generally incorporate the sources, transport, transformation, and removal processes affecting each species, including their dependence on climate. Although many uncertainties remain, such models provide the only tool available to improve estimates based on extrapolation.

### 4.4.1. Carbon Dioxide

Observational studies of the phenomena controlling the atmospheric $CO_2$ concentration are often impractical because of the time scales involved (decades to centuries), the vast extent and spatial heterogeneity of Earth's systems and the small perturbations to natural levels that fossil-fuel releases cause in very large reservoirs such as the oceans. Furthermore, many important variables cannot be measured directly. Mathematical models are an important means of contending with these limitations. They also provide a means of synthesizing the diverse data assembled to address the $CO_2$ issue and of interpreting these data in terms of changes in atmospheric $CO_2$ levels.

The simplest "model" for projecting future $CO_2$ concentrations is the assumption that the airborne fraction will remain constant, although this is clearly not the case for scenarios where the growth in fossil-fuel emission rates departs from historical exponential trends. More comprehensive models of the carbon cycle have been developed to explicitly represent the chemical and transport processes. Because the most likely carbon sink is the ocean, a number of carbon-cycle models have been developed that express the time-varying concentration of $CO_2$ in terms of oceanic transport process. Models of the oceanic component of the carbon cycle must represent (1) gas exchange at the air/sea interface, (2) the buffering effect by which a major part of dissolved $CO_2$ is converted to carbonate and bicarbonate ions, and (3) the vertical mixing and transport of carbon through the stably stratified thermocline and the deep ocean. In addition, the models must account for the organic carbon produced by marine biota and the $CaCO_3$ formed in the shells of marine organisms—particularly, they must account for the rain of organic carbon fecal pellets and $CaCO_3$ shells that fall from the surface photic zone to the deep sea. Sediment dissolution and accumulation are also potentially important on longer time scales and should be included if possible.

Classical box models treat the dissolved constituents of the oceanic carbon cycle as well-mixed in reservoirs that exchange carbon flux in both directions at their boundaries at rates proportional to the total carbon concentration in adjacent boxes (Bacastow and Keeling, 1973; Keeling, 1973; Bjorkstrom, 1979; Broecker and Peng, 1982). This type of transport is essentially diffusive. When many "boxes" are stacked vertically beneath a surface, mixed-layer box, they constitute a *box-diffusion model* whose vertical eddy-exchange coefficients are related to the vertical eddy diffusivity of a "deep ocean" with a continuous, vertical, concentration profile. Oeschger et al. (1975) demonstrated this limiting relationship between disaggregated and continuous models, but did not include advective transport, a downward transport path from high latitudes that ventilates the deep ocean, or the effects of marine biology. The Oeschger one-dimensional ocean, topped by a mixed layer that exchanges carbon with the atmosphere via a gas exchange coefficient, and incorporating the buffering effect of surface carbon chemistry, was nonetheless a milestone carbon-cycle model in its time, and it has proven quite useful for projecting future $CO_2$ levels (Siegenthaler and Oeschger, 1978; Kraus et al., 1989).

One result of the glacial-interglacial $CO_2$ changes that were evident in the Vostok ice-core record was the realization that a mechanism must be included that exchanges surface and deep-ocean carbon more rapidly than thermocline mixing (Sarmiento et al., 1988). A "fast track" to the deep sea is also needed on anthropogenic time scales if the model must (1) account for oceans that absorb a terrestrial biosphere carbon source of order 1 to 2 PgC/yr in addition to 50% of the fossil-fuel $CO_2$ emissions, and (2) still match the $CO_2$ concentration history of the fossil-fuel era. This requirement has led to a class of carbon-cycle box models that include coupling with the deep sea (e.g., see Kagan and Ryabchenko, 1981; Byutner and Zakharova, 1983; Sundquist and Broecker, 1985). These couplings are generally represented by high-latitude surface "outcrops," where carbon can be carried away directly to the deep sea (Sarmiento and Togweiller, 1984; Lapenis and Kolomietsev, 1987; Volk and Liu, 1988). Indeed, enhanced mixing with the deep sea along inclined, constant-density surfaces (isopycnals) can "rapidly" sequester excess atmospheric $CO_2$ (Siegenthaler, 1984), although the physical basis for such models is still problematical.

Predictions of future $CO_2$ levels are particularly sensitive to the oceanic carbon-cycle model when projected emission levels *decrease* with time. For a hypothetical scenario in which $CO_2$ emissions decrease from present levels of more than 5 PgC/yr to 1 PgC/yr over the next 50 years and remain constant thereafter, Kraus et al. (1989) estimated $CO_2$ levels in the year 2100 to be between 440 and 370 ppmv depending on whether the calculation was done with a constant airborne fraction, a box-diffusion model, or an outcrop model (marine biology and circulation feedbacks were excluded from this study). In a general sense, all models predict that a drastic rollback of global fossil-fuel emissions is needed almost immediately to maintain the atmospheric $CO_2$ level at close to its present amount (350 ppmv), although the details can be crucial for evaluating policy options. A parallel problem exists with respect to the time-dependent response of climate to varying $CO_2$ levels. Here too, the role of ocean transport and mixing can be quite important in slowing the observed projected global warming (Bretherton et al., 1990).

That atmospheric $CO_2$ levels depend on a combination of time-varying $CO_2$ emissions and ocean transport processes, and perhaps on other poorly characterized processes, underscores the need for better models of the oceanic carbon cycle. This work is continuing on several fronts.

One approach is to construct multi-ocean box models disaggregated by ocean basin in the spirit of Broecker and Peng's (1982) "Pandora" model. These models typically have exchange coefficients calibrated to recover observed ocean-tracer fields with little obvious connection to the physical transport and circulation of the ocean.

So-called inverse box models include dynamical constraints on the evaluation of exchange coefficients. As a result, these models must be viewed as interpreters of oceanographic data derived from steady and transient tracers rather than as wholly prognostic models. Such models, therefore, have not yet reached the stage where they can be applied with confidence (Moore and Bjorkstrom, 1986).

An alternate approach to box models is to compute concentration fields interior to the ocean using specified ocean currents and diffusivities in one, two, and three dimensions. One-dimensional (vertical) upwelling-diffusion models for temperature (Munk, 1966), dissolved oxygen (Wyrtki, 1962), and radiocarbon (Craig, 1969) are the classical "deep-sea recipes" of oceanography. Carbon-cycle models utilizing upwelling and diffusion have subsequently been extended to include polar sea outcrops, the full nutrient and $CaCO_3$ cycles, and carbon pumps (Hoffert et al., 1981; Volk and Hoffert, 1985; Shaffer, 1989a, 1989b). Models of this type account for internal sources and sinks due to marine biology, sediment dissolution, and lysocline depth; they are well-adapted to time-dependent, carbon-cycle calculations (Kheshgi, 1989). The advantage of such models is that they link depth-dependence of concentration fields to oceanographic transport processes and provide a picture of coupled marine biology and chemistry effects with relatively little computational investment. They also permit testing of hypothesis and submodels that can subsequently be incorporated in models resolved in two and three dimensions.

The long-term goal of this modeling of the oceanic carbon cycle is the design of fully three-dimensional, time-varying models capable of predicting chemical and biological properties of the ocean interactively in terms of the three-dimensional circulation of the oceans. Preliminary steps in this direction (Maier-Reimer and Hasselmann, 1987; Bacastow and Maier-Reimer, 1990) are encouraging, because they show internal concentration distributions that are in qualitative agreement with oceanographic observations. As is usual for such models, considerable tuning is needed, and all properties cannot be matched simultaneously. Moreover, three-dimensional models of the oceanic carbon cycle are computationally intensive; at present, questionable numerical approximations must be made to complete the calculations in reasonable time. Moreover, the circulation of the oceans is not well characterized observationally, and must be derived from numerical oceanic circulation models, which are themselves in a relatively primitive state of development (see Section 5.2.3). It is possible that the situation would improve if additional supercomputer and parallel-processing resources were devoted to the problem. However, computers are not enough—more integrated oceanic observations and modeling studies on subgrid-scale processes, such as turbulence and deep convection, are needed before these advanced models become fully credible.

An additional uncertainty in these models is the degree to which the photosynthesis-respiration cycle of the global biosphere is in a dynamic equilibrium with respect to net carbon flux. If the biosphere is a net source or sink of atmospheric $CO_2$ (see Section 4.2.3), there are implications for carbon-cycle models that impact our ability to predict future $CO_2$ variations. For example, for fossil-fuel emissions of about 5.6 PgC/yr and an atmospheric buildup of about 3 PgC/yr (1.4 ppmv/yr), a biospheric source implies that >2.6 PgC/yr is being sequestered by the oceans; a sink implies <2.6 PgC/yr. Given the difficulty of transporting carbon in the oceans down through the "thermocline bottleneck" of the first kilometer of ocean depth (where density stratification inhibits vertical mixing), most ocean modelers would prefer a biospheric sink (or at least a neutral biosphere).

Projections of future emission trends vary over a wide range due to the problems associated with carbon-cycle models and with forecasting future energy patterns (Bolin et al., 1986). Figure 4.10a displays five scenarios for future fossil-fuel $CO_2$ emissions for the period from 1989 to 2050 from several different sources. Three scenarios are included from those developed recently for the international report of the Intergovernmental Panel on Climate Change (IPCC, 1990). The "most probable" case scenario developed by Edmonds et al. (1986) is also included, along with their estimated uncertainty range for this scenario. The fifth scenario, developed by Legasov et al. (in Budyko and Izrael, 1987), assumes a

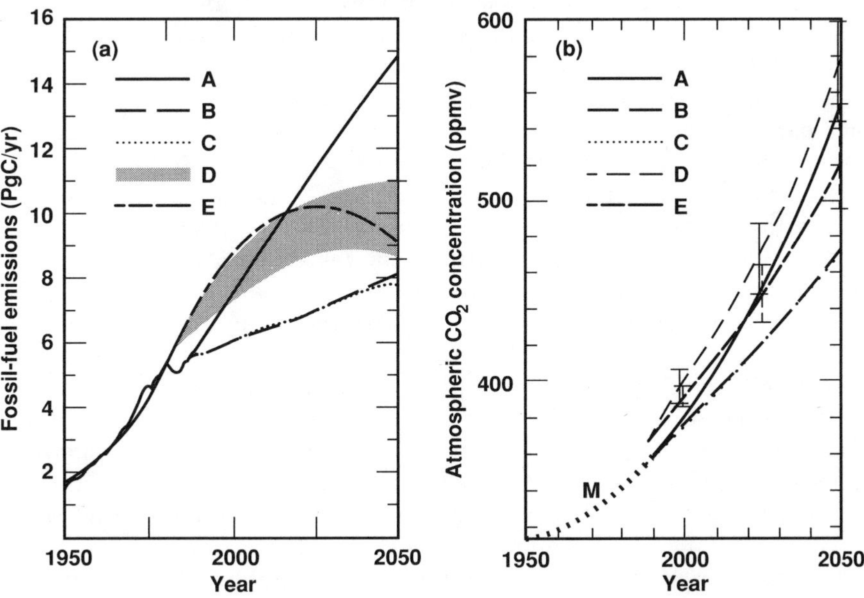


**Figure 4.10.** (a) Future scenarios for fossil-fuel $CO_2$ emissions based on: (A, B, C) the IPCC (1990), (D) Edmonds et al. (1986), and (E) Legasov et al. in Budyko and Izrael (1987); and (b) resulting atmospheric $CO_2$ concentration for these scenarios, with M indicating results of atmospheric monitoring. Vertical bars indicate estimated errors mainly due to uncertainty arising from changes in biomass.

gradual transition from fossil fuels to alternative energy sources by 2100; by 2050, there is a fifty-fifty ratio between the energy sources. The latter two scenarios assume that the total amount of fossil fuel burned by 2050 will be about 500 to 700 PgC; this is, on average, about double the total that would be generated at the current rate of emissions. Although representative examples of current estimates are included in Fig. 4.10a, other published scenarios extend over a wider range than those presented here.

Figure 4.10b shows the increasing atmospheric $CO_2$ concentration as derived from carbon-cycle models for these scenarios. The $CO_2$ concentrations for the IPCC scenarios were simulated using an atmosphere-ocean carbon-cycle model similar to that of Oeschger et al. (1975), as derived by Emanuel (1989). Estimates of the atmospheric $CO_2$ concentrations obtained from averaging the Legasov et al. (in Budyko and Izrael, 1987) and the Edmonds et al. (1986) scenarios are derived using the Byutner and Zakharova carbon-cycle model. Vertical bars in Fig. 4.10b indicate the differences between results due to varying biotic source intensity from –1 PgC/yr to 1 PgC/yr, along with different assumptions in ocean-mixing parameters. In general, the atmospheric $CO_2$ content will likely continue to increase through at least the middle of the twenty-first century, reaching values almost double the pre-industrial concentration.

The atmospheric $CO_2$ concentrations for an earlier, but similar, version of the Edmonds et al. (1986) scenario have also been evaluated using two different models of ocean-atmosphere interactions (one is the so-called box diffusion model, while the other assumes that a "blocking" layer separates the upper, mixed layer of the ocean from the deep layers, as discussed in Trabalka, 1985). The uncertainty range is based primarily on different assumptions of the future biotic source, ranging from close to zero growth to an increase of about 2.7 PgC/yr. Similar results for this scenario have also been obtained with the model of Byutner and Zakharova, as discussed in Budyko and Izrael (1987), and with the three-dimensional ocean circulation model of Maier-Reimer and Hasselmann (1987).

A significant fraction of anthropogenic carbon emissions is destined to remain in the atmosphere for a long time. If the entire recoverable fossil-fuel carbon reserve were rapidly burned and dumped to the atmosphere, the atmospheric $CO_2$ concentration would rise instantaneously to some eight times its current level. A more realistic estimate of the potential peak value for continuing fossil-fuel burning at current levels or greater is a factor of four increase over the next hundred to several hundred years, with $CO_2$ reaching twice pre-industrial levels sometime in the middle to latter decades of the next century.

Carbon dioxide from fossil-fuel burning will in time be redistributed to other reservoirs. However, if a significant fraction of the fossil reserve is burned, a factor of two above current $CO_2$ levels would very likely persist after 10 thousand years (Sundquist, 1985). Purging this residual fossil-fuel carbon from the atmosphere by sequestering it in carbonaceous rocks would take even longer, some 1 to 10 million years. This time is longer than the human species has inhabited the Earth.

### 4.4.2. Other Greenhouse Gases

There have been many attempts to estimate future emissions and concentrations of the non-$CO_2$ greenhouse gases (e.g., Wuebbles et al., 1984; WMO, 1985, 1989; Mintzer, 1987; IPCC, 1990). Although the development of scenarios for future concentrations of $CH_4$, CFCs, $N_2O$, and other gases is useful for theoretical studies evaluating the potential ranges of effects on ozone and climate, such projections are fraught with uncertainties that greatly limit their usefulness.

The factors controlling the increase in atmospheric concentrations of $CH_4$ and $N_2O$ are still poorly understood; there are too many uncertainties associated with their budgets to yet establish the cause of their increase. For $CH_4$, the sources are diverse, making it difficult to establish globally integrated emissions. In addition, any decrease in the global tropospheric OH concentration over recent decades would be contributing to the net increase in the atmospheric $CH_4$ concentration. Although recent measurements suggest that the increase in $CH_4$ may be slowing down (WMO, 1989), there are also several reasons indicating that its concentration will continue to increase. Many of the sources of $CH_4$ are related to human activities; thus, emissions are likely to continue to increase as human population increases. In addition, $CH_4$-producing microbial organisms work faster as temperature increases, suggesting that future climate change could add to emissions of $CH_4$. Finally, there is the possibility of a massive release of $CH_4$ from boreal deposits of $CH_4$ trapped in metastable hydrate formations, both beneath terrestrial permafrost and in sediments of the extensive, shallow seas around the Arctic basin (Revelle, 1983; Cicerone and Oremland, 1988). Although not well quantified, it is thought that the estimated reserves in these deposits are many times the present atmospheric levels of $CH_4$ and that global warming could lead to release of a significant increase in emissions from these deposits over the next century.

For $N_2O$, the recent evidence concerning the overestimate of combustion sources brings into question the actual cause of its increasing concentration. Generally, scenarios for future assessment studies have assumed that $CH_4$ and $N_2O$ concentrations will continue to increase at their current rates. A much better understanding of the budgets of these gases is needed to determine improved estimates of future concentrations.

Projections of future emissions and the atmospheric concentrations of CFC-11, CFC-12, and other halocarbons have been limited by uncertainties about the impact of international regulatory actions. Although the Montreal Protocol signed in 1987 called for a 50% decrease in the production of these gases by 1990, the possibility of limited participation in the Protocol restrictions by developing countries has led to scenarios assuming essentially constant emissions over coming decades

(Watson et al., 1988; WMO, 1989). However, there has been recent agreement on a total phaseout of CFC-11 and CFC-12 production as a result of the reevaluation of the Montreal Protocol by UNEP in June 1990. Although many replacement compounds have been suggested by the chemical industry and have been examined for potential environmental effects by the science community (AFEAS, 1989), the extent of future emissions of these compounds remains uncertain. Most of the compounds being considered have smaller potential effects on ozone and climate than the compounds they are replacing (Connell and Wuebbles, 1989; WMO, 1989).

Although reduced CFC emissions may limit the extent of further decreases in stratospheric ozone, it should be recognized that, because of their long atmospheric lifetimes, existing concentrations of these compounds will continue in the atmosphere for many decades. Based on the previous discussion about the Antarctic ozone hole and the recent trends in global ozone, the concentration of stratospheric chlorine must be reduced below current levels.

Computer models of global atmospheric chemical, radiative, and dynamical processes have been used in many studies of past variations in atmospheric composition and of the possible range of changes in composition that may occur in the future atmosphere (e.g., see Karol and Kiselyev, 1984, 1988; WMO, 1985, 1989; Karol, 1986; Karol et al., 1986; Karol and Babanova, 1987; Wuebbles and Kinnison, 1988; Wuebbles et al., 1989). Many of these studies concern the possibility of significant decreases in stratospheric ozone if current anthropogenic emissions of CFCs and other trace gases continue at current rates. Model calculations assuming continued growth in the atmospheric concentrations of $CH_4$, CO, and perhaps $NO_x$ suggest that tropospheric ozone concentrations could continue to increase, particularly in some areas of the Northern Hemisphere (Isaksen and Hov, 1987; Thompson et al., 1989a, 1989b).

## 4.5. THE DIRECT RADIATIVE INFLUENCE

The contribution of a gas to the greenhouse effect depends on the wavelength at which the gas absorbs radiation, the concentration of the gas, the strength of the absorption per molecule (line strength), and whether or not other gases absorb at the same wavelengths (e.g., see Ramanathan et al., 1985; Wang et al., 1986; Ramanathan et al., 1987; Mitchell, 1989). Gases absorb and emit radiation at wavelengths that correspond to transitions between discrete energy levels. Absorption at infrared (greenhouse) wavelengths occurs for triatomic or larger molecules, where vibrational and rotational energy transitions occur at appropriate wavelengths. Although each transition is associated with a discrete wavelength, the interval over which absorption occurs is "broadened" either by addition or removal of energy due to molecular collision (pressure broadening) or by the Doppler frequency shift due to the random velocities of molecules (Doppler broadening). If absorption is strong, there may be complete absorption (saturation) around the central wavelength of the spectral line.

As shown in Fig. 4.11, at a number of wavelengths the high concentrations of water vapor and $CO_2$ almost completely absorb the radiation emitted at the surface before it can be lost to space. Increases in the concentrations of these species would, therefore, lead only to increased absorption in the wings of the absorption lines, with the result that the net trapping of infrared radiation due to these gases would increase logarithmically rather than linearly with concentration. Because the atmospheric temperature changes with altitude, additional concentrations of these gases also change the effective altitude of emission, thereby changing the infrared flux and further enhancing the greenhouse effect of these gases. Gases absorbing at similar wavelengths as $CO_2$ and $H_2O$ will contribute little to the greenhouse effect, unless they have comparable

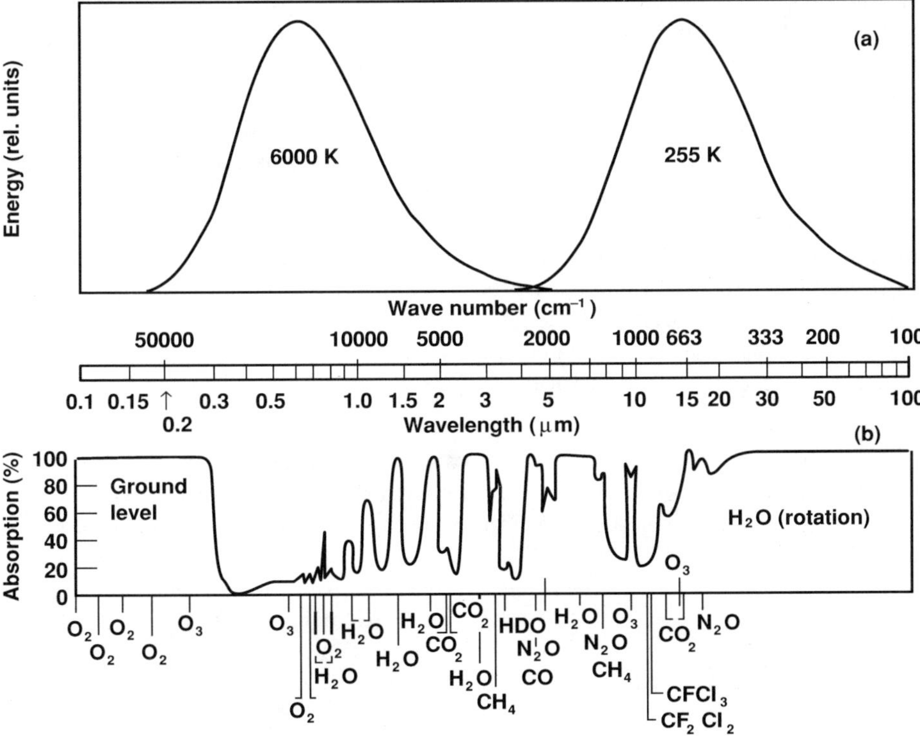


**Figure 4.11.** (a) Spectral distribution of long-wave blackbody emission at 6000 K and 255 K, corresponding to the mean emitting temperatures of the Sun and Earth, respectively, and (b) percentage of atmospheric absorption for radiation passing from the top of the atmosphere to the surface. Notice the comparatively weak absorption of the solar spectrum and the region of weak absorption from 8 to 12 μm in the long-wave spectrum. (From MacCracken and Luther, 1985.)

concentrations. However, Fig. 4.11 shows that there is a region from about 8 to 12 µm where absorption by $CO_2$ and $H_2O$ is weak; this is referred to as the window region. Most of the non-$CO_2$ gases with the potential to lead to climate change, including $CH_4$, $N_2O$, $O_3$, and the CFCs, have absorption lines in the "window" region. Some of these gases, such as $CH_4$ and $N_2O$, have absorption lines that lead to saturation of the line cores and emission from the pressure-broadened Lorentz line wings; the net result is that their absorption increases approximately as the square root of their concentration. Gases with little overlap, such as CFC-11 and CFC-12, exhibit absorption that increases linearly with concentration.

The net result of these varying radiative characteristics is that comparable increases in concentration of different greenhouse gases have vastly different effects on radiative forcing, as is shown in Table 4.8. The instantaneous radiative forcing of the surface-troposphere system is the change in the net radiative flux at the tropopause; equivalently, it can be defined as the surface temperature change with zero climate feedbacks (the importance of such feedbacks on the actual surface temperature change is discussed in Chapter 5). The addition of one molecule of $CH_4$ has about 21 times the effect on climate as the addition of one $CO_2$ molecule; one CFC-11 molecule is about 12,400 times as effective as one molecule of $CO_2$. These differences occur because $CH_4$ and CFC-11 absorb in the window region, whereas the $CO_2$ molecule competes not only with $H_2O$, but also with many other $CO_2$ molecules (i.e., the 15-µm region of $CO_2$ absorption is saturated). Although added $CO_2$ has the least effect on radiative forcing of the gases considered on a per-molecule basis, it is still the primary gas of concern for climate change due to the larger absolute change in concentration.

**Table 4.8.** Radiative forcing of various infrared-absorbing gases relative to $CO_2$ (therefore, $CO_2$ = 1) for each 1 ppbv (or molecule) added to the atmosphere. Based on Ramanathan et al. (1985), Wuebbles (1989), and IPCC (1990).

| Gas | Relative effect on radiative forcing |
|---|---|
| $CO_2$ | 1 |
| $CH_4$ | 21 |
| $N_2O$ | 206 |
| CFC-11 | 12,400 |
| CFC-12 | 15,800 |
| HCFC-22 | 10,700 |
| $CH_3CCl_3$ | 2,730 |

Ozone plays an important dual role in affecting climate. Although the climatic effect of $CO_2$ and the other trace constituents considered above depends primarily on their concentration in the troposphere, the climatic effect of ozone depends on its distribution throughout the troposphere and stratosphere. Ozone and molecular oxygen are the primary absorbers of solar radiation in the atmosphere; absorption of solar radiation by ozone is responsible for the increase in temperature with altitude in the stratosphere. Ozone is also an important absorber of infrared radiation. It is the balance between these radiative processes and the local changes in ozone with altitude (see Fig. 4.12; Lacis et al., 1990) that determines the net effect of ozone on climate. Increases in ozone

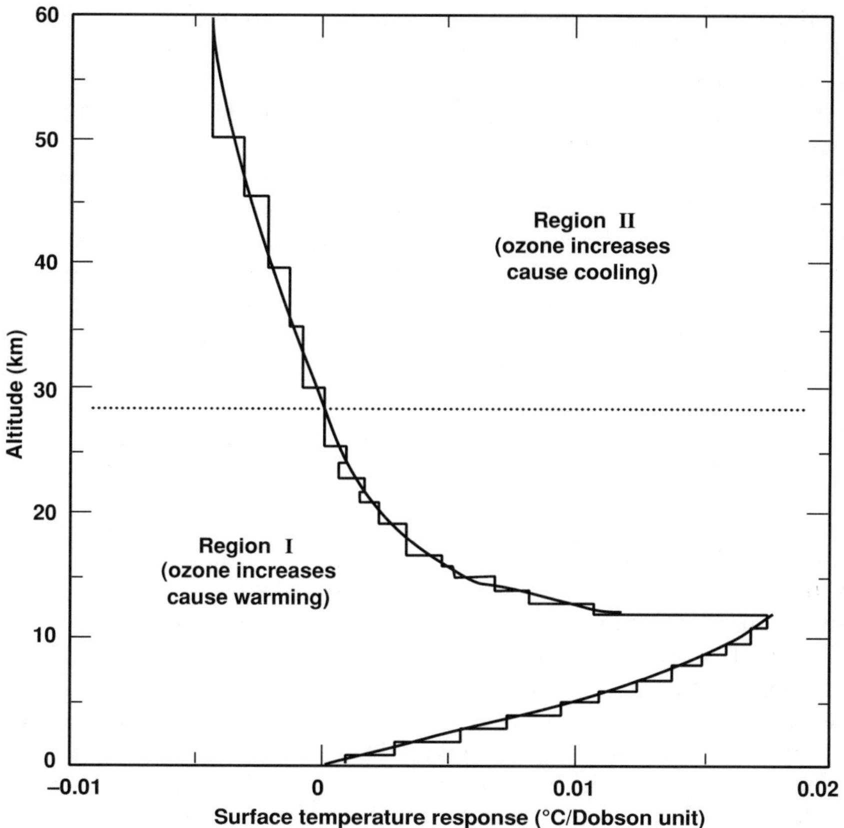


**Figure 4.12.** Radiative forcing sensitivity of global surface temperature to changes in vertical ozone distribution. The smoothed line is a least squares fit to one-dimensional model radiative-convective equilibrium results computed for ozone increments of 10 Dobson units (1 Dobson unit = 0.001 atm • cm of $O_3$) added to each atmospheric layer. Ozone increases in region I (below 30 km) warm the surface, whereas ozone increases in region II (above 30 km) cool the surface. No climate feedback effects are included in the radiative forcing. (Lacis et al., 1990.)

above about 30 km tend to decrease surface temperature; increases in ozone below 30 km tend to increase surface temperature.

Several research studies using global atmospheric models have attempted to examine the relative effects of other trace gases in affecting radiative forcing and climate as compared to the effects from the increasing carbon dioxide concentration (Lacis et al., 1981; Ramanathan et al., 1985, 1987; Karol et al., 1986; Wang et al., 1986; Wigley, 1987; Hansen et al., 1988). Each of these studies indicates that the combined effects of the other trace gases are comparable to the effects from the projected $CO_2$ emissions both for recent observed changes in species abundances and for reasonable assumptions about future changes in their concentrations. For example, model calculations by Lacis et al. (1981), Karol et al. (1986), and Hansen et al. (1988) suggest that observed changes in concentrations of $CH_4$, CFC-11, CFC-12, $N_2O$, $O_3$, and other gases during the 1970s and 1980s produced a radiative forcing that was 70 to 100% of that expected from the $CO_2$ increase during this period. In other words, the non-$CO_2$ trace gases provided about half of the total change in greenhouse forcing during these decades. Results of the Hansen et al. (1988) calculations are shown in Fig. 4.13; note that the radiative forcing is only the direct radiative effect on climate and does not account for the additional temperature change expected from the climate feedback effects of clouds, water vapor, sea ice, etc. Nor do these results account for the delay of the expected temperature change resulting from ocean-atmosphere interactions. These calculations indicate that by the 1980s, the effects of the other greenhouse gases on radiative forcing have together become as large as the effect from the increasing $CO_2$ concentration.

Evaluations of future scenarios, such as those by Ramanathan et al. (1985, 1987), Wang et al. (1986), and Hansen et al. (1988), suggest that these gases could effectively double, or more than double, the climatic effect from the increasing $CO_2$ concentration alone. It is not surprising, given the significant uncertainties concerning future trace constituent concentrations, that published scenarios for future changes in radiative forcing extend over a wide range of values. Recent evaluation of the "most probable" range in radiative forcing from $CO_2$ and other trace-gas concentration scenarios as developed by Ramanathan et al. (1987), Hansen et al. (1988), and Byutner and Zakharova (1989) are shown in Fig. 4.14. This range encompasses the best current estimates of changes in radiative forcing from existing evaluations of future changes in trace gas concentrations. Figure 4.14 also shows the change in equilibrium surface temperature calculated for this range of scenarios if the climatic feedback processes are assumed to increase the global surface temperature by 3°C for an assumed doubling of the atmospheric $CO_2$ concentration from 300 to 600 ppmv (see Chapters 5 and 6). If the actual sensitivity to a doubling of the $CO_2$ concentration is larger or smaller, then the scale of the derived temperature change would also change accordingly. For the assumed sensitivity, the derived change in surface temperature for this range of scenarios is substantial. The timing of such changes in temperature would

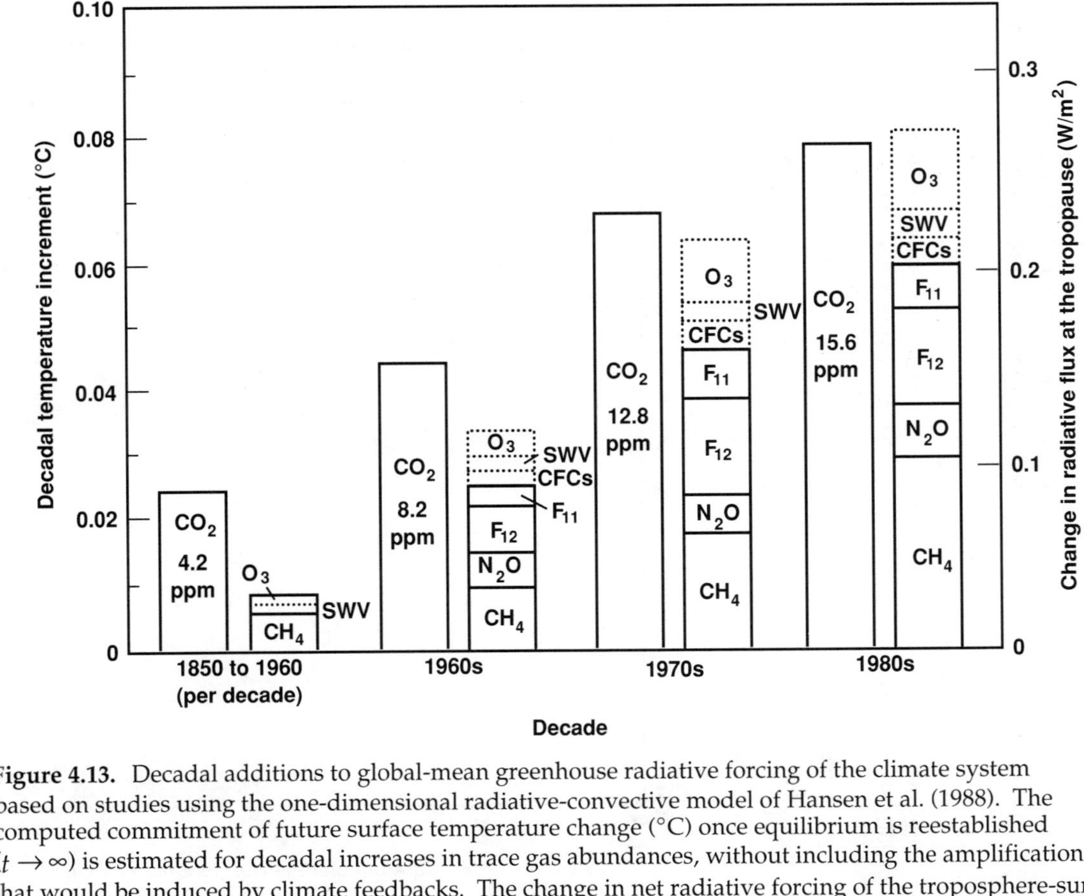


**Figure 4.13.** Decadal additions to global-mean greenhouse radiative forcing of the climate system based on studies using the one-dimensional radiative-convective model of Hansen et al. (1988). The computed commitment of future surface temperature change (°C) once equilibrium is reestablished ($t \rightarrow \infty$) is estimated for decadal increases in trace gas abundances, without including the amplification that would be induced by climate feedbacks. The change in net radiative forcing of the troposphere-surface system is indicated on the right ordinate. SWV indicates the contribution of stratospheric water vapor.

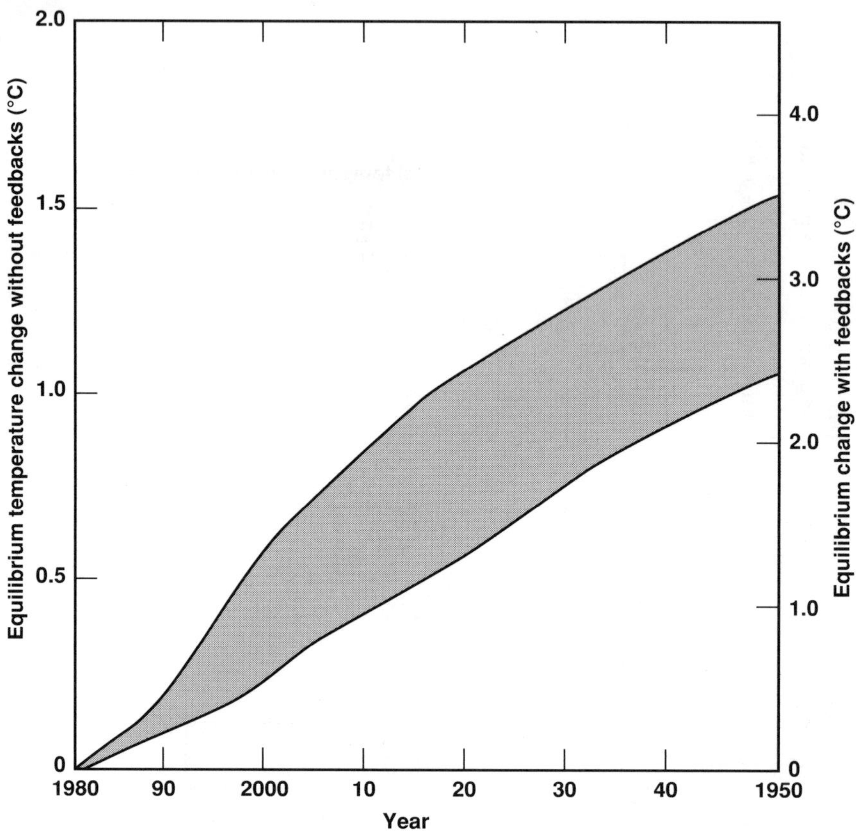


**Figure 4.14.** Range in global-mean greenhouse forcing for future trace gas concentration scenarios (based primarily on Ramanathan et al., 1987; Hansen et al., 1988; and Byutner and Zakharova, 1989), expressed in terms of the computed temperature change at equilibrium with no climatic feedback (assuming that 4 $W/m^2$ is equivalent to a 1.2°C temperature increase). The equilibrium surface temperature change with feedbacks is also shown on the right ordinate, assuming that an equivalent doubling of the $CO_2$ concentration from 300 to 600 ppmv produces a 3°C warming when feedbacks are included.

depend on the extent of the climate system's thermal inertia, which is due mainly to the large effective heat capacity of the ocean (see Section 5.4).

## 4.6. LESSONS AND OPPORTUNITIES FOR THE FUTURE

The composition of the atmosphere is of primary importance in determining the Earth's climate. The changes in composition over the Earth's history and over the past 100 years presage changes for the future.

(1) *What have we learned about the changes in atmospheric composition that have been occurring?*

The increases in atmospheric concentrations of $CO_2$ and other greenhouse gases have been conclusively documented by an extensive observational record. Even though the emissions and budgets of these gases in the atmosphere-ocean-land-biosphere system are somewhat uncertain, there are no reasons to indicate that, given present trends in human activities, the increasing concentration trends will do anything but continue upward.

(2) *What have we learned about recent changes in the $CO_2$ concentration?*

The concentration of $CO_2$ has increased about 25% over the last 150 years as a result of rapid industrial development and agricultural expansion. Projections of accelerating development of the fossil-fuel resource, particularly coal, to raise the global standard of living suggest that the $CO_2$ concentration could reach double its pre-industrial value by the middle to late decades of the next century.

(3) *What have we learned about recent changes in the concentrations of non-$CO_2$ trace gases?*

The concentration of $CH_4$ has more than doubled during the same period, and its rate of increase seems likely to continue as the production of food and the use of natural gas continue to increase. Although its absolute concentration is less than for $CO_2$, on a per-molecule basis, its effect on atmospheric radiation is about 30 times larger. The concentrations of CFCs have been increasing at relatively rapid rates. International action to reduce CFC emissions seems possible as substitute compounds are developed that will have much less effect on stratospheric ozone and somewhat less effect on the greenhouse effect. The $N_2O$ concentration is also increasing, probably due in large part to the increasing use of fertilizers. Uncertainties are so large, however, that forecasts of concentrations are quite unreliable.

(4) *What have we learned about the effects of these emissions on the atmospheric ozone concentration?*

The extensive emissions of $CO_2$ and other trace gases are also altering the natural concentrations of stratospheric and tropospheric ozone and of other compounds. These changes in ozone concentration will also not only enhance the greenhouse effect, but may also increase the downward flux of UV radiation at the surface and reduce the chemical cleansing capacity of the atmosphere.

(5) *What have we learned about the consequences of these composition changes on the atmospheric radiation balance?*

The increasing concentrations of $CO_2$ and other trace gases and the perturbing of the concentrations of tropospheric and stratospheric ozone will increase the trapping of infrared radiation. The radiative effect of the non-$CO_2$ gases has provided about a 50% increment to the trapping of infrared radiation over the last 150 years.

## Prospects for Future Climate

Although there has been a significant advance in understanding, much remains to be done:

(1) Continue study of the sources, sinks, and transformation processes affecting the concentrations of $CO_2$ and other greenhouse gases of both natural and anthropogenic origin.

(2) Extend drilling and study of ice cores and conduct isotopic and other analyses to reconstruct past concentrations of $CO_2$ and other trace gases.

(3) Expand study of the feedbacks between atmospheric composition and climate, including representations of atmospheric chemistry and radiative fluxes.

# CHAPTER 5. THEORETICAL ESTIMATES OF GREENHOUSE-GAS-INDUCED CLIMATE CHANGE

## 5.1. INTRODUCTION

If the Earth's atmosphere were composed of only its two major constituents, nitrogen ($N_2$, 78% by volume) and oxygen ($O_2$, 21%), and if the planetary albedo remained at its present value, the Earth's average surface temperature would be about –18°C, which is the radiative-equilibrium temperature at which the amount of radiation emitted to space by the Earth equals the amount of radiation absorbed from the Sun. The fact that the Earth's global-mean surface temperature is a life-supporting 15°C is a consequence of the greenhouse effect of the atmosphere's minor constituents, mainly water vapor ($H_2O$, about 0.2%) and carbon dioxide ($CO_2$, now up to about 0.035%).

The concentration of $CO_2$ and other greenhouse gases started to increase at the beginning of the Industrial Age, about 150 years ago, as described in Chapter 4. As these concentrations increase, less of the temperature-dependent infrared radiation emitted by the Earth can escape through the atmosphere to space, but the amount of solar radiation absorbed by the Earth remains almost unchanged. It is to be expected that this *enhancement* of the greenhouse effect will result in a higher average temperature as the climate system seeks to restore the balance between the infrared cooling and solar heating. However, it is also to be expected that the temperature change at any location may be higher, lower, or even of opposite sign from the average temperature change, and that there may be concomitant changes in other climatic elements such as precipitation and soil moisture.

How can the potential greenhouse-gas-induced changes in climate be estimated? The classical approach in science is to perform controlled experiments in a laboratory. Although some aspects of the behavior of the atmosphere or ocean have been studied by the laboratory method, the Earth's climate system is just too complex to miniaturize *in toto* in a laboratory. Alternatively, the climatic history of the Earth can be considered a model. Chapter 6 describes use of this empirical method, in which the seasonal and regional patterns of past warmer climates are used to gain insights about future climatic conditions.

This chapter focuses on the theoretical approach to projecting greenhouse-gas-induced climate change. With this approach, the behavior of the Earth's climate system (see Fig. 5.1) is calculated from the fundamental physical laws of nature. These physical laws are expressed mathematically to form a mathematical climate model that is then used to perform simulations on a computer. The advantage of the theoretical method is that it can be used to simulate, in a physically consistent manner, not only the present climate, but also how that climate would change in response to increases in the concentrations of greenhouse gases. A limitation of the theoretical method at present is its inability to construct a model that reproduces the behavior of the actual climate system with the desired accuracy.

Most simulations of greenhouse-gas-induced climate change have been performed to determine the change in the equilibrium climate of the Earth resulting from a doubling of the $CO_2$ concentration, this because the effective $CO_2$ concentration[1] is projected to double sometime within the next century. These simulations of equilibrium climate change have not been concerned with the time required for the climate change to reach its equilibrium and, in fact, have generally tried to minimize that time in an effort to economize on computer time.

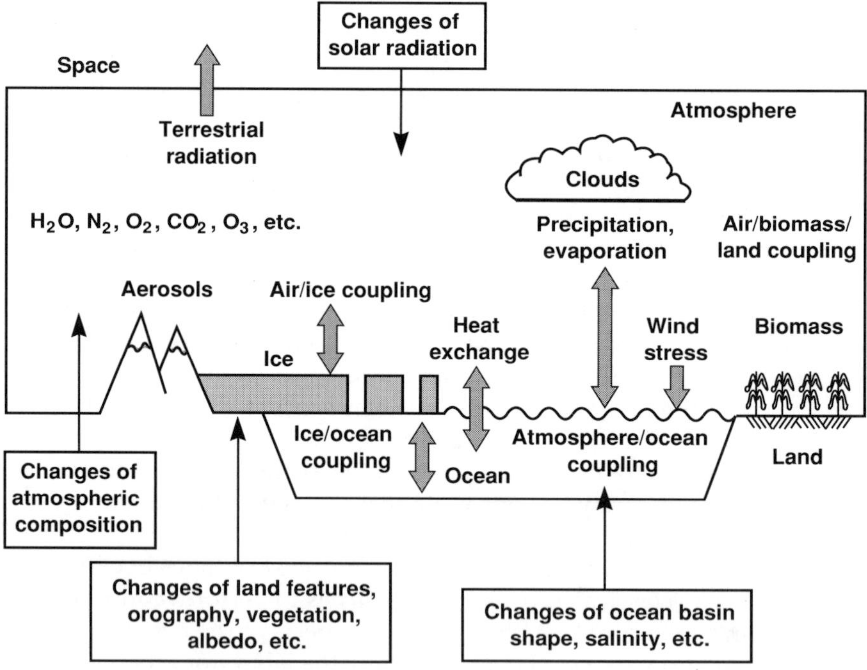


**Figure 5.1.** Schematic illustration of the components of the Earth's climate system, together with some physical processes responsible for climate and climate change. (From Gates, 1979.)

---

[1] The effective $CO_2$ concentration is the actual $CO_2$ concentration plus the concentration of the other greenhouse gases expressed in terms of their equivalent $CO_2$-radiative influence (see Section 4.5).

More recently, concern has been focused on the detection of a greenhouse-gas-induced climate change. Because such detection depends on the evolving response of the Earth's climate system to the actual, slowly increasing concentrations of greenhouse gases, simulations of nonequilibrium, time-dependent climate change have recently been performed with mathematical climate models.

This chapter presents results from simulations of both equilibrium and nonequilibrium climate change. Before doing so, however, an overview of the formulation and use of mathematical climate models is provided.

## 5.2. MATHEMATICAL CLIMATE MODELS

Three different types of climate model have been used to simulate greenhouse-gas-induced climate change: energy balance models, radiative-convective models, and general circulation models. Each of these types of climate model is described below.

### 5.2.1. Energy Balance Models

Energy balance models, developed first in the Soviet Union by Budyko (1968, 1969) and then independently in the United States by Sellers (1969), predict only the surface temperature, most often in terms of its globally or zonally averaged value. In an energy balance model, meridional energy fluxes induced by the atmospheric and oceanic circulations are parameterized,[2] generally in terms of temperature gradients, and losses of energy from the planet are parameterized based on the surface temperature. Increasing the concentration of a greenhouse gas results in a change in the infrared (longwave) radiative energy flux that initiates a positive energy imbalance. As a result, the surface temperature begins to increase, and the energy fluxes that depend on this temperature begin to change. Ultimately a new, warmer surface temperature is attained at which the energy fluxes at the selected level again sum to zero.

Estimates from energy balance models of the global-average surface temperature warming that would be induced by a doubling of the $CO_2$ concentration cover a wide range. This wide range is largely due to the fact that the energy fluxes in this type of climate model must be specified solely in terms of the surface temperature. This requires that assumptions be made about the behavior of the climate system; several alternative assumptions can and have been made (Schlesinger, 1985, 1988). These different assumptions have led to the different estimates of the greenhouse-gas-induced warming. For example, Budyko (1974) took into account the dependence of albedo on the snow-ice area and obtained a temperature

---

[2] Parameterization is the parametric treatment of physical processes whose characteristic size is smaller than the smallest size resolved by a model or whose complexity cannot be fully represented by the limited physical representations inherent in a model.

increase of 3°C due to a $CO_2$ doubling. Later, Mokhov (1981) used a similar model with improved parameters and obtained a 2°C warming when the above-mentioned dependence was absent and a 3.5°C warming when it was present, in good agreement with present estimates from more detailed models.

### 5.2.2. Radiative-Convective Models

Radiative-convective models, developed first by Manabe and co-workers in the 1960s, determine the vertical distribution of atmospheric temperature and the surface temperature, virtually always in terms of their globally averaged values (Manabe and Wetherald, 1967). Increasing the concentration of a greenhouse gas in a radiative-convective model results in a change in the vertical profile of the longwave radiative energy flux. This initiates a warming of the surface and lower atmosphere (troposphere) and a cooling of the higher atmosphere (stratosphere). The warming of the troposphere is such that the vertical profile of air density becomes unstable; that is, more dense air begins to overlie less dense air. When this happens in the real atmosphere, convection ensues, causing horizontally small-scale vertical overturning motions to occur that tend to restore the vertical density profile to its neutral value. These convective motions are not explicitly resolved in a radiative-convective model. Instead, they must be parameterized. For convection, this is done by restoring the lapse rate at which temperature decreases with increasing altitude to a prescribed value whenever the simulated value exceeds the prescribed value. This parameterized convection redistributes the surface heating throughout the troposphere, but not into the stratosphere, because vertical motion in the stratosphere is strongly inhibited by the stable density profile.

In addition to parameterizing convection, radiative-convective models often parameterize other processes such as those that govern the amount of water vapor in the atmosphere, the temperature lapse rate, the horizontal extent and optical properties of clouds, and the reflectivity (albedo) of the Earth's surface due to changes in snow, ice, and surface vegetation cover.

Estimates from radiative-convective models of the global-average surface temperature warming induced by a doubling of the $CO_2$ concentration range from 0.48 to 4.2°C. (For example, the model by Karol and Rozanov, 1982, calculates 2.5°C.) This wide range results because of the different representations of physical processes that are included in the models and how these processes are represented. By analyzing these radiative-convective model simulations, it is possible to estimate quantitatively the contribution of each process to the warming, that is, to determine whether the process enhances (*positive feedback*) or diminishes (*negative feedback*) the warming induced by the radiative flux change that is caused solely by the increase in greenhouse-gas concentration (Schlesinger, 1985, 1988). However, a high degree of confidence cannot be placed in these feedback estimates because radiative-convective models are not models of the global climate system and, of necessity, prescribe the behavior of much of that system. These feedbacks, and thus the greenhouse-gas-induced climate change itself, can be predicted credibly only by a physically based, global, general circulation model that includes comprehensive treatments of both dynamical and thermodynamical processes.

### 5.2.3. General Circulation Models

General circulation models (GCMs) of the atmosphere are the only type of climate model that determines, in addition to temperature, an ensemble of other climatic quantities, including precipitation, soil moisture, and cloud cover. Furthermore, atmospheric GCMs are the only type of climate model that is formulated to determine the geographical distributions of these climatic quantities. Consequently, atmospheric GCMs are the only type of climate model that can be used to estimate the regional distribution of greenhouse-gas-induced climate change. To use an atmospheric GCM for this purpose, however, it must be coupled to a model of the ocean so that the geographical distributions of sea surface temperature and sea ice extent can be determined.

Because of the reliance on GCMs for present-day estimates of greenhouse-gas-induced equilibrium climate change, it is helpful to understand the formulation of this type of model and the limitations imposed thereon even by today's supercomputers.

#### *5.2.3.1. Formulation of Atmospheric GCMs*

Atmospheric general circulation models are based on the fundamental laws of nature, namely, conservation of energy, conservation of mass, and Newton's second law of motion. These laws govern the time rates of change of temperature, surface pressure, and wind velocity; these quantities are therefore called prognostic variables. The large-scale motion pattern of the atmosphere, or atmospheric general circulation, is determined in large part by the release of latent heat[3] when water vapor condenses into liquid (or sublimates into ice). Accordingly, the concentration of water vapor is a prognostic variable of an atmospheric GCM governed by a water vapor conservation law. Because the atmosphere is heated by the underlying surface, the temperature of the ground, the amount of water in the soil, and the mass of snow on the ground are prognostic variables governed, respectively, by energy, water, and snow

---

[3] The evaporation of liquid water at the Earth's surface requires the input of energy. The evaporated water vapor therefore has a higher energy level than its parent liquid water. Eventually, the water vapor can condense back into liquid at a place and time remote from where it was evaporated. When this occurs, the atmosphere is given the heat that was acquired when the water evaporated at the surface. This remote heating of the atmosphere is therefore called latent heating.

conservation laws. Atmospheric GCMs also have diagnostic variables, that is, quantities whose magnitudes rather than time rates of change are determined by the governing laws. A summary of the prognostic and diagnostic variables in atmospheric GCMs is given in Table 5.1.

The governing laws are expressed mathematically as partial differential equations that describe the behavior of the atmosphere continuously in both space and time. These equations are so complex, however, that there are no analytic methods available to obtain their solution. Accordingly, it is necessary to obtain their solution by numerical methods in which the behavior of the atmosphere is determined discretely in both space and time.

The numerical methods subdivide the atmosphere vertically into layers (Fig. 5.2) and horizontally into either grid boxes (Fig. 5.3), as in finite difference models, or into a number of prescribed mathematical functions, as in spectral models. The evolution in time of the prognostic and diagnostic variables is also determined by advancing the solution in discrete time steps.

The atmospheric GCMs that have been used to simulate $CO_2$-induced climate change have from 2 to 11 vertical layers and horizontal resolutions (the distance between grid points) from about 500 to 1000 km (Fig. 5.4). These horizontal resolutions are chosen not on the basis of physical grounds (clearly, they are so coarse that they also impose severe limitations for impact analyses), but rather by the limitation of present-day supercomputers. For example, a simulation of $CO_2$-induced climate change with an atmospheric GCM/mixed-layer ocean model having a

**Table 5.1.** The principal prognostic and diagnostic variables in atmospheric general circulation models.

| Prognostic variables | Diagnostic variables |
|---|---|
| Surface pressure | Vertical velocity |
| Temperature | Geopotential height |
| Horizontal velocity components[a] | Density |
| Water vapor concentration | Clouds |
| Soil temperature | Surface albedo[b] |
| Soil moisture | |
| Snow mass | |

[a] In spectral models (see text), the vertical component of vorticity and the horizontal divergence replace the horizontal velocity components (which give velocity and direction) as the prognostic variables, and the latter are determined diagnostically from the former.

[b] Due to changes in the presence of snow and sea ice.

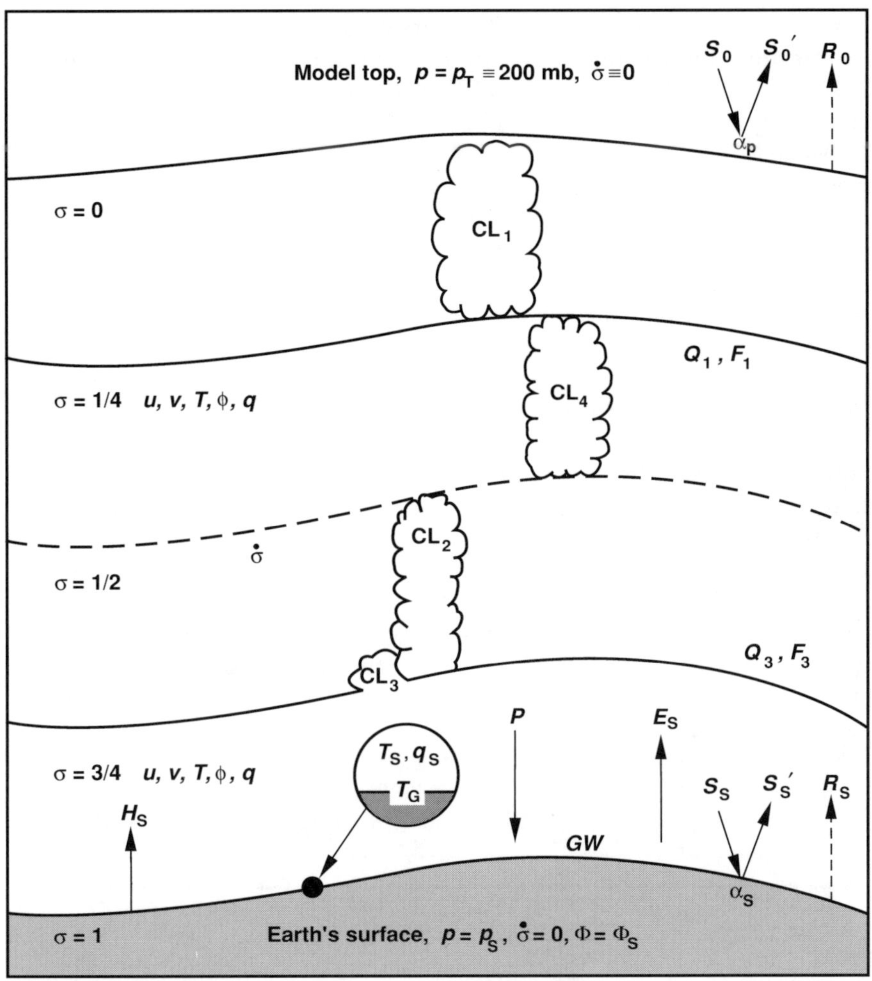


**Figure 5.2.** The vertical structure and principal variables of the Oregon State University (OSU) two-layer atmospheric general circulation model. Here, $\sigma = (p - p_T)/(p_s - p_T)$, where $p$ is pressure, $p_T$ the (constant) pressure at the model top, and $p_s$ the (variable) pressure at the Earth's surface; $\dot{\sigma}$ is the vertical velocity in the $\sigma$ coordinate system; $u$ and $v$ are the eastward and northward velocity components, $T$ the temperature, $\phi$ the geopotential height, $\Phi_s$ the geopotential height of the Earth's surface relative to mean sea level, and $q$ the water vapor mixing ratio; $S$ and $R$ are the solar and terrestrial radiation at the top of the model atmosphere (subscript 0) or at the Earth's surface (subscript s), and the prime denotes the reflected component of solar radiation; $\alpha_s$ and $\alpha_p$ are the surface and planetary albedos, $Q$ and $F$ the diabatic heating and friction, $H_s$ the surface sensible heat flux, $P$ the precipitation rate, $E_s$ the surface evaporation rate, and $GW$ the ground wetness. $CL_1$, $CL_2$, $CL_3$, and $CL_4$ denote the model's cloud types. (From Schlesinger and Gates, 1980.)

500-km resolution requires about 300 hours on a contemporary supercomputer. Increasing the resolution tenfold to 50 km would require about 40 years on this same machine! This thousandfold increase in required computer time for a tenfold increase in horizontal resolution occurs because, in addition to the resulting hundredfold increase in the number of grid boxes, there is a multiplicative factor of ten due to the decrease in the time step of a factor of ten that is required to prevent computational instability.

The limitation on the horizontal resolution of atmospheric GCMs imposed by present-day supercomputers has a major influence on which physical processes can be resolved. In particular, the horizontal scale of the physical processes occurring in the Earth's climate system range from a millionth of a meter for the condensation of water vapor onto condensation nuclei (aerosol particles and dust) to 40 million meters for atmospheric waves that span the entire circumference of the Earth. Thus, the physical processes range over 14 powers of ten in horizontal size. With present supercomputers, atmospheric GCMs resolve physical processes with a minimum horizontal scale of about 1000 km; thus, they cover only 2 of the 14 powers of ten. While a computer 1000 times faster than present-day supercomputers may be available by the end of this century, allowing resolution to be decreased to about 50 km, a computer fast enough to permit the resolution of the complete range of physical processes in the climate system remains much further in the future.

Although all of the processes in the climate system cannot be resolved, their effects must be represented. The reason for this, as noted above, is that the general circulation of the atmosphere is determined in

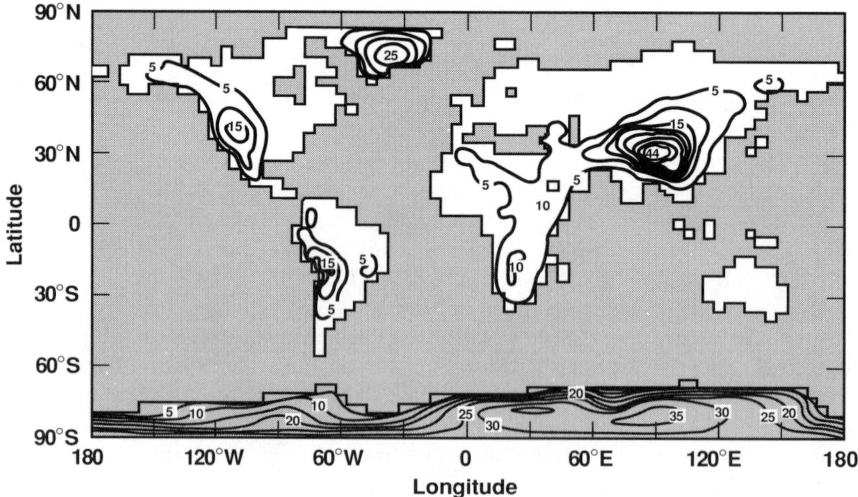


**Figure 5.3.** The distribution of land and ocean as resolved by the 4° latitude × 5° longitude grid of the OSU atmospheric GCM. For clarity, the continental grid points are not shown. The contours show the surface elevation of the continents ($10^2$ m). (After Schlesinger and Gates, 1981.)

large part by the release of latent heat that occurs when water vapor condenses into liquid (or sublimes into ice), and this happens on the smallest unresolved scales. The approach taken is to parameterize, that is, to represent the effects of the unresolved or subgrid-scale physical processes on the scales resolved by the GCM using information available only on the resolved scales. The subgrid-scale processes that are parameterized in atmospheric GCMs are shown in Table 5.2.

To simulate climate and climate change with an atmospheric GCM, it is also necessary to prescribe certain parameters and boundary conditions, as listed in Table 5.3. It is also necessary to include the ocean and sea ice components of the climate system (Fig. 5.1); see the following section.

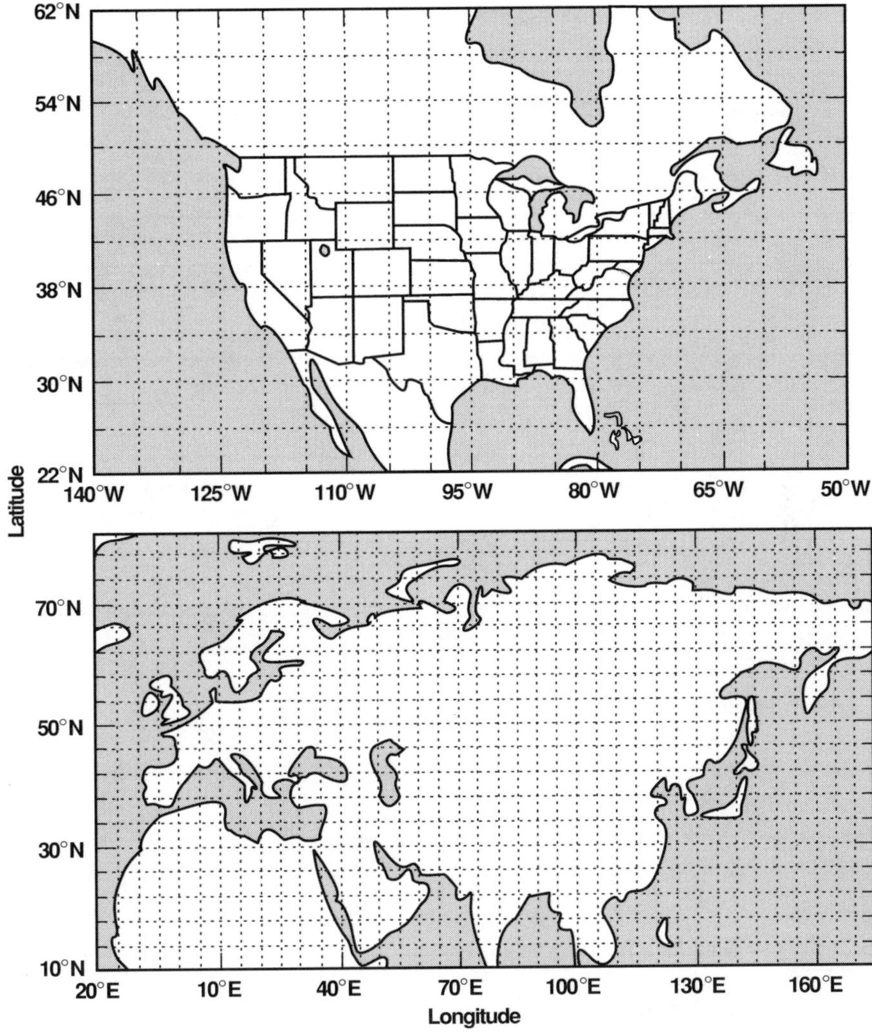


**Figure 5.4.** Grid points of the 4° latitude × 5° longitude resolution OSU atmospheric GCM for: (a) North America, and (b) Eurasia. The grid points are located at the intersections of the dotted lines.

**Table 5.2.** Subgrid-scale processes that are parameterized in atmospheric general circulation models.

- Turbulent transfer of heat, moisture, and momentum between the Earth's surface and the atmosphere.
- Turbulent transfer of heat, moisture, and momentum within the atmosphere by dry and moist (cumulus) convection.
- Condensation of water vapor.
- Transfer of solar and terrestrial radiation.
- Formation of clouds and their radiative interaction.
- Formation and dissipation of snow.
- Physics of soil heat and moisture.

**Table 5.3.** Prescribed parameters and boundary conditions in atmospheric general circulation models.

- Radius, surface gravity, and rotation speed of the Earth.
- Solar constant and orbital parameters of the Earth.
- Total mass of the atmosphere and, in models that do not treat atmospheric chemistry, its composition.
- Thermodynamic and radiation constants of the atmospheric gases and clouds.
- Surface albedo[a] and soil type (to define hydrological characteristics)
- Surface elevation.

[a] Except as changed with the occurrence of snow and sea ice.

#### 5.2.3.2. *Ocean and Sea Ice Models*

To evaluate the performance of an atmospheric GCM in simulating the present climate, it is possible to treat the sea surface temperature (SST) and sea ice extent as given boundary conditions rather than as predicted quantities. To simulate a greenhouse-gas-induced climate change, however, the SST and sea ice extent must be determined by the model.

Consequently, atmospheric GCMs have been coupled with different ocean and sea ice models, the comprehensiveness of such models again being limited by the speed of the fastest computers available.

The earliest simulations by atmospheric GCMs of greenhouse-gas-induced climate change were performed primarily to estimate the change in the global-mean surface air temperature, this because of the wide range in the estimates of this quantity obtained by energy balance and radiative-convective models. The first such GCM simulation was performed 15 years ago by Manabe and Wetherald (1975) at a time when the fastest computer was about 10% the speed of a single processor of a present-day supercomputer. Accordingly, the treatment of the ocean in this and other such early simulations was quite rudimentary. In particular, the ocean was considered to be infinitely thin, but to act as an infinite source of water, with the existence of sea ice being predicted whenever the SST was less than the prescribed freezing temperature of seawater. This type of ocean model is called a swamp ocean model because of its similarity to perpetually wet land. Such an infinitely shallow ocean is always in thermal equilibrium; therefore, the model must be run for only about a year to permit the atmospheric component of the system to achieve a new equilibrium in response to the change in the $CO_2$ concentration.

Because a swamp ocean has zero heat capacity, however, it cannot be used to simulate the annual or day-night (diurnal) cycle; such variations would cause the ocean to freeze where it was night, and to boil where it was day. To avoid this, the geographical distribution of the annually averaged solar radiation (insolation) was prescribed. The simulation results were then interpreted as the annual-mean climate, even though the actual annual mean is the result of both the diurnal and annual cycles that the model cannot represent.

The results of simulations of greenhouse-gas-induced simulations using atmospheric GCM/swamp ocean models have been reviewed by Schlesinger (1983, 1984), Schlesinger (1984), and Schlesinger and Mitchell (1985, 1987). These models yielded a global-mean surface air temperature warming induced by a $CO_2$ doubling of 1.3 to 3.9°C, with a concomitant increase in global-mean precipitation of 2.7 to 7.8% of the unperturbed value. Such a $CO_2$-induced warming represents a large fraction of the 5°C increase of the global average surface air temperature between the last ice age 18,000 years ago and the present warm (interglacial) period, a natural warming that took about 10,000 years to occur.

To simulate the annual cycle of climate and its greenhouse-gas-induced change requires an improved treatment of the ocean. In recognition of supercomputer limitations, the most commonly used ocean model includes only the upper part of the ocean, that is, the oceanic mixed layer wherein temperature is virtually uniform with depth. In such a mixed-layer ocean model, the depth of the mixed layer is prescribed as a function of geographical location and time of year, most often as a constant value between 50 and 70 m, chosen so that the simulated annual cycle of SST averaged globally over the ocean is in agreement with the corresponding observed mean value (Manabe and Stouffer, 1979, 1980). At any location, however, the simulated annual cycle of SST in models with a constant mixed layer depth need not agree with observations, particularly where the annual cycle of mixed layer depth is large.

There are two additional limitations of mixed-layer ocean models. First, in such models there is no vertical exchange of heat between the mixed layer and the underlying ocean. The advantage of this is that atmospheric GCM/mixed-layer ocean models will achieve equilibrium in about 20 simulated years. Mixed-layer ocean models cannot, however, be used to simulate the time evolution of the response of the climate system to the slowly increasing rise in greenhouse gas concentrations because of the missing coupling to the deep ocean.

Secondly, mixed-layer ocean models are nondynamic models of the ocean; that is, they do not simulate ocean currents. In some models, the effects of these currents on SST have been prescribed; as a result, atmospheric GCM/mixed-layer ocean models do not simulate the effects on SST of greenhouse-gas-induced changes on oceanic heat fluxes.

To simulate the time evolution of the nonequilibrium climate change resulting from increasing greenhouse-gas concentrations requires use of a dynamic model of the entire ocean that extends vertically from the sea surface to the ocean floor and includes horizontal and vertical heat transports. Such a comprehensive oceanic GCM is the oceanic counterpart to the atmospheric GCM. In oceanic GCMs, the prognostic variables are the temperature, horizontal currents, and salinity; the diagnostic variables include density, pressure, and the vertical velocity. Subgrid-scale processes that must be parameterized in ocean GCMs include the turbulent transfers of heat, momentum, and salt in both the vertical and horizontal directions, and also parameters and boundary conditions similar to those in Table 5.2 that must be prescribed. The solution of the governing equations is obtained numerically in a manner similar to that used for atmospheric GCMs. The ocean is subdivided vertically into layers and horizontally into grid boxes, and the predicted quantities are determined as a function of time by numerical integration.

In both mixed-layer ocean models and oceanic GCMs, the thickness of sea ice is predicted based on a thermodynamic energy budget that includes seawater freezing, the accumulation of snowfall, and the melting and sublimation of ice. The transport of sea ice by the upper ocean currents has been included only in some oceanic GCMs.

An oceanic GCM can be coupled to an atmospheric GCM to form an atmosphere-ocean GCM. Although this type of model is the most physically comprehensive climate model, it requires about 1000 simulated years to achieve equilibrium, hence, about 50 times more computer time than the atmospheric GCM/mixed-layer ocean model. For this reason, there have been no atmosphere-ocean GCM simulations of the greenhouse-gas-induced changes in equilibrium climate with the annual cycle included. However, atmosphere-ocean GCM simulations of climate change have been performed using annually averaged insolation, because a method to accelerate the deep ocean to equilibrium has been developed that is successful for this idealized case (Bryan, 1984). Simulations now being made are starting to use coupled atmosphere-ocean models in which, rather than simply doubling the $CO_2$ concentration, the greenhouse gas concentrations are slowly increased. However, these time-dependent simulations can be run only for periods of about 100 years, again because of the large amount of computer time required by these models.

## 5.3. EQUILIBRIUM CLIMATE CHANGE

The purpose of equilibrium studies is to determine what the change in the climate would be if the greenhouse gas concentrations increased to some constant values and the climate system reached a new equilibrium for these higher levels, regardless of the time required to attain that new equilibrium climate. Because the increase in the concentrations of the greenhouse gases $CO_2$, methane ($CH_4$), chlorofluorocarbons (CFCs), and nitrous oxide ($N_2O$) can be considered roughly equivalent to an appropriate increase in the concentration of $CO_2$ alone (see Section 4.5), studies of equilibrium climate change have been performed only for increased $CO_2$ concentrations.

In an equilibrium study, a control simulation is made with a fixed $CO_2$ concentration (typically about 300 ppmv), and an experiment simulation is made with another fixed $CO_2$ concentration. Both the control and the experiment are run sufficiently long to achieve their respective equilibrium climates, as illustrated in Fig. 5.5.

Equilibrium simulations have been performed for elevated $CO_2$ concentrations that are double ($2 \times CO_2$), quadruple ($4 \times CO_2$) and 10 times ($10 \times CO_2$) their control values. The $2 \times CO_2$ simulations have been made because, as noted in Chapter 4, it is projected that the $CO_2$ level will reach twice the preindustrial value sometime in the next century. The $4 \times CO_2$ and $10 \times CO_2$ simulations with GCMs have been made not because such large increases are foreseen, but rather to increase the statistical significance of the results (and because such elevated levels have probably occurred earlier in the Earth's history, as explained in Chapter 2).

In GCM simulations, as in nature, there is an inherent variability for any time-averaged quantity, even though the climate is unchanged. This natural variability constitutes noise against which the signal of the difference between the experiment and control must be contrasted. If the ratio of the signal to the noise is sufficiently large, then it can be said that the experiment climate is not just another realization (sample) of the control climate, but is in fact a climate different from that of the control. To increase the signal-to-noise ratio, the noise can be reduced by averaging over longer time periods—this requires extending the length of the simulations. Alternatively, the magnitude of the signal can be increased by increasing the magnitude of the forcing, for example, by quadrupling the $CO_2$ level.

The annual cycles of the equilibrium changes in climate induced by a doubling of the $CO_2$ concentration have been simulated by the atmospheric GCM/mixed-layer ocean (AGC/ MLO) models of (1) the Geophysical Fluid Dynamics Laboratory (GFDL; Wetherald and Manabe, 1986), (2) the Goddard Institute for Space Studies (GISS; Hansen et al., 1984), (3) the National Center for Atmospheric Research (NCAR; Washington and Meehl, 1984), (4) Oregon State University (OSU; Schlesinger and Zhao, 1989), and (5) the United Kingdom Meteorological Office (UKMO; Wilson and Mitchell, 1987). The equilibrium changes in the surface air temperatures, precipitation rate, and soil moisture simulated by these five AGC/MLO models for a $CO_2$ doubling are presented in the following subsections.

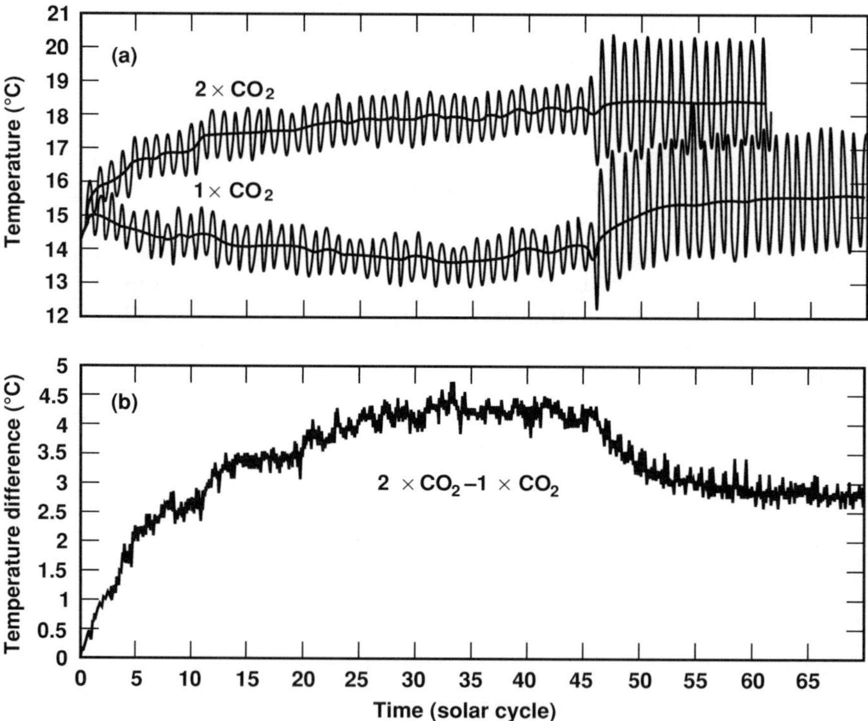


**Figure 5.5.** Time evolution of the global-mean surface air temperature for the $1 \times CO_2$ and $2 \times CO_2$ simulations with the OSU atmospheric GCM coupled to a mixed-layer ocean model (a), and their difference (b). A 12-month running mean is also shown in both panels. An accelerated annual cycle is used out to year 45 to accelerate the adjustment to a new equilibrium, after which a full annual cycle is used; the new equilibrium cycles show a larger seasonal oscillation and a slight adjustment to a new equilibrium. (From Schlesinger and Zhao, 1989.)

### 5.3.1. Surface Air Temperature

Table 5.4 presents the changes simulated by the GCMs in the global-mean equilibrium surface air temperature, $\Delta T_{eq}$, and in the global-mean precipitation rate per unit temperature change, $\Delta P/\Delta T_{eq}$. The results in this table indicate that the simulated $CO_2$-induced changes in the annual-mean, global-mean surface air temperature range from 2.8 to 5.2°C, and the corresponding changes in $\Delta P/\Delta T_{eq}$ range from 2.0%/°C to 2.9%/°C. The change in precipitation, $\Delta P$, increases with the size of $\Delta T_{eq}$ as a result of the Clausius-Clapeyron relation between the saturation vapor pressure of water vapor and temperature, which leads to a higher vapor pressure (i.e., more water vapor in the atmosphere) when the climate is warmer.

**Table 5.4.** Changes in the equilibrium global-mean surface air temperature ($\Delta T_{eq}$) and precipitation rate per unit temperature change ($\Delta P/\Delta T_{eq}$) simulated by atmospheric GCM/mixed-layer ocean models for a $CO_2$ doubling.

| Model/citation | $\Delta T_{eq}$ (°C)[1] | $\Delta P/\Delta T_{eq}$ (% of $1 \times CO_2$ value/°C) |
|---|---|---|
| GFDL/Wetherald and Manabe (1986) | 4.0 | 2.2 |
| GISS/Hansen et al. (1984) | 4.2 | 2.6 |
| NCAR/Washington and Meehl (1984) | 3.5 | 2.0 |
| OSU/Schlesinger and Zhao (1989) | 2.8 | 2.8 |
| UKMO/Wilson and Mitchell (1987) | 5.2 | 2.9 |

A partial explanation for the range of values of $\Delta T_{eq}$, and thus of $\Delta P$, is provided by Fig. 5.6, which displays the $CO_2$-induced warming plotted versus the annual-mean, global-mean surface air temperature of the simulated $1 \times CO_2$ climate. This figure shows that, in general, the warmer the simulated $1 \times CO_2$ control climate, the smaller the simulated $CO_2$-induced warming. This is due in part to the fact that there is a decrease in the positive ice-albedo feedback[4] with increasing $1 \times CO_2$ temperature,

[4] An initial warming in high latitudes is amplified by melting snow and/or sea ice, which results in a large decrease in surface albedo, an increase in the absorbed solar radiation, and an enhancement of the warming.

this as a result of there being less sea ice and snow. From these results, it appears that the $CO_2$-induced changes in temperature and precipitation simulated by these models would be in better agreement if their $1 \times CO_2$ global-mean surface air temperatures were in better agreement. Furthermore, it is tempting to conclude that the resulting model-derived responses (sensitivities) of temperature and precipitation of about 4.5°C and 12% would be correct if the $1 \times CO_2$ surface air temperature were in agreement with the observed temperature. However, while this agreement is a necessary condition for the common sensitivities of the models to be in agreement with the sensitivities of nature, it is not a sufficient condition.

That there is considerable residual uncertainty in the climate sensitivity is evidenced in Fig. 5.6 by the simulation results from the UKMO model. With a treatment of clouds similar to that of the other four models, the UKMO model simulated the largest warming with a value of 5.2°C. However, when it was assumed that the conversion rate from cloud water to precipitation is higher in ice clouds than in water clouds, the sensitivity of the UKMO model decreased to 2.7°C, this as a

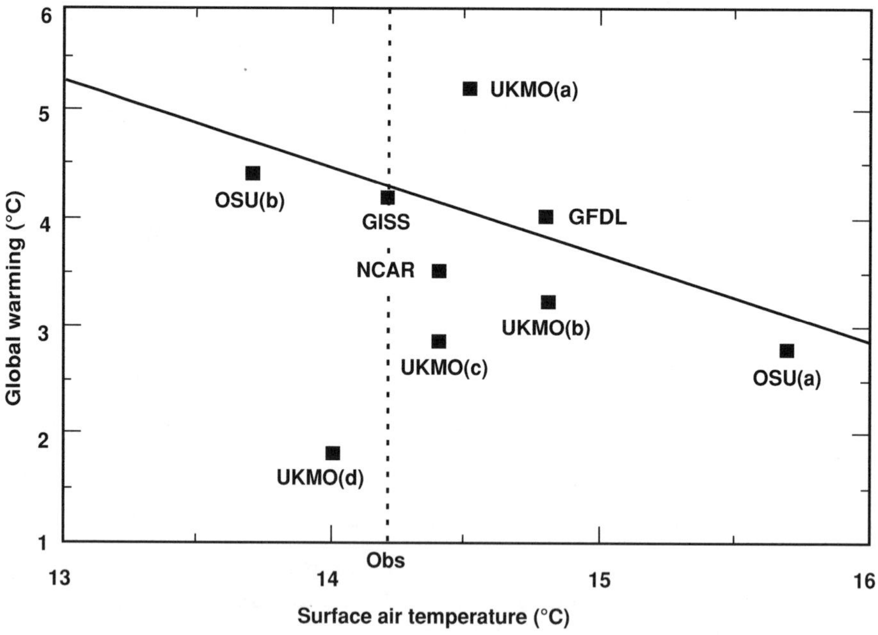


**Figure 5.6.** Increases in the global-mean surface air temperature as simulated for a $CO_2$ doubling by five GCMs versus their simulated $1 \times CO_2$ global-mean surface air temperature. For some models, different approximations of physical processes have given different results. "Obs" indicates the observed global-mean surface air temperature based on the data of Jenne (1975). (Adapted from Cess and Potter, 1988, including results from Mitchell et al., 1989.)

result of cloud-cover feedback.[5,6] When, in addition, the radiative properties of clouds were assumed to depend on the predicted cloud water, the resulting cloud optical depth feedback reduces the warming even further, to 1.9°C (Mitchell et al., 1989). These results indicate that the effects of clouds produce a large uncertainty in the simulated temperature sensitivity of the Earth to increased concentrations of greenhouse gases and that refinement of these preliminary estimates requires further study.

To establish the correctness of the models' temperature and precipitation sensitivities, even if they simulated the present climate correctly, is not a simple task. Validation of climate sensitivity requires the simulation of at least one climate different from that of the present, for example, that of the Wisconsin Ice Age 18,000 years before the present, and the comparison of such a simulated paleoclimate with observations. This requirement for model validation is a fundamental and inherently difficult problem in climate modeling and simulation (Schlesinger and Mitchell, 1985, 1987). This notwithstanding, it is of interest here to present and compare additional results of the $CO_2$-induced changes in climate simulated by these models.

The time-latitude distributions of the zonal-mean surface air temperature changes simulated by the five models for doubled $CO_2$ are presented in Fig. 5.7. This figure shows that the $CO_2$-induced temperature changes increase from the tropics, where the values range from about 2°C for the NCAR and OSU models to about 4°C for the GISS and UKMO models,[7] toward the poles, and that the seasonal variations of the $CO_2$-induced surface temperature changes are small between 50°S and 30°N and are large in the regions poleward of 50° latitude in both hemispheres. In the Northern Hemisphere, a warming minimum of about 2°C in summer is simulated near the pole by all five models. The models also simulate a warming maximum in fall with values that range from 8°C for the NCAR model to 16°C for the GFDL and UKMO models. This

---

[5] The direct radiative forcing resulting from a doubling of the $CO_2$ would increase the Earth's surface air temperature by about 1.2°C. That the projected temperature increase is larger than this value is the result of feedbacks in the climate system, with positive feedbacks enhancing the warming and negative feedbacks diminishing it. Because such positive and negative feedbacks strongly determine the magnitude of the warming, the warming is said to be $CO_2$-induced.

[6] When the troposphere warms as a result of increased $CO_2$, clouds located where temperatures lie between 0°C and about –15°C change from ice clouds to water clouds. Because of this and the assumed higher cloud water-to-precipitation conversion rate for ice clouds than for water clouds, the cloudiness increases. Because the solar effect of water clouds dominates their infrared effect, there is a reduction in the net radiation at the top of the atmosphere. This reduces the $CO_2$-induced warming.

[7] Versions (a) of the OSU and UKMO models are displayed (refer to Fig. 5.6).

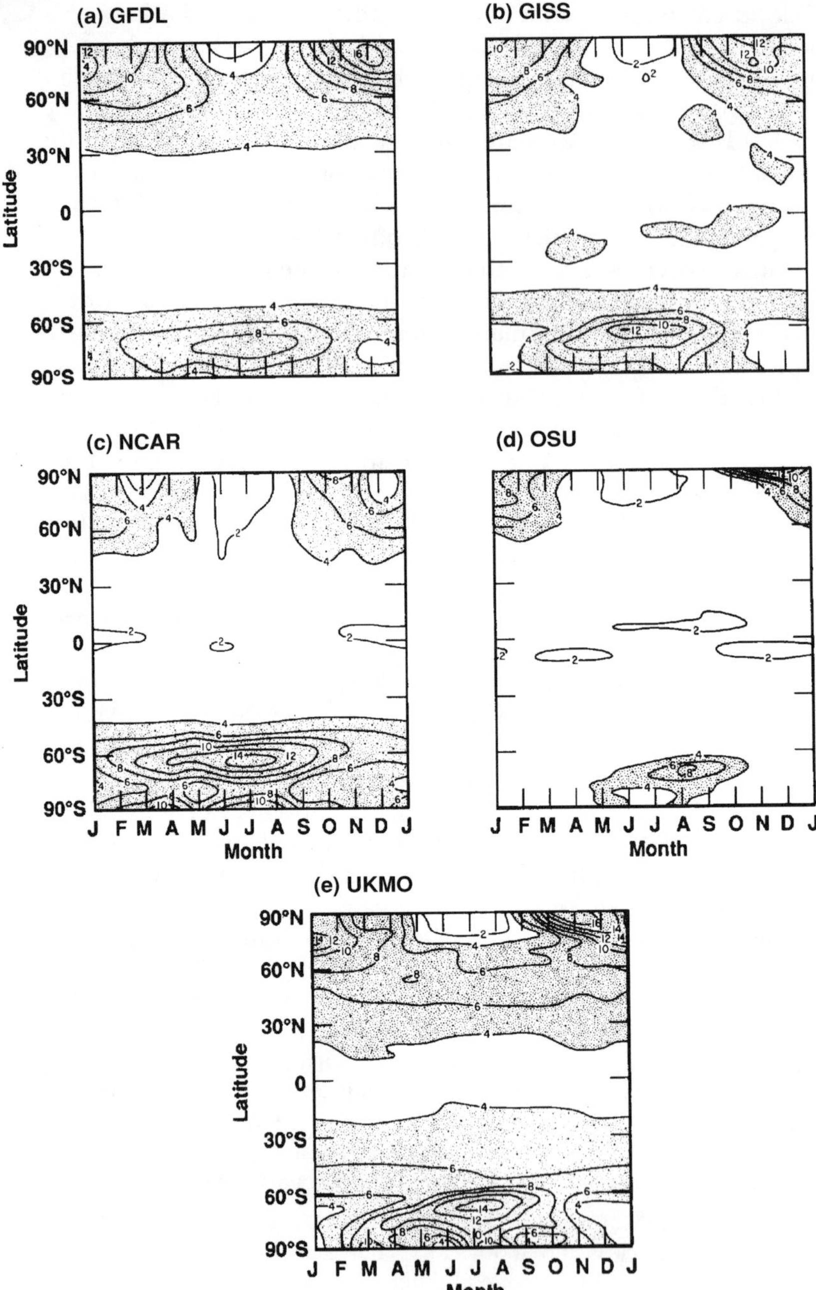


**Figure 5.7.** Time-latitude distribution of the change in zonal-mean surface air temperature (°C), $2 \times CO_2$ minus $1 \times CO_2$, as simulated by: (a) the GFDL model; (b) the GISS model; (c) the NCAR model; (d) the OSU model; and (e) the UKMO model. Shading indicates temperature increases larger than 4°C.

maximum extends into winter in all the simulations except that of NCAR, which instead exhibits a relative warming minimum near the pole. The NCAR model also simulates another polar warming minimum in spring that is not found in the other simulations. In the Southern Hemisphere, all five models simulate a maximum warming in winter and a minimum warming in summer. The winter warming maximum occurs near the Antarctic coast in all five simulations and ranges from 8°C in the GDFL and OSU simulations to 14°C in the NCAR and UKMO simulations.

The geographical distributions of the $2 \times CO_2$ minus $1 \times CO_2$ surface air temperature changes simulated for December-January-February (DJF) and June-July-August (JJA) by the five models are presented in Figs. 5.8 and 5.9, respectively. These figures show that all five models simulate a $CO_2$-induced surface air temperature warming virtually everywhere. In general, the warming is a minimum in the tropics during both seasons, at least over the ocean, and increases toward the winter pole. The tropical maritime warming minimum ranges from about 2°C in the NCAR and OSU simulations to about 4°C in the GFDL, GISS, and UKMO simulations.

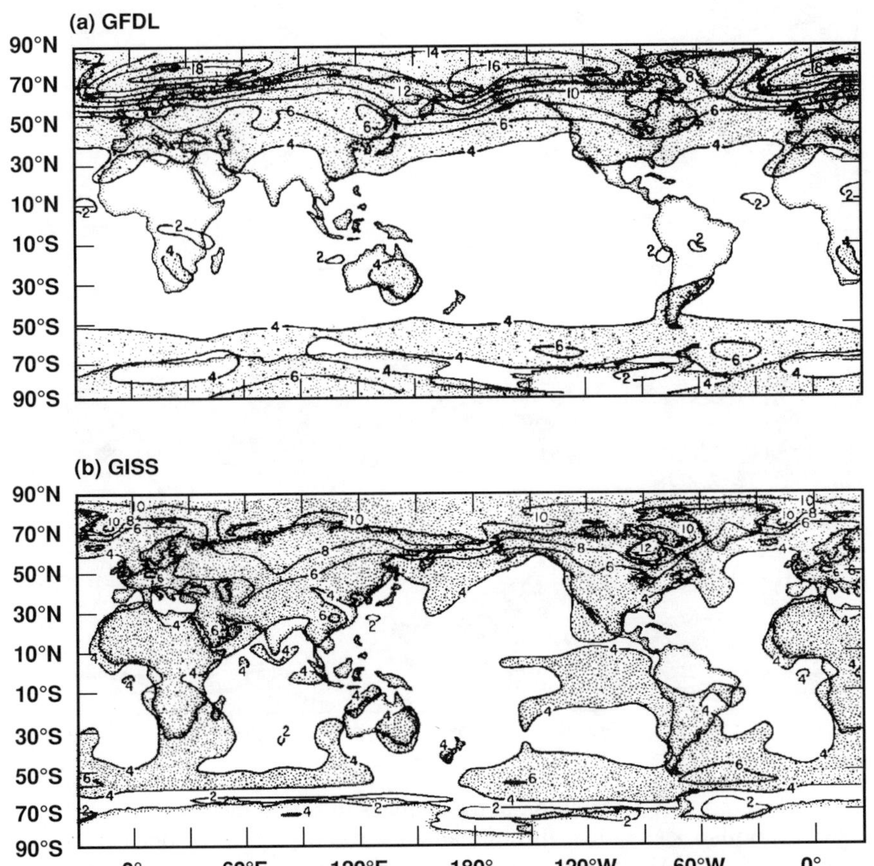

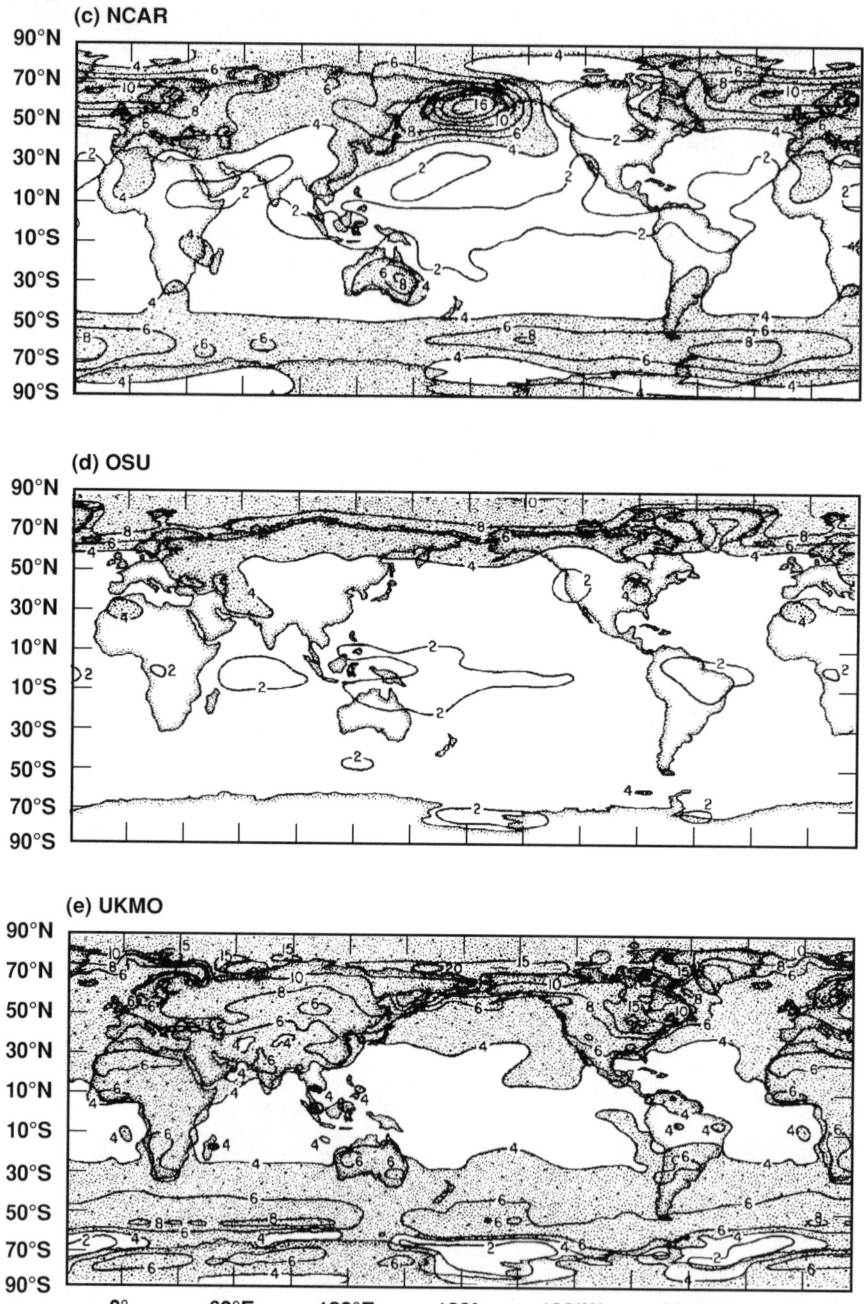


**Figure 5.8.** Geographical distribution of the changes in surface air temperature (°C), $2 \times CO_2$ minus $1 \times CO_2$, for DJF simulated by: (a) the GFDL model; (b) the GISS model; (c) the NCAR model; (d) the OSU model; and (e) the UKMO model. Shading indicates temperature increases larger than 4°C.

Maximum warming in DJF occurs in the Arctic in all the simulations except that of NCAR, which instead exhibits a warming maximum near 65°N. The maximum warming in JJA occurs around the Antarctic coast in all five simulations. The locations of the wintertime warming maxima in both hemispheres coincide with the locations where the $1 \times CO_2$ sea ice extent retreats in the $2 \times CO_2$ simulation. The magnitude of the wintertime warming maxima in the Northern Hemisphere ranges from 10°C in the GISS and OSU simulations to 20°C in the UKMO simulation; in the Southern Hemisphere, it ranges from 10°C in the OSU simulation to 20°C in the UKMO simulation. In JJA, there is a warming minimum in the Arctic of about 2°C in all five simulations.

Figures 5.8 and 5.9 also show that, although there are similarities in the $CO_2$-induced regional temperature changes simulated by the models, there are marked differences in both their magnitude and seasonality.

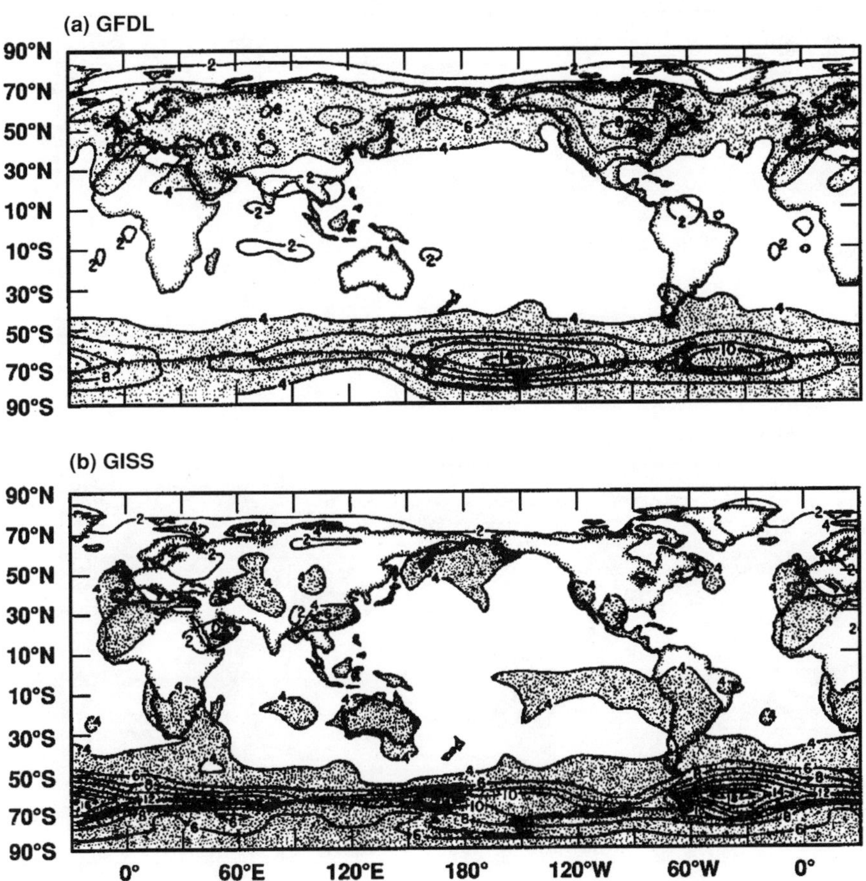

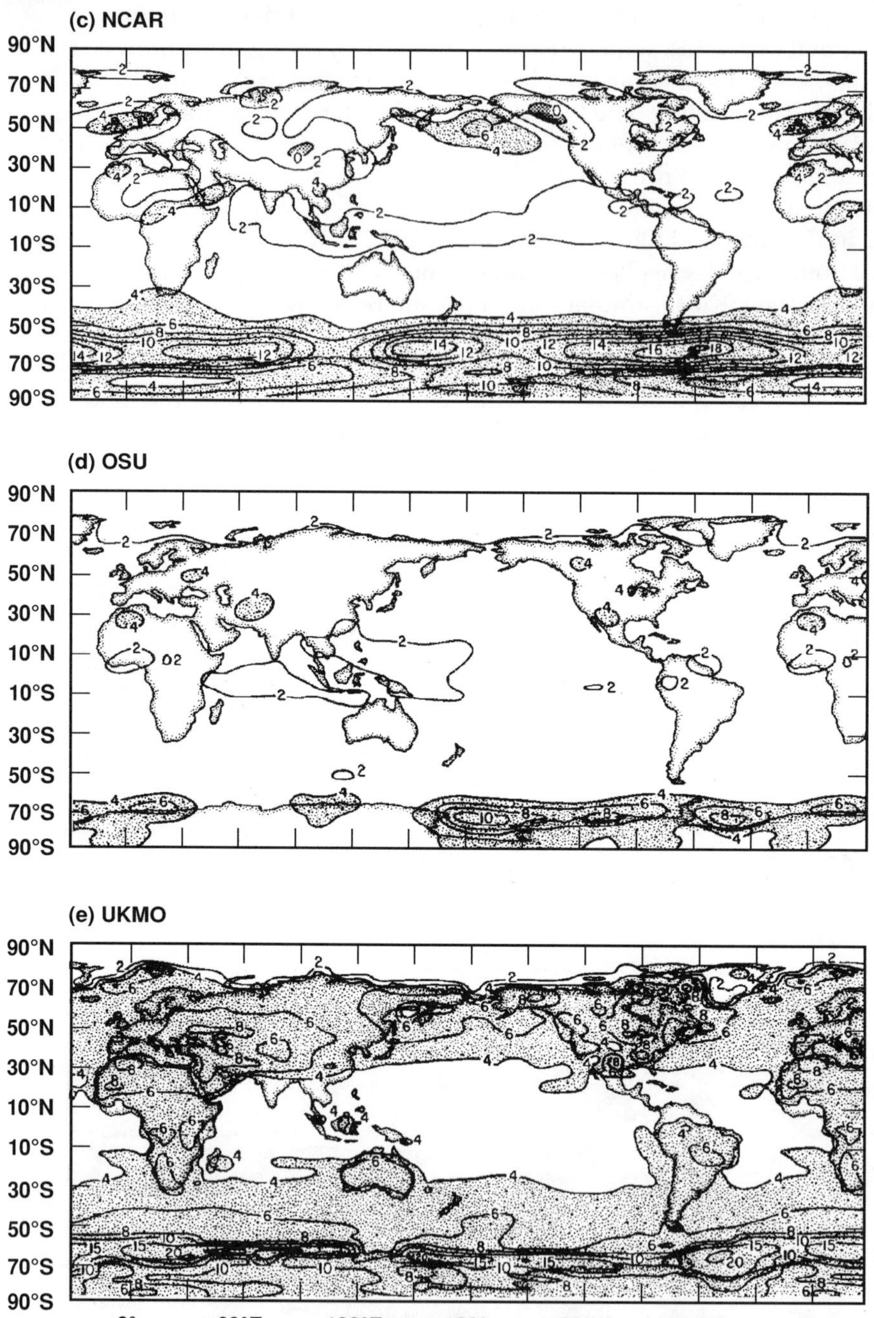


**Figure 5.9.** Geographical distribution of the changes in the surface air temperature (°C), 2 × $CO_2$ minus 1 × $CO_2$, for JJA simulated by: (a) the GFDL model; (b) the GISS model; (c) the NCAR model; (d) the OSU model; and (e) the UKMO model. Shading indicates temperature increases larger than 4°C.

For example, in North America the wintertime warming generally increases with latitude in the GFDL, GISS, and UKMO simulations, with values of 4°C in the south and 10°C in the north, while both the NCAR and OSU simulations exhibit a warming minimum of 2°C centered over Canada and over the Pacific northwest, respectively. Also, the $CO_2$-induced warming in summer compared to that in winter is simulated to be smaller by the GISS and NCAR models, and to be comparable by the GFDL, OSU and UKMO models. Further similarities and differences can be seen for the $CO_2$-induced temperature changes simulated for the other continents.

The level of quantitative agreement among the models' geographical distributions of $CO_2$-induced temperature changes for DJF and JJA is presented in Tables 5.5 and 5.6, respectively, in terms of a pattern-correlation coefficient. This coefficient has been calculated so that the differences among the models' simulated global-mean temperature changes are eliminated. Thus, for example, if the geographical distributions of two of the models are the same except that one is a constant multiple of the other, the pattern-correlation coefficient will be unity. On the other hand, if the patterns relative to their respective global means are opposite, the coefficient will be –1. Tables 5.5 and 5.6 compare the patterns pairwise among the models, thus giving 10 comparisons. These

**Table 5.5.** Pattern correlation between pairs of model calculations by different groups for the DJF temperature changes resulting from a doubling of the $CO_2$ concentration.

| | GFDL | GISS | OSU | NCAR | UKMO |
|---|---|---|---|---|---|
| GFDL | 1.000 | 0.718 | 0.804 | 0.520 | 0.765 |
| GISS | 0.718 | 1.000 | 0.714 | 0.340 | 0.820 |
| OSU | 0.804 | 0.714 | 1.000 | 0.380 | 0.749 |
| NCAR | 0.520 | 0.340 | 0.380 | 1.000 | 0.445 |
| UKMO | 0.765 | 0.820 | 0.749 | 0.445 | 1.000 |

**Table 5.6.** Pattern correlation between pairs of model calculations by different groups for the JJA temperature changes resulting from a doubling of the $CO_2$ concentration.

| | GFDL | GISS | OSU | NCAR | UKMO |
|---|---|---|---|---|---|
| GFDL | 1.000 | 0.618 | 0.679 | 0.578 | 0.715 |
| GISS | 0.618 | 1.000 | 0.509 | 0.696 | 0.730 |
| OSU | 0.679 | 0.509 | 1.000 | 0.370 | 0.604 |
| NCAR | 0.578 | 0.696 | 0.370 | 1.000 | 0.667 |
| UKMO | 0.715 | 0.730 | 0.604 | 0.667 | 1.000 |

comparisons are all positive, with values between 0.340 and 0.820. The patterns of temperature change simulated by the GISS and UKMO models are most similar for both DJF (0.820) and JJA (0.730), while the patterns simulated by the NCAR and GISS models are least similar for DJF (0.340), and the NCAR and OSU models are least similar for JJA (0.370).

### 5.3.2. Precipitation Rate

The geographical distributions of the $2 \times CO_2$ minus $1 \times CO_2$ precipitation rate changes simulated for DJF and JJA by the five models are presented in Figs. 5.10 and 5.11. These figures show that both positive and negative changes in precipitation rate are simulated by all five models, and that the largest changes generally occur between 30°S and 30°N. The precipitation changes poleward of these latitudes are generally positive in both seasons over both the ocean and land. However, the GFDL model simulates a decreased precipitation rate in JJA over most of North America and Europe, and over much of Asia. With this exception,

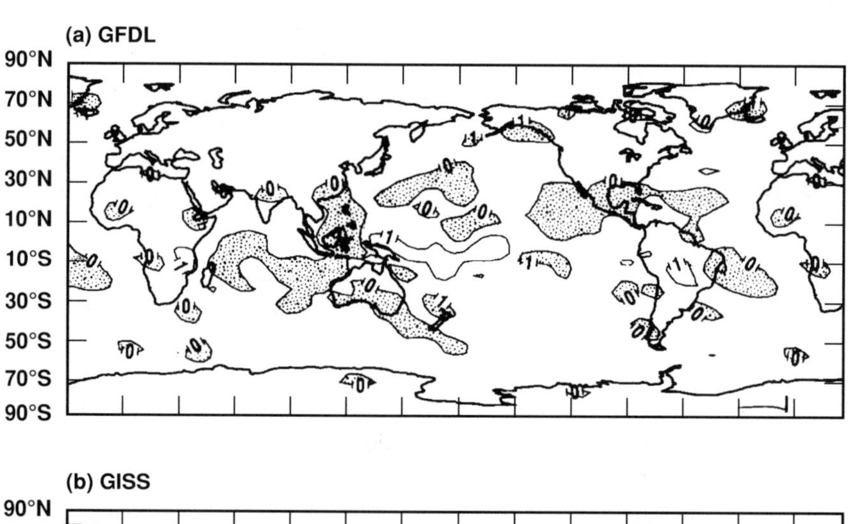


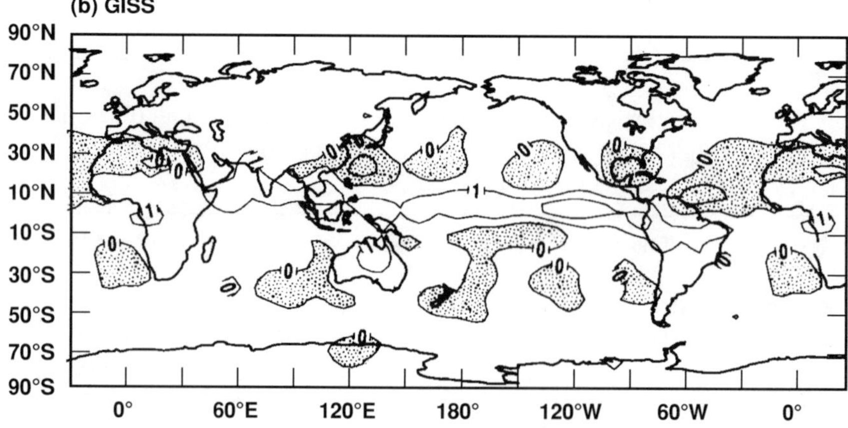

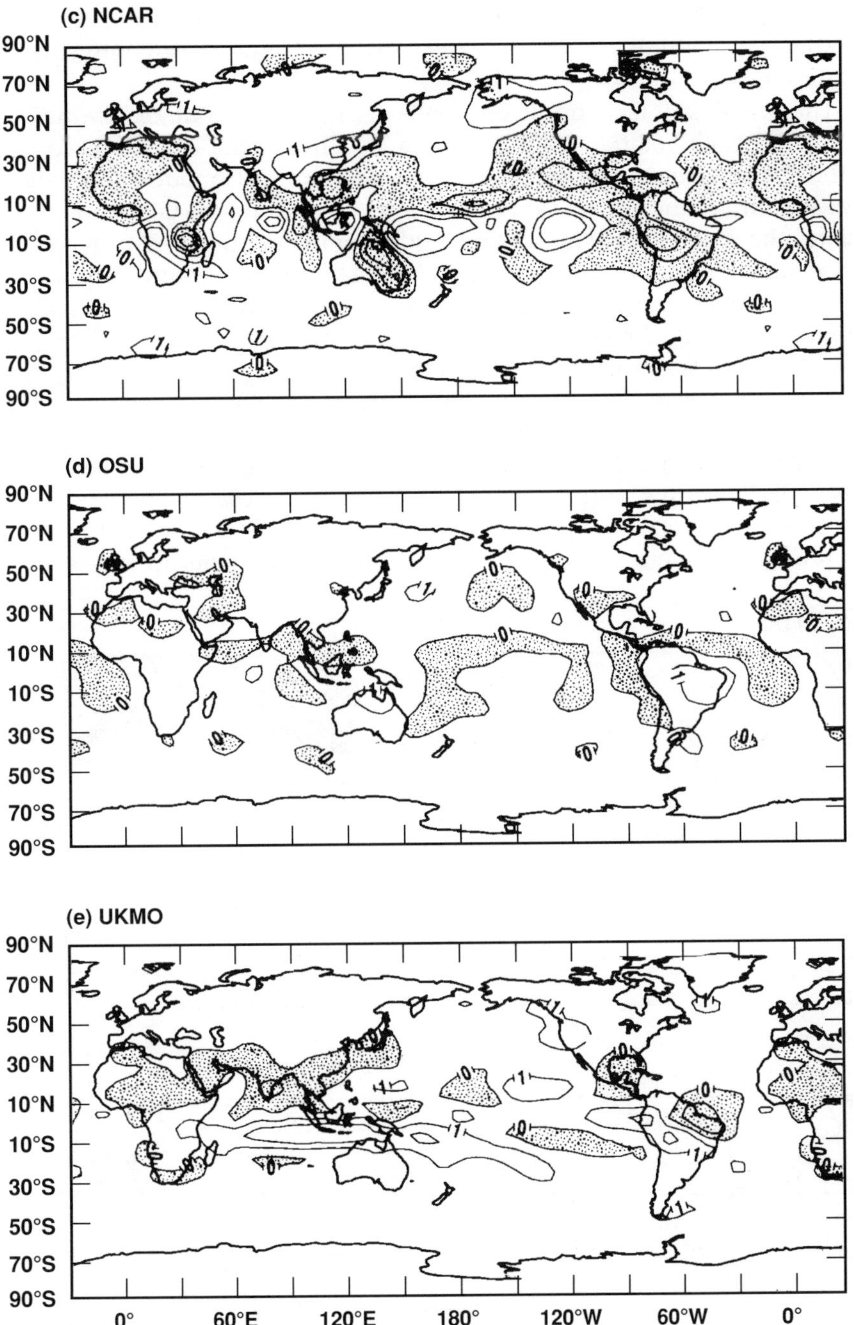


**Figure 5.10.** Geographical distribution of the changes in precipitation rate (mm/day), $2 \times CO_2$ minus $1 \times CO_2$, for DJF simulated by: (a) the GFDL model; (b) the GISS model; (c) the NCAR model; (d) the OSU model; and (e) the UKMO model. These fields have been smoothed with a 9-point filter. Light shading indicates precipitation decreases, and heavy shading indicates precipitation increases larger than 5 mm/day.

the models generally simulate precipitation increases of less than 1 mm/day during both seasons over the Northern Hemisphere continents and Antarctica.

The precipitation changes between 30°S and 30°N simulated by the models show both qualitative and quantitative differences. For example, over the Pacific Ocean in DJF the GFDL, NCAR, and OSU models simulate increases south of the equator and decreases to the north, the GISS model simulates the reverse, and the UKMO model simulates an increase on the equator and decreases to the north and south. As another example, in JJA the GFDL model simulates a large increase in the Indian monsoon precipitation, the OSU model simulates a smaller increase, but the GISS and NCAR models simulate a small decrease.

The pattern-correlation coefficients between the models' geographical distributions of $CO_2$-induced precipitation changes are presented in Tables 5.7 and 5.8 for DJF and JJA, respectively. These tables show negative as well as positive values, which means that at least somewhere over the Earth the models depict precipitation changes of opposite sign. In all cases, however, the magnitudes of the correlation coefficients are not

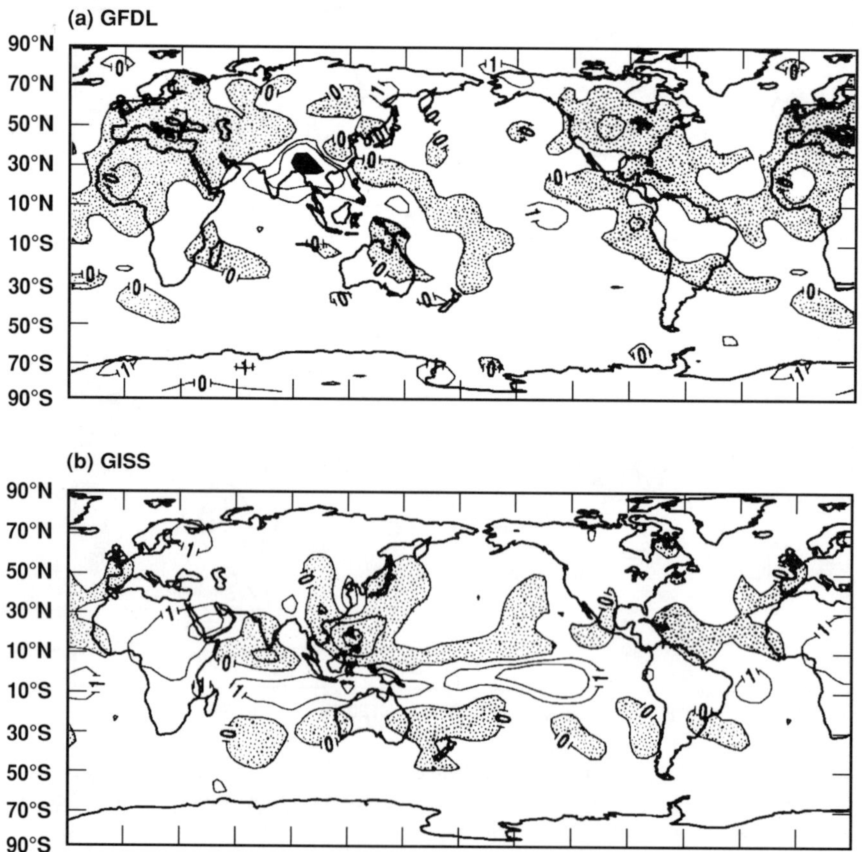

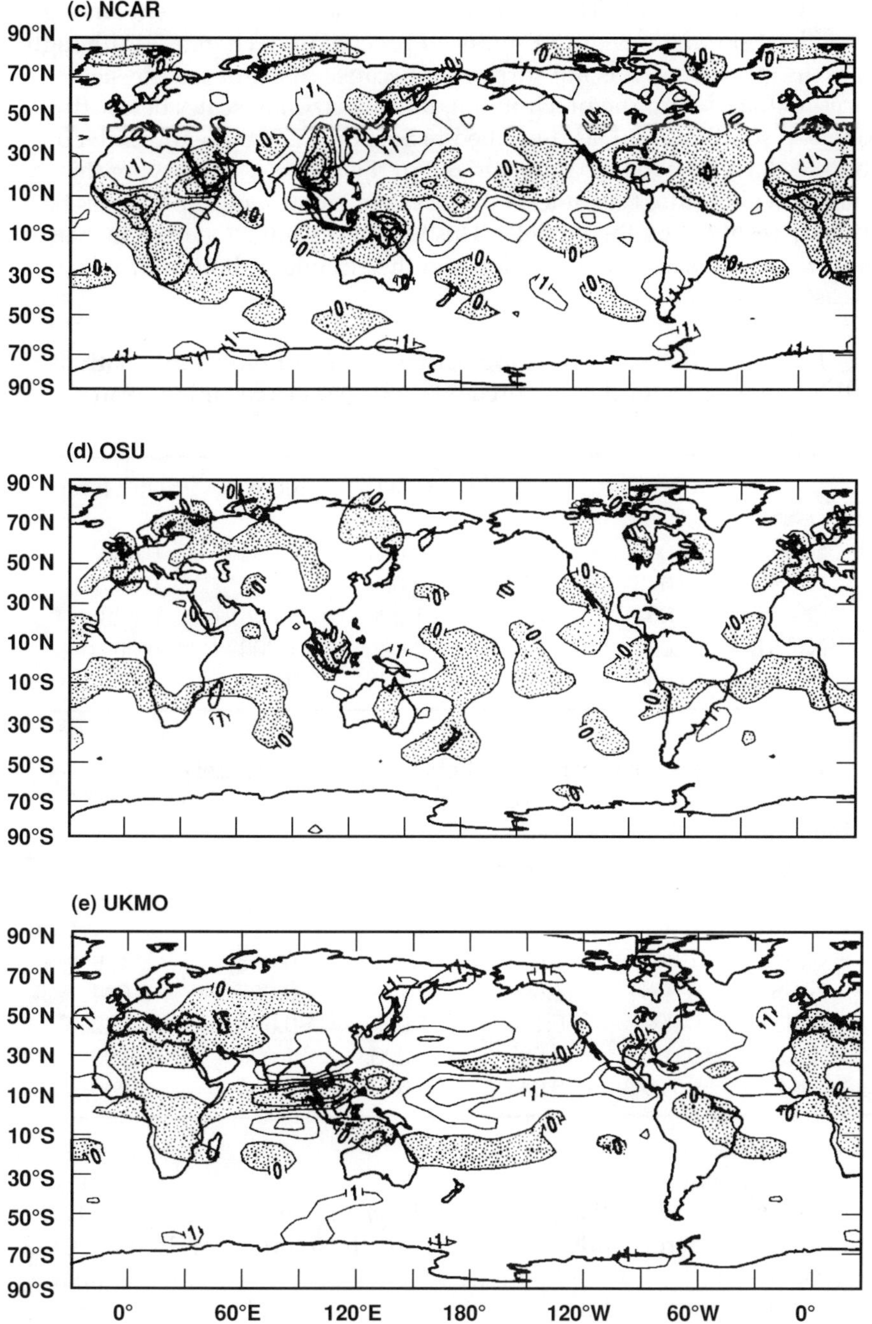


**Figure 5.11.** Geographical distribution of the precipitation rate change (mm/day), $2 \times CO_2$ minus $1 \times CO_2$, for JJA simulated by: (a) the GFDL model; (b) the GISS model; (c) the NCAR model; (d) the OSU model; and (e) the UKMO model. These fields have been smoothed with a 9-point filter. Light shading indicates precipitation decreases, and heavy shading indicates precipitation increases larger than 5 mm/day.

greater than about 0.13. This means that there is little agreement among the models' simulations of $CO_2$-induced precipitation changes. Although this is disappointing, it is not surprising because many of the physical processes that determine precipitation have horizontal scales below the models' resolutions and, as described in Section 5.2.3, these unresolved processes have had to be parameterized in the models. Improving this situation will require that at least mesoscale meteorological processes be explicitly resolved by the models. This will necessitate a tenfold increase in the models' horizontal resolution and the attendant thousandfold increase in computer time.

**Table 5.7.** Pattern correlation between pairs of model calculations by different groups for the DJF precipitation changes resulting from a doubling of the $CO_2$ concentration.

| | GFDL | GISS | OSU | NCAR | UKMO |
|---|---|---|---|---|---|
| GFDL | 1.000 | 0.048 | –0.054 | 0.094 | 0.060 |
| GISS | 0.048 | 1.000 | –0.069 | –0.115 | 0.092 |
| OSU | –0.054 | –0.069 | 1.000 | 0.025 | 0.028 |
| NCAR | 0.094 | –0.115 | 0.025 | 1.000 | –0.009 |
| UKMO | 0.060 | 0.092 | 0.028 | –0.009 | 1.000 |

**Table 5.8.** Pattern correlation between pairs of model calculations by different groups for the JJA precipitation changes resulting from a doubling of the $CO_2$ concentration.

| | GFDL | GISS | OSU | NCAR | UKMO |
|---|---|---|---|---|---|
| GFDL | 1.000 | –0.008 | 0.099 | 0.083 | 0.129 |
| GISS | –0.008 | 1.000 | –0.104 | 0.007 | –0.044 |
| OSU | 0.099 | –0.104 | 1.000 | –0.090 | 0.042 |
| NCAR | 0.083 | 0.007 | –0.090 | 1.000 | –0.100 |
| UKMO | 0.129 | –0.044 | 0.042 | –0.100 | 1.000 |

### 5.3.3. Soil Moisture

The time-latitude distributions of the change in zonal-mean soil moisture over ice-free land simulated by the five models for doubled $CO_2$ are presented in Fig. 5.12. This figure shows that all five models simulate increased soil moisture during winter in the Northern Hemisphere from about 45° to 70°N. In the Northern Hemisphere summer, the GISS and NCAR models simulate a minimum increase in soil moisture, while the GFDL, OSU, and UKMO models simulate decreased soil moisture from about 30° to 70°N.

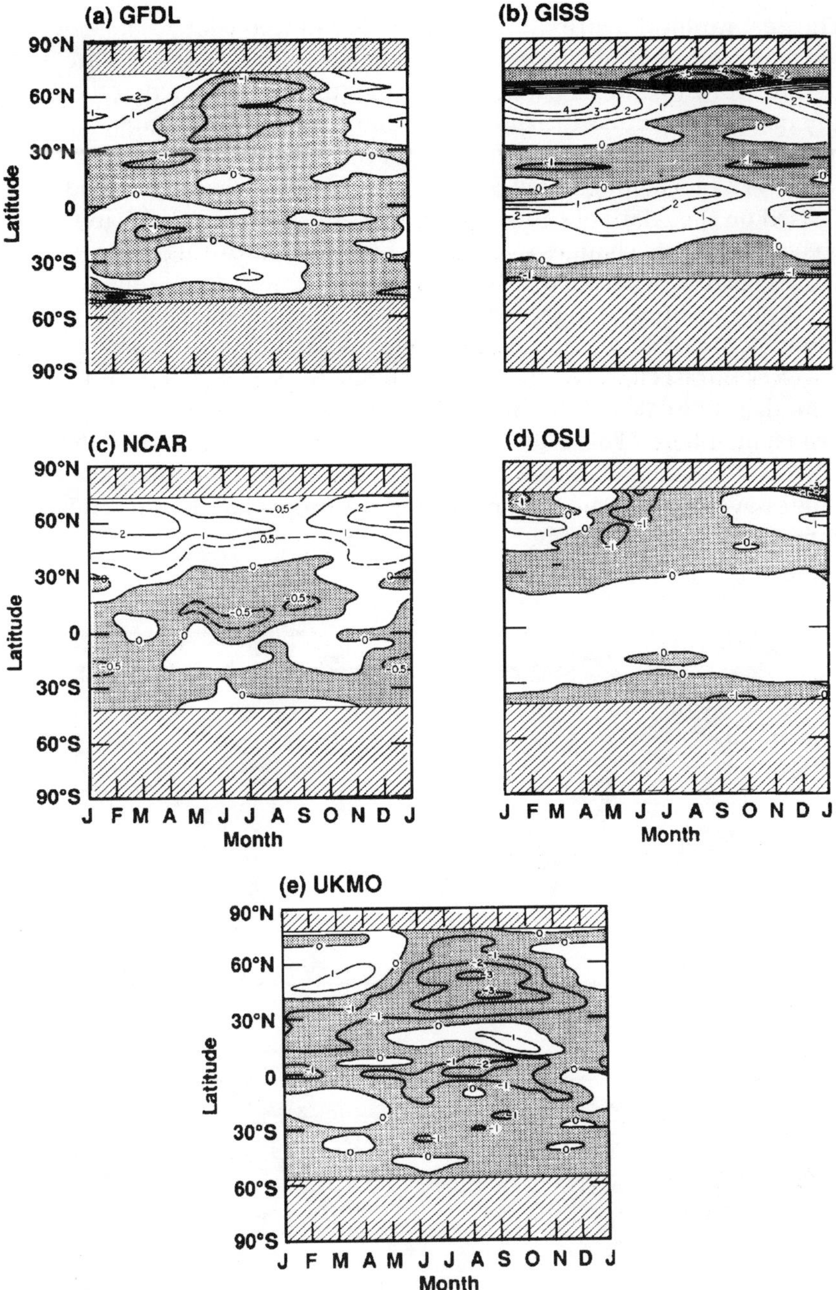


**Figure 5.12.** Time-latitude distribution of the zonal-mean soil water change (cm), $2 \times CO_2$ minus $1 \times CO_2$, over ice-free land only, simulated by: (a) the GFDL model; (b) the GISS model; (c) the NCAR model; (d) the OSU model; and (e) the UKMO model. Stipple indicates a decrease in soil water, hatching indicates latitudes where there is no ice-free land.

The geographical distributions of the $CO_2$-induced changes in soil moisture over ice-free land simulated by the models for DJF and JJA are presented in Fig. 5.13 and 5.14, respectively. Figure 5.13 shows that there are many qualitative similarities among the simulated soil-moisture changes for DJF, particularly over the Northern Hemisphere continents where there is a moistening of much of the soil. This is what one would expect based on the relatively good agreement among the models' simulated precipitation rate changes over these same regions during DJF (Fig. 5.10). Figure 5.14 shows that there is less agreement among the simulated soil-moisture changes for JJA. In particular, the GISS and NCAR models simulate regions of both increased and decreased soil moisture over Eurasia and North America, while the GFDL, OSU, and UKMO models simulate a desiccation virtually everywhere in the Northern Hemisphere. This is also what one would expect based on the relatively poor agreement among the models' simulated precipitation rate changes over these same regions during JJA (Fig. 5.11). Because it is the summer soil-moisture changes that will strongly determine the

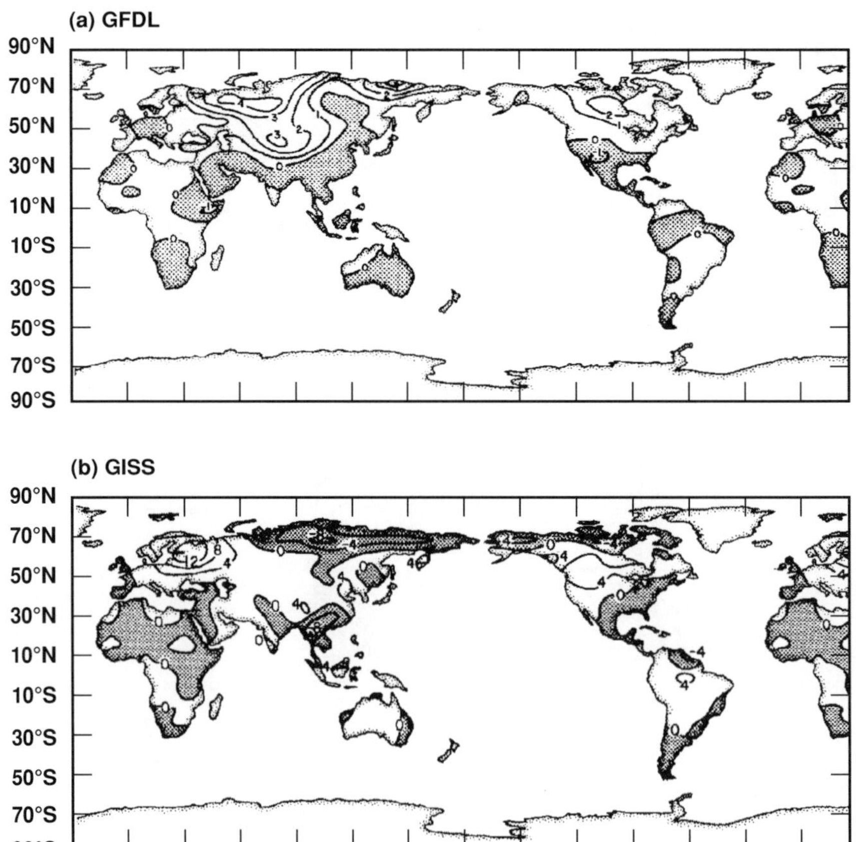

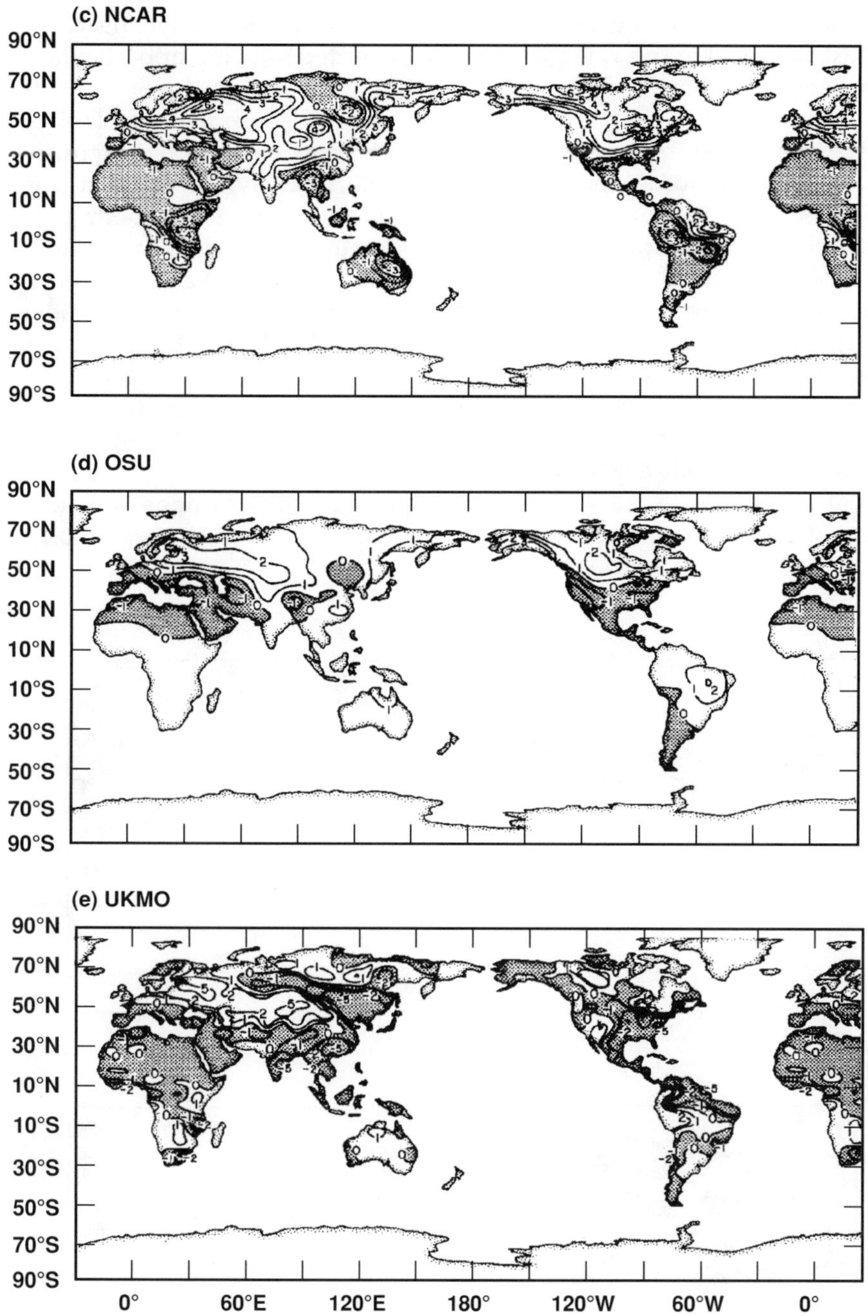


**Figure 5.13.** Geographical distribution of the soil water change (cm), $2 \times CO_2$ minus $1 \times CO_2$, for DJF simulated by: (a) the GFDL model; (b) the GISS model; (c) the NCAR model; (d) the OSU model; and (e) the UKMO model. Shading indicates a decrease in soil water.

agricultural impact of a greenhouse-gas-induced climate change, the causes of the differences in this quantity must be regarded as being of particular importance.

One way to assess the credibility of the models' simulations of $CO_2$-induced soil moisture changes is to evaluate their performance in simulating soil moisture for the present climate. Unfortunately, the soil-moisture content is a quantity for which measurements are incomplete in both geographical and temporal distributions. However, such data have been gathered for some parts of the Soviet Union (see Chapter 6) and have been compared with the soil moisture simulated by two models for four regions, as presented in Fig. 5.15 (Vinnikov and Yeserkepova, 1989). This figure shows that the models simulate the annual cycle of soil moisture reasonably well, but underestimate the annual-mean soil moisture, particularly in the northernmost region (55° to 60°N, 22.5° to 37.5°E), where the observed soil moisture exceeds the field capacity as a result of proximity of the ground water table to the surface. The importance of this underestimated annual-mean soil moisture on the simulated, $CO_2$-induced change in soil moisture must be determined.

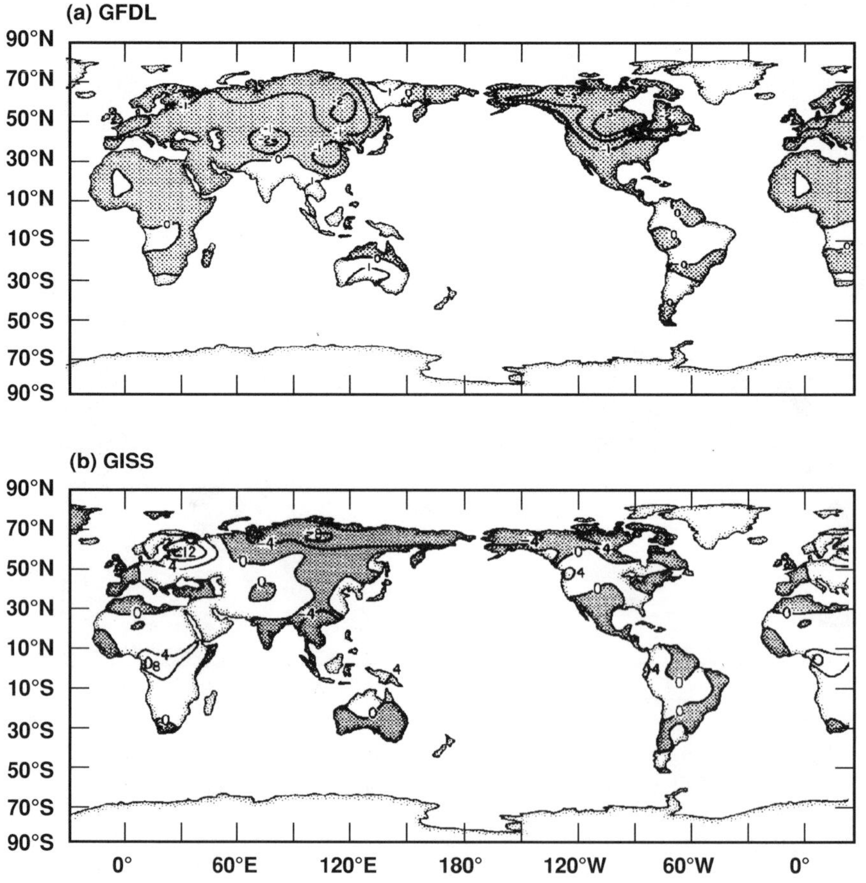

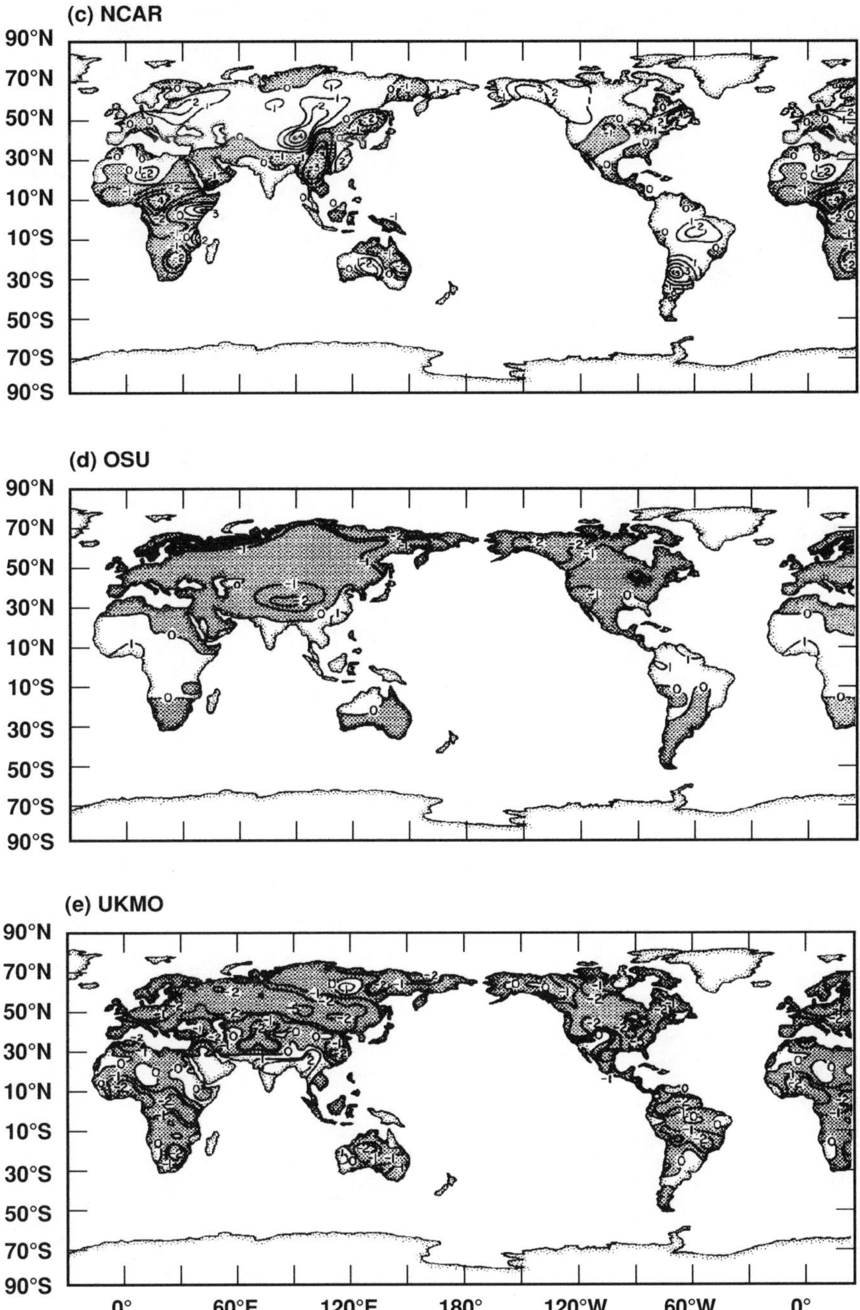


**Figure 5.14.** Geographical distribution of the soil water change (cm), 2 × $CO_2$ minus 1 × $CO_2$, for JJA simulated by: (a) the GFDL model; (b) the GISS model; (c) the NCAR model; (d) the OSU model; and (e) the UKMO model. Shading indicates a decrease in soil water.

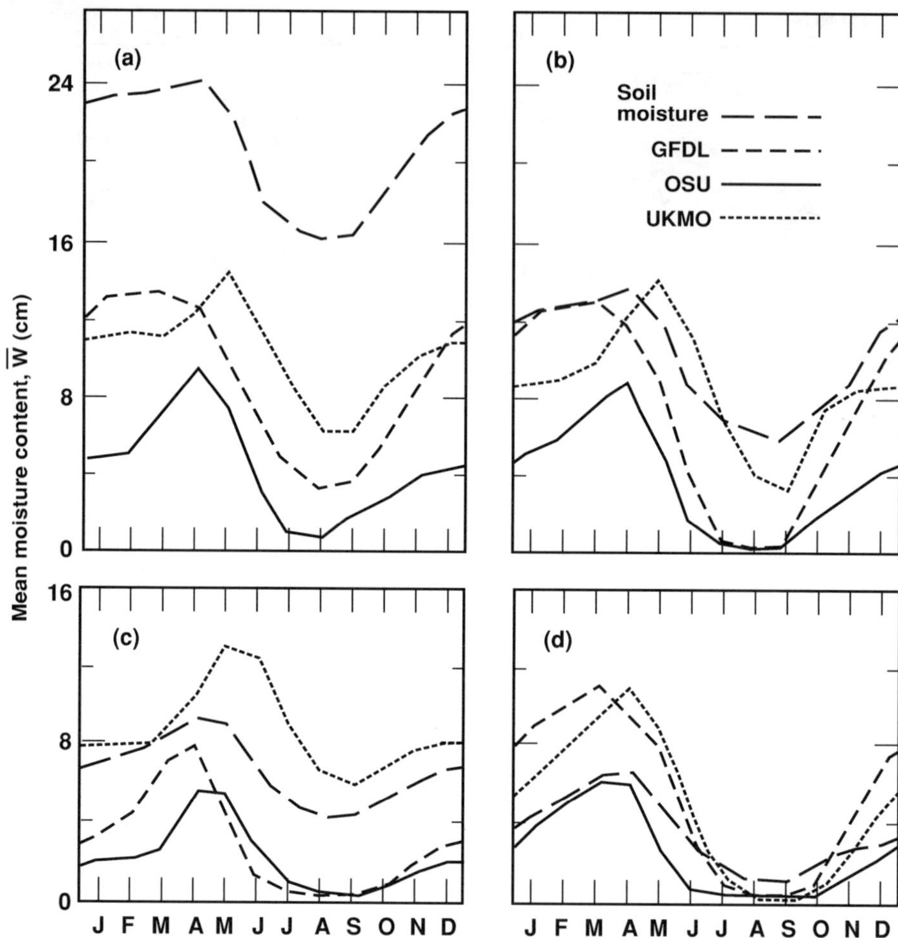


**Figure 5.15.** Comparison of annual changes of soil moisture simulated by the GFDL, the OSU, and the UKMO models with soil moisture measurements for (a) 55° to 60°N, 25° to 50°E; (b) 50° to 55°N; 25° to 60°E; (c) 50° to 55°N, 60° to 110°E; and (d) 45° to 50°N, 50° to 75°E. (From Vinnikov and Yeserkepova, 1990.).

## 5.4. TIME-DEPENDENT CLIMATE CHANGE

In the preceding section, results have been presented from the most sophisticated climate models that have been used to estimate the changes in the equilibrium climate of the Earth if the effective $CO_2$ concentration were to become twice as large as its pre-industrial value. This was a reasonable first question to have investigated because such an effective $CO_2$ doubling is projected to occur sometime during the next century (Chapter 4). Having found that the estimated changes in equilibrium climate are large, it is now natural to ask:

(1) How much should the temperature have warmed during the past 100 years as a result of the increases in greenhouse gas concentrations during that time?

(2) Has such a warming been detected in the observational record?

(3) How much should the climate in the future change due to the projected increases in concentrations of greenhouse gases?

These questions concern the time-dependent changes in climate. To answer them we must address two other questions; namely, how much time is required by the climate system to evolve from one climatic state to another, and are the patterns of climate change during this evolution different from the patterns determined in equilibrium simulations?

**5.4.1. Response Time of the Climate System**

As discussed in Chapter 4, the $CO_2$ concentration has increased from about 290 ppmv in 1880 to 351 ppmv in 1988, and the concentrations of the other greenhouse gases, $CH_4$, CFCs 11 and 12, and $N_2O$, have also increased (Chapter 4). The expected change in the equilibrium surface air temperature from 1880 to 1988, $\Delta T_{eq}$ (1988), induced by these increases in greenhouse gas concentrations can be estimated from the known radiative properties of these gases and from the equilibrium warming of the climate system induced by a $CO_2$ doubling, $\Delta T_{eq}(2 \times CO_2)$. If it is assumed that the latter is 4°C based on results from equilibrium simulations by the atmospheric GCM/mixed-layer ocean models (Fig. 5.6), then $\Delta T_{eq}$ (1988) is about 1.6°C (Fig. 5.16). Observations show, however, that the global-mean surface air temperature of the Earth has increased by only about 0.4°C to 0.5°C during this time period (see Chapter 3). One explanation for this discrepancy might be that our climate models are about three times as sensitive as nature. Another explanation might be that natural variability is counter-balancing at least some of the greenhouse-gas-induced effect. Of special interest in this regard is the 100-year control simulation of Hansen et al. (1988), which shows a temperature decrease of nearly 0.3°C midway through the simulation (Fig. 5.17). This change in temperature was apparently caused by a change in tropical cloudiness without any external changes to the model (Barnett, 1990). The importance of these results in understanding variations in the modern climate record (e.g., the decadal fluctuations from the slowly increasing trend shown in Fig. 5.16) is the possibility that the variations could be produced, at least in part, simply by the chaotic behavior of the climate (Lorenz, 1984).

A third, and very important, reason for the difference between model equilibrium estimates and observed temperatures is that the climate system is not expected to be in equilibrium with the present concentrations of greenhouse gases due to the transport into the ocean of the surface heating induced by the greenhouse gases. This heat transport would reduce the greenhouse-gas-induced surface warming at any time to below its equilibrium value and, thereby, would result in a delay in reaching the equilibrium warming.

Estimates of this time delay due to the ocean have been made with a variety of simplified climate/ocean models with results that range from about 10 years to 100 years for the e-folding time, $\tau_e$, of the climate system, that is, the characteristic time required for the evolving temperature change to reach 63% of the equilibrium value (Hoffert et al., 1980; Bryan et al., 1982; Hansen et al., 1984; Bryan et al., 1984; Bryan and Spelman, 1985). If $\tau_e$ is only 10 years, then the climate system is close to being in equilibrium with the instantaneous greenhouse gas concentrations, and the discrepancy between the observed and model-predicted warmings would mean that our climate models are about three times more sensitive than nature. However, if $\tau_e$ is as large as 100 years, then the discrepancy between the

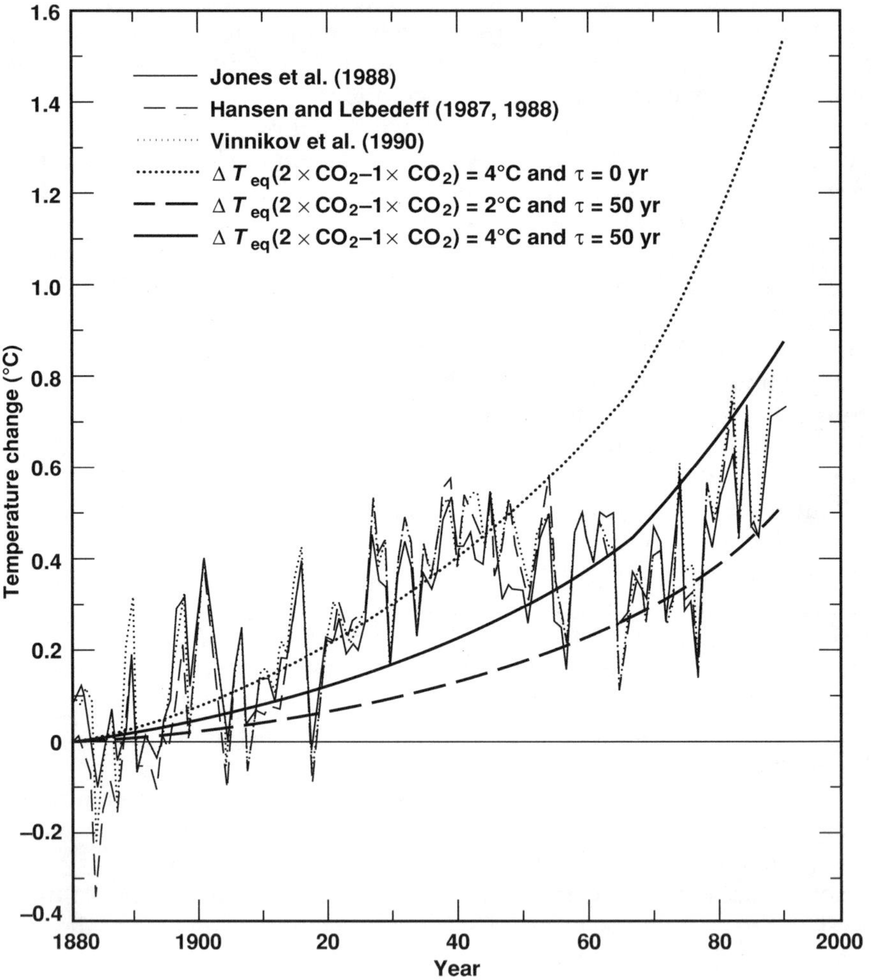


**Figure 5.16.** The change in global-mean surface air temperature expected from the increase in the greenhouse gas concentrations from 1880 to 1988 compared with the corresponding equilibrium temperature change, $\Delta T_{eq}$, and the observed temperature change as determined by Jones et al. (1986), Hansen and Lebedeff (1987, 1988), and Vinnikov et al. (1990).

observed and model-predicted warmings could be explained as being due to the thermal inertia of the ocean that results from the downward transport of heat into its interior.

Obtaining an accurate estimate of $\tau_e$ requires a physically based model of the atmosphere-ocean system, that is, a global atmosphere-ocean general circulation model in which the annual cycle of solar insolation is included. Schlesinger et al. (1985) used such a coupled atmosphere-ocean general circulation model to estimate $\tau_e$ by performing two 20-year simulations, a $1 \times CO_2$ simulation with a $CO_2$ concentration of 326 ppmv, and a $2 \times CO_2$ simulation with the $CO_2$ concentration doubled. By using energy balance climate/multi-box ocean models to reproduce and extend the temperature evolutions of these simulations, Schlesinger et al. (1985) and Schlesinger and Jiang (1990) estimated that $\tau_e$ was about 50 years.

The effect of a 50-year climatic response time on the temperature change expected due to the increase in greenhouse gases from 1880 to 1988 is presented in Fig. 5.16, together with the equilibrium and observed temperature evolutions during this time period. This figure shows that the expected warming in 1988 relative to 1880 is 0.87°C, while the equilibrium warming is 1.57°C. Thus, the transport of heat into the ocean reduces the expected warming to somewhat more than half the equilibrium warming. The expected warming, however, is larger than the observed warming. To reconcile this difference would require a decrease in the assumed value of $\Delta T_{eq}(2 \times CO_2)$ from 4°C to about 2°C. Such a decrease in $\Delta T_{eq}(2 \times CO_2)$ is a possibility that cannot now be excluded because the feedback effect of changes in the optical depth of the clouds has been included in only one model simulation of $CO_2$-induced climate change (Mitchell et al., 1989), and that simulation showed a large decrease in the resultant value of $\Delta T_{eq}(2 \times CO_2)$ (Fig. 5.6).

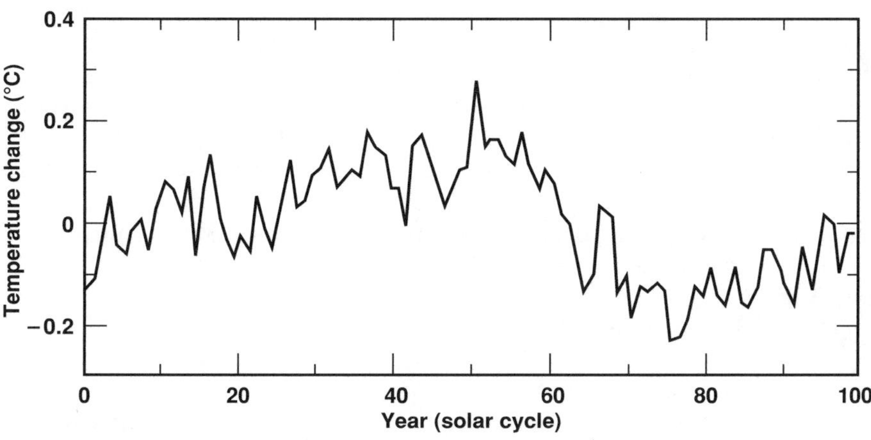


**Figure 5.17.** Departures of the annual global mean surface air temperature from the 100-year mean during the 100-year control simulation of the GISS GCM. (From Hansen et al., 1988.)

Figure 5.16 also illustrates an important consequence of the delay in the response of the climate system to increasing greenhouse gas concentrations. In particular, if concentrations were to remain at their present levels, the Earth's surface temperature would continue to increase toward its as-yet unattained equilibrium value. This demonstrates the important role of the ocean in the greenhouse-gas-induced climate change issue: Because the ocean transports heat into its interior, the present warming is relatively "small" and either just within or slightly above the natural variation of climate; however, because of this, when the warming becomes unequivocally evident, continued future warming is inevitable, even if the greenhouse gas concentrations are prevented from increasing further.

### 5.4.2. Response Similitude of the Climate System

In all of the GCM studies discussed above, the $CO_2$ concentration in the experiment simulation was taken to be constant in time with a value twice that of the control simulation. Recently, three GCM simulations have been performed in which the $CO_2$ concentration in the experiment was taken to increase with time. In the study performed with the GISS model by Hansen et al. (1988), the concentrations of $CO_2$, $CH_4$, CFCs 11 and 12, and $N_2O$ were taken to increase from 1958 to the mid-1980s according to their observed or estimated increases, and three scenarios for the future increases in these gases from the mid-1980s to 2050 were investigated. In the study performed with the NCAR model by Washington and Meehl (1989), the $CO_2$ concentration alone was increased by 1% per year over a 30-year period, whereas, in the study performed with the GFDL model by Stouffer et al. (1989), the $CO_2$ concentration alone was increased by 1% per year over a 100-year period. In the nonequilibrium study by Hansen et al. (1988), the atmospheric GCM was coupled to a mixed-layer ocean with horizontal heat transport prescribed and vertical heat transport to the deeper ocean predicted, but only by diffusive heat transfer. In the nonequilibrium studies by Washington and Meehl (1989) and Stouffer et al. (1989), the atmospheric GCM was coupled to an oceanic GCM in which both the horizontal and vertical transports of heat were predicted, the former due to advection and diffusion, and the latter due to these processes as well as convection.

The time evolutions of the global-mean temperature change from 1958 to 2025 simulated in the study by Hansen et al. (1988) are presented in Fig. 5.18 for three scenarios:

(1) A, which assumes that the annual growth rates in the emissions of $CO_2$, CFCs 11 and 12, $CH_4$, and $N_2O$ essentially remain at their present values of 1.5%, 3%, 1.5%, and 0.4% to 0.9%, respectively;

(2) B, which assumes that the annual growth rates in the emissions decrease in time to zero in 2010 for $CO_2$ and CFCs 11 and 12, to 0.5% per year in 2000 for $CH_4$, and 0.5% per year in 2010 for $N_2O$;

(3) C, which assumes that the rate of increase of annual emissions for all these gases decreases to zero in 2000.

Figure 5.18 shows that the projected increases in global-mean surface air temperature in 2025 relative to 1958 are about 1.8°C, 1.3°C, and 0.6°C for scenarios A, B, and C, respectively. For the middle scenario B, the temperature rise projected from 1989 to 2025 is about 0.9°C, which is about double the observed temperature rise of 0.4 to 0.5°C from the late nineteenth century to the present.

The geographical distributions of the July surface air temperature changes relative to July 1958 simulated by Hansen et al. (1988) for scenario B are presented in Fig. 5.19 for the decades of the 1980s, 1990s, and 2010s. This figure shows that in each of these periods there are regions where the temperature changes are projected to be quite small and in the 1980s (and for particular years) even negative (see Hansen et al., 1988); this is in contrast to the equilibrium surface air temperature changes induced by a $CO_2$ doubling that are everywhere strongly positive (Fig. 5.9). Similar results were found in the studies by Washington and Meehl (1989) and Stouffer et al. (1989). These nonequilibrium simulation results reveal that our expectations of what a greenhouse-gas-induced climate change should look like, based on simply scaling the equilibrium response for a

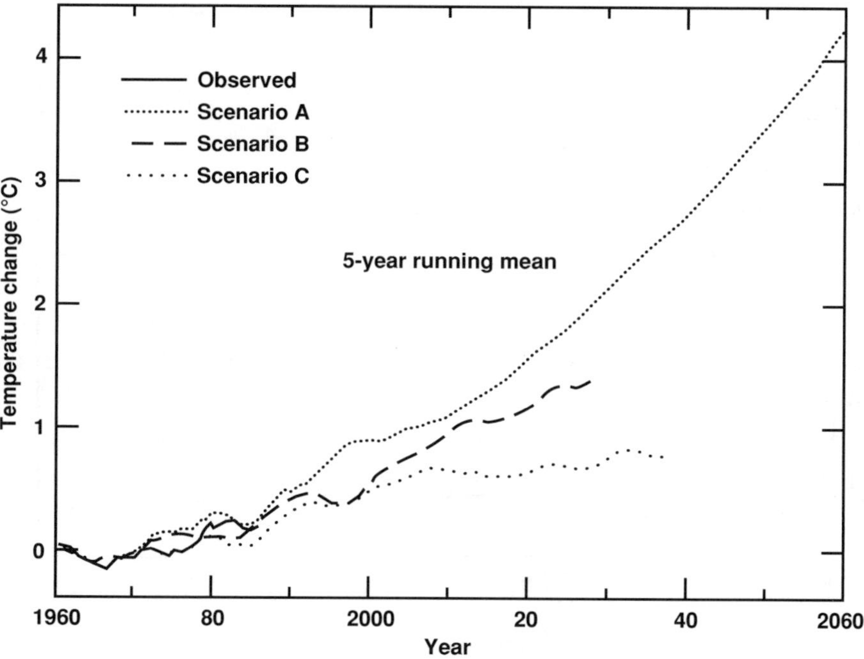


**Figure 5.18.** The change in 5-year running mean of global-mean surface air temperature starting in 1958 for scenarios A, B, and C by Hansen et al. (1988) and as observed to 1986.

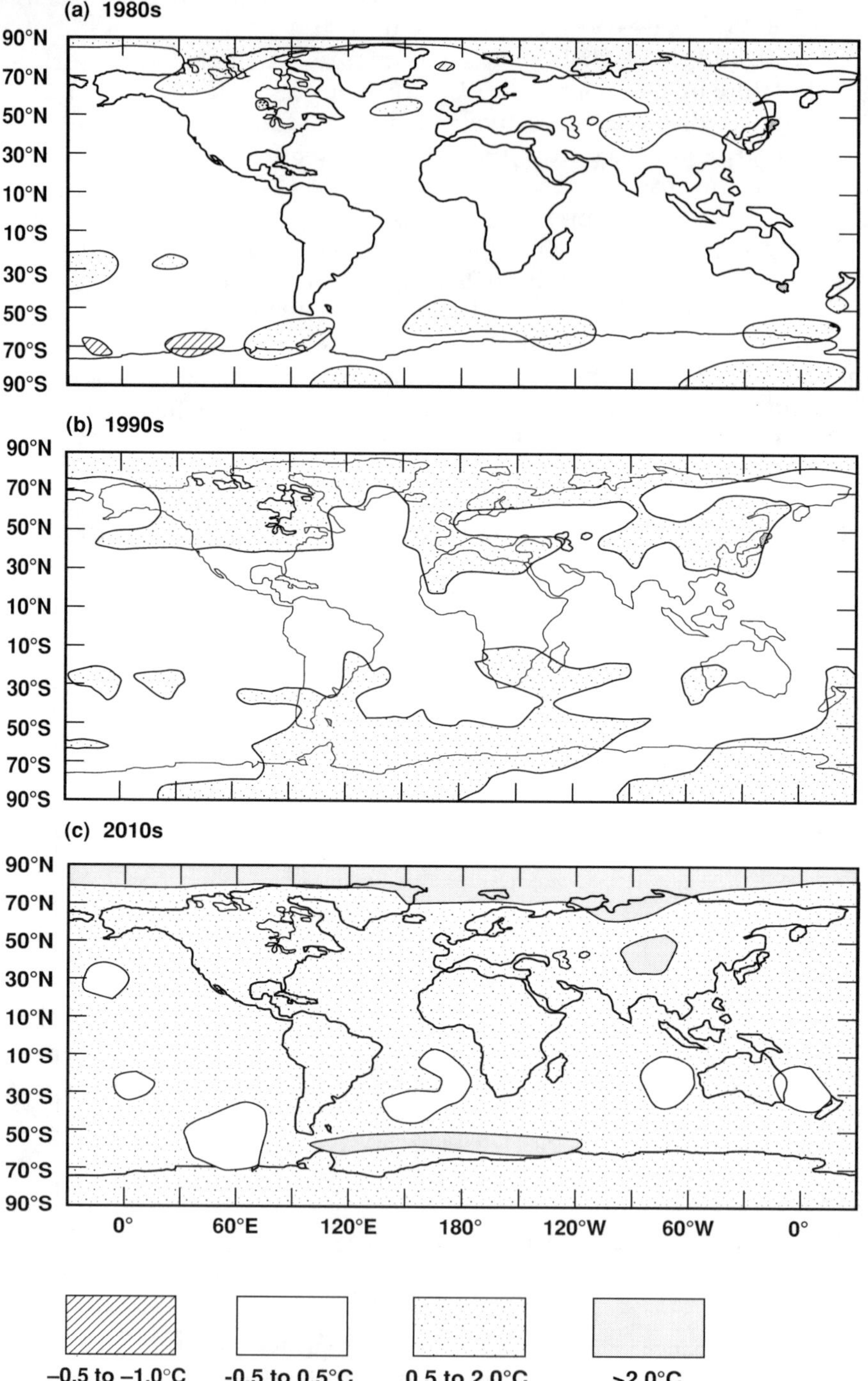


**Figure 5.19.** Geographical distribution of the change in July surface air temperature relative to the July case based on 1958 atmospheric concentrations as simulated by Hansen et al. (1988) for the decades of the (a) 1980s, (b) 1990s, and (c) 2010s.

$CO_2$ doubling to any smaller level of increase in the equivalent $CO_2$ concentration, are not correct. Instead, the actual patterns of climate change are comprised of the greenhouse-gas-induced signal superposed with the natural variability of climate that exists even in the absence of climate change. When the increases in the concentrations of the greenhouse gases are small, the greenhouse-gas-induced signal is small and can be dominated regionally by the natural variability and by the different time constants governing interactions of the atmosphere, upper ocean, and deep ocean. These factors can lead, for example, to negative regional temperature changes in any year, with the locations of these regions varying from year to year, and even to quite extensive regions where the temperature does not increase or even decreases. Eventually, however, the rise in greenhouse gas concentrations becomes so large that its signal begins to dominate the natural variability and the differing time constants of the ocean and atmosphere. How long it will be before the patterns of climate change take on their equilibrium greenhouse-gas-induced appearance is not yet certain.

This recent finding is of extreme importance to the effort to attribute the change in the observational record to the change in the greenhouse gas concentration. Such an attribution can be said to have been achieved only by discerning in the observations the patterns of greenhouse-gas-induced climate change simulated by the climate models. This has been attempted in the past by Barnett and Schlesinger (1987) using the "Fingerprint Detection Method," in which the evolution of the observed climate patterns was compared with a climate model simulation of the equilibrium climate change for a doubled $CO_2$ concentration. It is now evident, however, that, as a result of the evolution in greenhouse gas concentrations, any detection study must compare the observed climate patterns with the corresponding evolution of climate simulated by a climate model. This comparison should be made over the longest possible time in the past, consistent with there being adequate observations available. In addition, other characteristics of the simulated climate change should be compared with their observational counterparts. Of particular importance to agriculture are the amplitude of the diurnal cycle (nights are expected to be warmer than now due to an increase in the absorption and re-emission of infrared radiation from the surface) and the length of the frost-free season.

## 5.5. LESSONS AND OPPORTUNITIES FOR THE FUTURE

A hierarchy of climate models has been employed to simulate both the equilibrium and time-dependent climate changes induced by increasing concentrations of greenhouse gases in the Earth's atmosphere.

(1) *What have we learned about how best to represent the climate in numerical models?*

The most physically comprehensive model used to date to simulate equilibrium climate change consists of an atmospheric general circulation

(AGC) model coupled to a mixed-layer oceanic (MLO) model in which the effects of horizontal heat transport are prescribed and there is no vertical heat exchange with the underlying ocean. Because of the absence of this vertical heat exchange, such an AGC/MLO model cannot be used to simulate nonequilibrium climate change. The most sophisticated model used for this latter purpose consists of an atmospheric general circulation model coupled to an oceanic general circulation model. Although this type of model determines both the horizontal and vertical heat transports in a self-consistent manner, it has not been used to simulate equilibrium climate change because of the large computing time required to simulate the 1000 years or more necessary to reach equilibrium.

(2) *What have we learned about the sensitivity of the climate to an increasing concentration of carbon dioxide?*

The AGC/MLO model simulations of the equilibrium changes in climatic conditions induced by a doubling of the $CO_2$ concentration show an increase in both the global-mean surface air temperature and precipitation rate, with the increase of the latter being dependent on the increase of the former. The magnitude of the increase in surface air temperature simulated by the models has recently ranged from 2.8 to 5.2°C, with the size of the warming generally decreasing as the simulated surface air temperature for the present climate increases. This indicates that the temperature and precipitation changes simulated by the models would likely be in better agreement if the models simulated the present climate better, at least in terms of the global-mean surface air temperature. However, the most recent simulations show that the simulated 5.2°C warming decreases to 2.7°C when the effect on cloudiness of the change of phase from ice to water is included, and an additional decrease to 1.9°C results when the change in the optical properties of clouds is incorporated (Mitchell et al., 1989). Consequently, the largest uncertainty in the temperature sensitivity of the Earth is due to clouds, namely, to how their coverage (Cess et al., 1989), phase, and optical properties will change in response to a greenhouse-gas-induced warming.

(3) *What have we learned about the potential geographical distribution of future climate change?*

The geographical distributions of the equilibrium surface air temperature changes simulated by the AGC/MLO models for a $CO_2$ doubling show a warming virtually everywhere for both winter and summer. In general, the warming is a minimum in the tropics during both seasons and increases toward the winter pole. The locations of the wintertime warming maxima in both hemispheres coincide with the locations where the sea ice extent retreats in the doubled $CO_2$ simulation. Despite these similarities, there are differences in the magnitude and seasonality of the regional temperature changes simulated by the models. The geographical distributions of the change in precipitation rate simulated by the models display both positive and negative values, with the largest changes occurring between 30°S and 30°N latitudes. The precipitation changes poleward of these latitudes are generally positive in both seasons. The models generally simulate precipitation changes of less than 1 mm/day in magnitude over the Northern Hemisphere continental areas and Antarctica during both seasons.

The agreement among the model simulations of the regional changes in precipitation is considerably less than that for the changes in surface air temperature. The changes in soil moisture simulated by the models show many qualitative similarities for winter, particularly over the Northern Hemisphere continents where there is a moistening over most areas. There is less agreement in the soil moisture changes simulated for summer, with three of the models displaying a desiccation virtually everywhere in the Northern Hemisphere and two of the models a moistening in some regions and a drying elsewhere.

(4) *What have we learned about the rate of climate change to be expected?*

Simulations of nonequilibrium climate change from the steadily increasing concentrations of greenhouse gases have given a characteristic response time of the climate system that ranges from 10 to 100 years. One estimate of this characteristic time based on a simulation with a coupled atmosphere ocean general circulation model yields 50 years, this as a result of the transport into the ocean of the $CO_2$-induced heating. With a climate sensitivity of 4°C for a $CO_2$ doubling and a 50-year characteristic response time, the temperature change expected due to the increase in greenhouse gases from 1880 to 1988 is about 0.9°C. Although this is smaller than the expected equilibrium warming of about 1.6°C, it is larger than the observed warming of about 0.4 to 0.5°C. To reconcile this difference would require a decrease in the sensitivity to about 2°C. Such a decrease in sensitivity is a likely result of including changes in the phase and optical properties of clouds in atmospheric GCMs.

(5) *What have we learned about the validity of interpolating equilibrium climate change patterns to time-dependent conditions?*

Recently, two GCM simulations were performed in which the $CO_2$ concentration was taken to increase with time. These time-dependent, nonequilibrium studies show that our expectations of what a greenhouse-gas-induced climate change should look like, based on simply scaling the equilibrium response for a $CO_2$ doubling to any smaller level of increase in the equivalent $CO_2$ concentration, are not completely correct. Instead, the actual patterns of climate change are comprised of the greenhouse-gas-induced signal superposed on the natural variability of climate. This finding is of extreme importance for the effort to detect the greenhouse-gas-induced climate change in the observational record.

(6) *What have we learned about areas needing further research?*

Although much has been accomplished, there is much that can and needs to be done to improve the theoretical models currently available.

## Prospects for Future Climate

Based on the results of the model simulations of greenhouse-gas-induced climate change so far carried out, further research is needed in at least six areas:

(1) Improved and more comprehensive parameterizations of clouds, their optical properties, and their radiative interactions must be developed and tested.

(2) Interactive ocean/sea ice models capable of simulating bottom-water formation must be coupled to atmospheric general circulation models to study time-dependent climate change.

(3) Improved representations of soil hydrology must be included to accurately estimate changes in water resources.

(4) The horizontal resolution of GCMs must be refined by a factor of about 10 to better represent regional climatic features.

(5) Model simulations of the time-dependent increase in greenhouse gas concentrations are needed to estimate the evolving climatic signal.

(6) Detailed analyses of extended simulations are needed to estimate the potential changes in the frequency and intensity of extreme events such as droughts, floods, and severe storms.

# CHAPTER 6. EMPIRICAL METHODS FOR ESTIMATING FUTURE CLIMATIC CONDITIONS

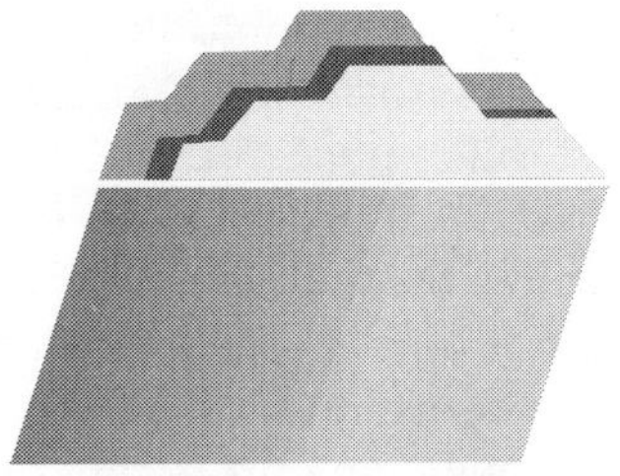

## 6.1. INTRODUCTION

About 15 years ago, Budyko (1974) suggested the possibility of applying an empirically based approach for forecasting future climate under the conditions of mankind's increasing impact on atmospheric composition. Budyko noted, "In spite of great progress made in modern studies on climate theory, the available numerical models for the meteorological regime are approximate, and their verification is limited by considerable restrictions. It is obvious that empirical data on climatic changes are very important both for constructing and for checking climate change theories."

Although there are also limitations when utilizing the empirical approach, the strengths of this approach can complement what can be learned from models, just as models can help to indicate how to interpret the empirical results. The greatest strength of the empirical, as compared to the theoretical approach of employing model simulations, is that the empirical estimates are drawn from a data set of past climatic conditions that actually took place. As such, the reconstructions may be able to provide indications of regional climatic relationships that should be incorporated when future projections are being developed. When reconstructions of past climatic conditions are accurate and thorough, they may be able to provide relatively reliable and self-consistent insights into the spatial patterns of these changes.

Limitations in applying the reconstructed paleoclimatic patterns can arise because of uncertainties in the temporal resolution of the proxy data, in extending limited areal samples to global scales, in interpreting the effects of changing land patterns and orography, in applying equilibrium climatic conditions to a nonequilibrium situation, and in determining the relative influences of the various factors that have influenced past climate. Just as for currently available climate models, the empirical method also is unable to estimate changes in the frequency of extreme events such as droughts, floods, and severe storms. Given these strengths and weaknesses, a dual approach involving both modeling and empirical methods, rather than pursuing either alone, seems more likely to bring rapid progress in understanding how the future climate may change.

## 6.2. THE EMPIRICAL APPROACH TO ESTIMATING FUTURE CLIMATE CHANGE

Since the early 1970s, many researchers have attempted to develop semi-empirical methods for gaining insights about forthcoming climate change by using data from a number of types of past climatic conditions. These efforts have ranged from using meteorological data collected since establishment of the global network of stations to using paleoclimatic evidence derived from warm epochs of the near and distant past. To utilize these data properly and to ensure consistency among the various conditions, experience derived from the study of the recent climate must also be used (see Budyko, 1980). By combining these data and analysis methods, an empirical approach is made possible that allows not only the checking of results of theoretical calculations of past climatic conditions, but also the development of additional indications of future conditions.

Application of these techniques by Soviet scientists to the problem of estimating the changing climatic conditions of the first half of the twenty-first century is described in a comprehensive monograph (Budyko and Izrael, 1987). The techniques rely on a complex method that draws from empirical data on climatic conditions of the warm epochs of the geological past, supplemented by inferences drawn from theoretical calculations. Thus, applying the empirical approach permits the derivation of estimates of the future climate that are nearly independent of conclusions based on theoretical (model) estimates. This creates an opportunity to compare these results with those derived from the model simulations of the forthcoming changes in climate, thus increasing confidence in areas of agreement and focusing research attention on areas of disagreements.

The premise underlying this approach for predicting anthropogenic climate change is based on associating the conditions of the climatic optimums of the Holocene, Eemian, and Pliocene with corresponding stages of the projected increase of mean global surface air temperature. Provided that certain assumptions are fulfilled in matching the value of the increased mean temperature for a certain epoch with the model-projected change in global mean temperature in the future, the empirical approach suggests that relationships leading to the regional variations in air temperature and other meteorological elements could be deduced and interpreted based on use of empirical data describing climatic conditions for past warm epochs.

Considerable care must be taken, of course, in making use of these spatial relationships, especially in accounting for possible large-scale differences that might, in some cases, result from different factors contributing to past climate changes than future changes and, in other cases, might result from the possible influences of changes in orography and geography on regional climatic conditions over time. Increasing evidence that past changes in climate were associated with changes in atmospheric composition, particularly the carbon dioxide ($CO_2$) concentration, does suggest that insights from past climatic periods may be helpful in predicting future conditions, subject to the uncertainties in the conditions that occurred and the degree to which the conversion from equilibrium to nonequilibrium situations can be made.

## 6.3. EMPIRICAL ESTIMATES OF CLIMATE SENSITIVITY

As a first step in the use of past conditions to project future conditions, the availability of paleoclimatic data on the climate and on atmospheric composition can be used to derive an empirical estimate of the overall sensitivity of climate to the increasing $CO_2$ concentration.

### 6.3.1. Mean Air Temperatures in the Past

Figure 6.1, from Budyko et al. (1985), plots the estimated $CO_2$-induced component of the departure of mean surface temperature over the last 100 million years from modern temperature versus the estimated atmospheric $CO_2$ concentration (ppmv) for corresponding epochs, here spaced logarithmically (base 10). Note that this $CO_2$-induced component

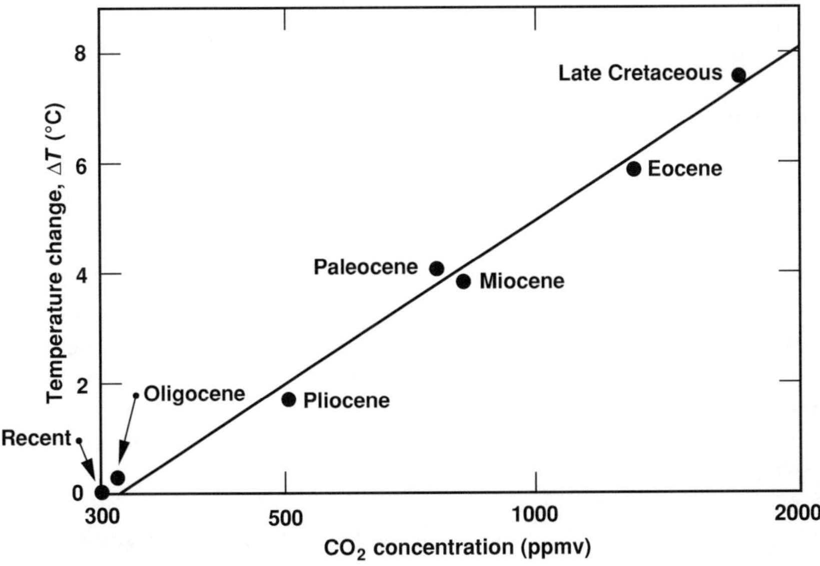


**Figure 6.1.** Scatter diagram showing estimates of $CO_2$-induced changes in the mean annual Northern Hemisphere temperature of the past geological epochs from the modern temperature ($\Delta T$) versus the atmospheric $CO_2$ concentration (in ppmv) spaced logarithmically (base 10), from Budyko et al. (1985). The estimated temperature changes shown are the part of the total temperature change for the particular epoch that is attributable to $CO_2$ changes after accounting for the estimated effects of changes in solar irradiance and planetary albedo.

was deduced after the estimated contributions of changes in solar irradiance and planetary albedo were subtracted from the total temperature change estimate derived from geological evidence (typically a 25 to 50% adjustment). Although this is an indirect technique, this plot shows a strong association between surface temperature departure and the logarithm of the atmospheric $CO_2$ concentration, especially when corrected for changes in solar irradiance and surface albedo. This association is in exceptionally good accord with theoretical studies indicating that the relationship between temperature change and the $CO_2$ concentration should be logarithmic (see Section 4.5). The slope of the best-fit line indicates that the sensitivity of the global average temperature to a doubling of the $CO_2$ concentration, denoted by $\Delta T_{eq}$ $(2 \times CO_2)$, is equal to about 3°C. In summaries of the results of theoretical studies (e.g., Chapter 5 and NRC, 1982), doubling of the $CO_2$ concentration was estimated to lead to an increase in mean global surface air temperature of 1.5 to 4.5°C. Such a wide range of estimates for $\Delta T_{eq}$ from models makes it difficult to accurately estimate the rate of the expected anthropogenic warming. Having available an empirical estimate of $\Delta T_{eq}$ near the center of this range not only generally supports the model-projected magnitude of the expected climate change, but also may be used to associate the expected changes in the mean surface air temperature with those of the available reconstructions of the warm epochs.

If this logarithmic dependence is formulated as an equation, the increase in mean surface air temperature $\Delta T$ at any time $t$ due to the growth of the atmospheric $CO_2$ concentration can be expressed in the form

$$\Delta T(t) = \Delta T_{eq} \frac{\log [c(t)/c_0]}{\log 2} \beta(t) \ , \qquad (6.1)$$

where $c$ is the $CO_2$ concentration at time $t$, $c_0$ is the $CO_2$ concentration in the pre-industrial epoch (i.e., early in the last century), and $\beta(t)$ is a parameter that accounts for the influence of the thermal inertia of the climate system on temperature change at time $t$. To use Eq. (6.1) to estimate future warming, the value of $c(t)/c_0$ must be increased to also take into account the effects of greenhouse gases other than $CO_2$.

Reconstruction of the $CO_2$ concentration in the nineteenth century shows that it was about 280 ppmv (see Chapter 4). Observations show that the modern concentration of $CO_2$ is near 350 ppmv, indicating that it has increased by about 25% compared to the pre-industrial value. From Eq. (6.1), if $\Delta T_{eq} = 3$ and climatic equilibrium were to be reached (i.e., $\beta = 1$), this 25% increase in $CO_2$ concentration would have led to an increase in the mean temperature of almost 1°C compared to the pre-industrial period. This value becomes about 50% higher when account is taken of the increase in methane ($CH_4$) and other trace gases (Dickinson and Cicerone, 1986).

At present, the actual increase in temperature should be somewhat less than the equilibrium value (i.e., $\beta < 1$), because there has not been time for equilibrium to become reestablished. Based on empirical and theoretical estimates, Byutner (in Budyko and Izrael, 1987) suggests a value of 0.7 for $\beta$ for cases of a slow and steady change in forcing; for the more rapid increase in concentrations in recent decades, a lower value is expected. Using a range of $\beta$ from 0.3 to 0.7 leads to an estimate that the temperature should have increased by about 0.5 to 1.0°C from preindustrial to present times, when the effects of all trace gases are included. As indicated in Chapter 3, the temperature over the past century is estimated to have increased by about 0.4 to 0.5°C, which is at the lower bound of the estimate indicated by using the empirical relationship.

### 6.3.2. Probable Mean Air Temperatures in the Future

Over the past two decades, a series of empirically based estimates has been made of future changes in global mean air temperature. Budyko (1972) used estimates of the growth in $CO_2$ concentration and the empirically derived climate sensitivity to project an increase in mean global surface air temperature at equilibrium of about 2.5°C from the middle of the twentieth century out to the second half of the twenty-first century (see Fig. 6.2a). This projected warming was associated approximately with a $CO_2$ doubling. If account is taken of the additional role of other greenhouse gases, the equilibrium changes would be expected sooner. However, this curve has remained a viable projection of the actual expected temperature change because the delay due to the oceans approximates the accelerating effect of the non-$CO_2$ trace gases. As indicated in Fig. 6.2b, the evolution of the climate over the succeeding 20-year period has been in quite good agreement with the original prediction.

Following Budyko's pioneering estimate, which was made at a time when many other climatologists were projecting a continued cooling, additional estimates of the potential $CO_2$ effect were made, both by individual scientists and as part of international workshops (see Table 6.1). It is of some interest to compare these empirically derived estimates of expected temperature changes with similar estimates from general circulation models (GCMs) reported in the "Declaration" adopted at the scientific conference in Villach, Austria (Bolin et al., 1986). In that Declaration, the most probable value for the equivalent increase in $CO_2$ concentration in the atmosphere (i.e., including consideration of the greenhouse effects of minor trace gases) corresponded to an effective doubling of the preindustrial $CO_2$ concentration in 2030. The equilibrium increase in mean global surface air temperature for a doubling of the $CO_2$ concentration was said to range from 1.5 to 4.5°C, with a mean value of 3.0°C. For comparison with the estimates shown in Table 6.1, the temperature change in 2025 can be estimated by lowering the equilibrium value by about 30 to 70% to take account of the moderating effects of the thermal inertia of the oceans on temperature increase (see Section 6.3.1).

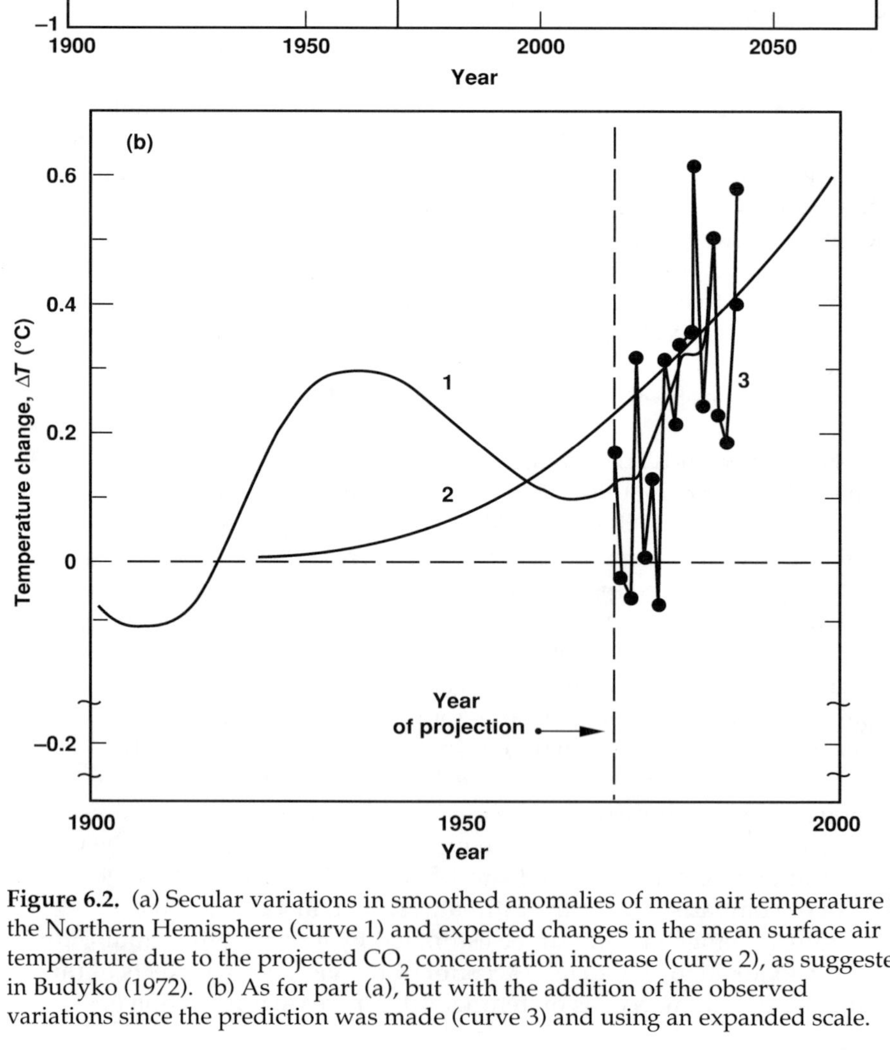


**Figure 6.2.** (a) Secular variations in smoothed anomalies of mean air temperature in the Northern Hemisphere (curve 1) and expected changes in the mean surface air temperature due to the projected $CO_2$ concentration increase (curve 2), as suggested in Budyko (1972). (b) As for part (a), but with the addition of the observed variations since the prediction was made (curve 3) and using an expanded scale.

Using the mean values of the indicated climate sensitivity (3°C) and thermal lag (0.5), the mean temperature increase, as projected by Bolin et al. (1986), is about 1.5°C in 2025, a value that is generally consistent with the range of estimates in Table 6.1.

For the near future, the simplest empirically based hypothesis assumes that the rate of increase in mean temperature over the next few decades will be close to its rate of increase over the last decade. A few factors are recognized that could cause this rate to become either higher or lower. For instance, natural variability could affect determination of the trend over this short period. It is also quite possible that the rate of growth in trace gas concentrations due to progress in technology will be different from the rates of increase in the recent past (see Chapter 4). It is also possible that the rate of increase of mean air temperature will change as a result of a change in inertia of the climate system.

Although there have been some attempts to estimate the expected departure of the increase in mean air temperature from the linear trend (e.g., Budyko and Izrael, 1987), there is an unavoidable inaccuracy in such calculations. As an initial approach, Budyko (1972) assumed that the total effect of all factors leading either to increases or decreases in the temperature growth rate would be small compared with the value of departures from the linear trend. Assuming that the rate of increase in mean temperature from the end of the twentieth century to 2070 would be constant seemed particularly appropriate at that time, when understanding of the potential for other factors to influence mean air temperature was more qualitative than quantitative. Additional information developed since 1972 is used in calculations by Budyko and Izrael (1987) to show that the time dependence of the expected air temperature rise is actually close to linear. Table 6.2 compares the results of calculating mean air temperature increase in the period 2000 to 2050 by using the most detailed calculation in Budyko and Izrael (1987) with similar results obtained by linear extrapolation of the trend from 1975 to

**Table 6.1.** Projections of mean increases in air temperature (°C) relative to the mean temperature at the end of the nineteenth century.

| Author | 1975 | 2000 | 2025 | 2050 |
|---|---|---|---|---|
| Budyko, 1972[a] | — | 0.7[a] | 1.4[a] | 2.0[a] |
| Kellogg, 1977–1978 | 0.5 | 1.2 | — | 4.0 |
| Budyko, 1979 | 0.5 | 0.9 to 1.3 | 1.8 to 2.5 | — |
| Flohn, 1981 | — | 1.0 | 1.5 | 2.8 |
| Budyko et al., 1978 | — | 0.9 | 1.8 | 2.8 |
| US-USSR, 1982 | — | 1 to 2 | 2 to 3 | 3 to 5 |
| Budyko and Izrael, 1987 | 0.5 | 1.3 | 2.5 | 3 to 4 |

[a] These values are relative to 1970; i.e., they are about 0.3 to 0.5°C less than the values estimated when comparing with temperatures at the end of the nineteenth century.

1985. Although the differences shown in Table 6.2 increase in time, as might be expected, the differences in the results obtained from these different methods lie within the limits of calculational accuracy.

The two major factors that Budyko and Izrael (1987) suggest are likely to contribute to errors in these estimates are (1) the changing rate of increase and inaccuracies in estimates of greenhouse gas concentrations, and (2) the lack of treatment of the effects on temperature of natural factors, primarily atmospheric transparency fluctuations, that might affect temperature anomalies by as much as a few tenths of a degree. Estimation of errors caused by the first factor will be quite small over the next few decades due to the considerable inertia in industrial activity that causes the growth of greenhouse gases in the atmosphere, but the errors in these estimates for the period 2010 to 2030 and beyond will grow rapidly. Relative errors associated with the second factor, however, will decrease for the more distant future because the temperature change due to "noise" induced by natural factors will decrease relative to the growing temperature increase due to greenhouse gases. When these factors are taken into account, the validity of climatic projections by trend extrapolation will reach a maximum in the comparatively near future as the $CO_2$ -induced changes exceed natural variability, possibly being most reliable for the time interval between the years 2000 and 2020.

The estimates of climatic conditions for later in the twenty-first century will then become less accurate as errors in the estimation of greenhouse gas concentrations increase; such estimates should be considered only as possible future changes of climate. An additional uncertainty is introduced by the inertial effect created by the oceans. If the value of $\beta$ is calibrated by applying Eq. (6.1) to the 0.5°C increase in global average temperature over the past 100 years due to the 50% increase in equivalent $CO_2$ concentration, then $\beta \approx 0.33$. If the calibrated equation is used to estimate future climatic warming, the results indicate that the warming (compared to the pre-industrial value) will not reach 1°C until the equivalent $CO_2$ concentration doubles, which may occur

**Table 6.2.** Projected increases in mean air temperature (°C) compared with that at the end of the nineteenth century.[a]

| | Year | | |
|---|---|---|---|
| Technique | 2000 | 2025 | 2050 |
| Adjusting for possible nonlinear factors | 1.3 | 2.5 | 3 to 4 |
| Linear extrapolation of trend from 1975 to 1985 | 1.25 | 2.0 | 2.75 |

[a] Using two empirical techniques and considering the perturbing effects of both $CO_2$ and other trace gases (from Budyko and Izrael, 1987). In subsequent studies by the same authors, projections of the changes in temperature for the years 2000, 2025, and 2050 have been estimated to be 1, 2, and 3 to 4°C , respectively.

around 2030. However, a gradual increase in β seems likely (see Chapter 5), suggesting that warming may occur somewhat more rapidly.

Thus, given projections of the future increases in greenhouse gas concentrations (e.g., see Chapter 4) and the logarithmic relationship established by paleoclimatic data [Eq. (6.1)], the empirical estimates of future climatic conditions suggest that there will be a progressing series of warmer and warmer climates.

## 6.4. ASSOCIATING PAST WARM EPOCHS WITH FUTURE CLIMATES

As described in Chapter 2, reconstructions of the temperature changes of past epochs suggest that the Holocene optimum, the Eemian interglacial, and the Pliocene optimum represent a series of successively warmer climates. If the estimates in Table 6.2 are valid, the temperature increase expected soon after the year 2000 corresponds approximately to the temperature increase in the Northern Hemisphere during the Holocene climatic optimum (5000 to 6000 years ago), the warming expected soon after 2025 corresponds approximately to the Northern Hemisphere warming experienced during the Eemian interglacial (125,000 years ago), and the warming expected in the middle of the next century corresponds approximately to that experienced during the Pliocene optimum (from 4.3 to 3.3 million years ago).

These associations in time are, of course, only approximations, in that natural fluctuations of climate make it necessary to consider average conditions over extended periods and not the climates of specific years. Natural variability and other factors can also affect the character of future spatial patterns, perhaps moderating or amplifying the strength of the warming tendencies derived from past warm epochs. The obscuring effects of natural climate variations should be less and less noticeable in the future, as compared to the pre-industrial temperatures.

In the book *Anthropogenic Climatic Changes* (Budyko and Izrael, 1987), it is proposed that these past periods be used, even if only generally, as indicators of future climatic conditions. In evaluating this hypothesis, it is important to consider whether differences in the nature and causes of past and future climate change may be important. For instance, due to slow changes of climate in the past, those climatic conditions should be close to equilibrium. However, the relatively rapid changes in recent greenhouse gas concentrations mean that the climate may be far from equilibrium (see Section 5.4), which may create differences in the regional climatic patterns for the future, as compared to past climates. Although the projected increases in the concentrations of greenhouse gases will likely be the major cause of the warming expected in the next century, the primary causes of past warming may have been different; therefore, the resulting climates may also have been somewhat different. In applying the empirical-analog forecasting approach, the problem of the causes of climate change during the Holocene, Eemian, and Pliocene warm epochs is of significant importance, although it is not the primary question to be resolved in order to substantiate the analog method of forecasting future climate. For the approach to be valid, the basic requirement is that the future changes in the main climate parameters (temperature and precipitation) as a consequence of the anthropogenic impact on the atmosphere be similar to changes occurring during the warm epochs of the past. This condition would be expected to be met if the causes of the past warmings were the same as the cause of the modern anthropogenic warming—namely, the increasing concentrations of greenhouse gases in the atmosphere. However, if past warmings occurred mainly due to some other factors, further studies are necessary to determine whether these warm epochs can be used as analogs for future climatic conditions.

There are conflicting opinions in evaluating this hypothesis. Climate model experiments performed several years ago (e.g., Manabe and Wetherald, 1980; Hansen et al., 1984), found that the latitudinal and seasonal responses of atmospheric temperature and other climate elements were very similar for the different radiative forcings caused by a 2% increase in the solar constant and a doubling of the $CO_2$ concentration. An analogous conclusion was reached by Budyko and Yefimova (1984), based on empirical studies of climate in nonequilibrium conditions. If this is generally the case, then past warm epochs would be expected to be very useful indicators of future climatic conditions, assuming that long-term equilibrium climatic conditions can be applied to the nonequilibrium changes we are experiencing.

There are, however, several reasons to believe that a similarity in response to different forcings may not always be the case. For example, Chapter 2 describes some of the effects of orbital forcing in changing the seasonal distribution of solar radiation and, therefore, the changes in hemispheric and seasonal climates in many regions over the globe. Model calculations with orbital forcing also give patterns of response that are different than their results for $CO_2$ forcing alone (e.g., Mitchell et al., 1988; Mitchell, 1990). In addition, as explained in Chapter 3, the two major warmings of the twentieth century have had different characteristics. The first of the warmings, which reached its maximum in the 1930s and can probably be attributed at least in part to increased atmospheric transparency due to a diminution in volcanic activity (Budyko, 1974), was particularly noticeable in the high latitudes of the Northern Hemisphere in both winter and summer. The second of the warmings, which has been continuing since starting strongly in the 1970s, has been more global in extent and has occurred mainly in winter. Its magnitude and latitudinal pattern seem generally consistent with the warming suggested by energy-balance models and past warm periods (Budyko and Izrael, 1987; Budyko, 1988).

Figure 6.3 shows the latitudinal distribution of temperature anomalies for Holocene, Eemian, and Pliocene warm epochs. The latitudinal distributions of the air temperature anomalies are generally in close agreement for the three epochs.[1] These results suggest that these epochs

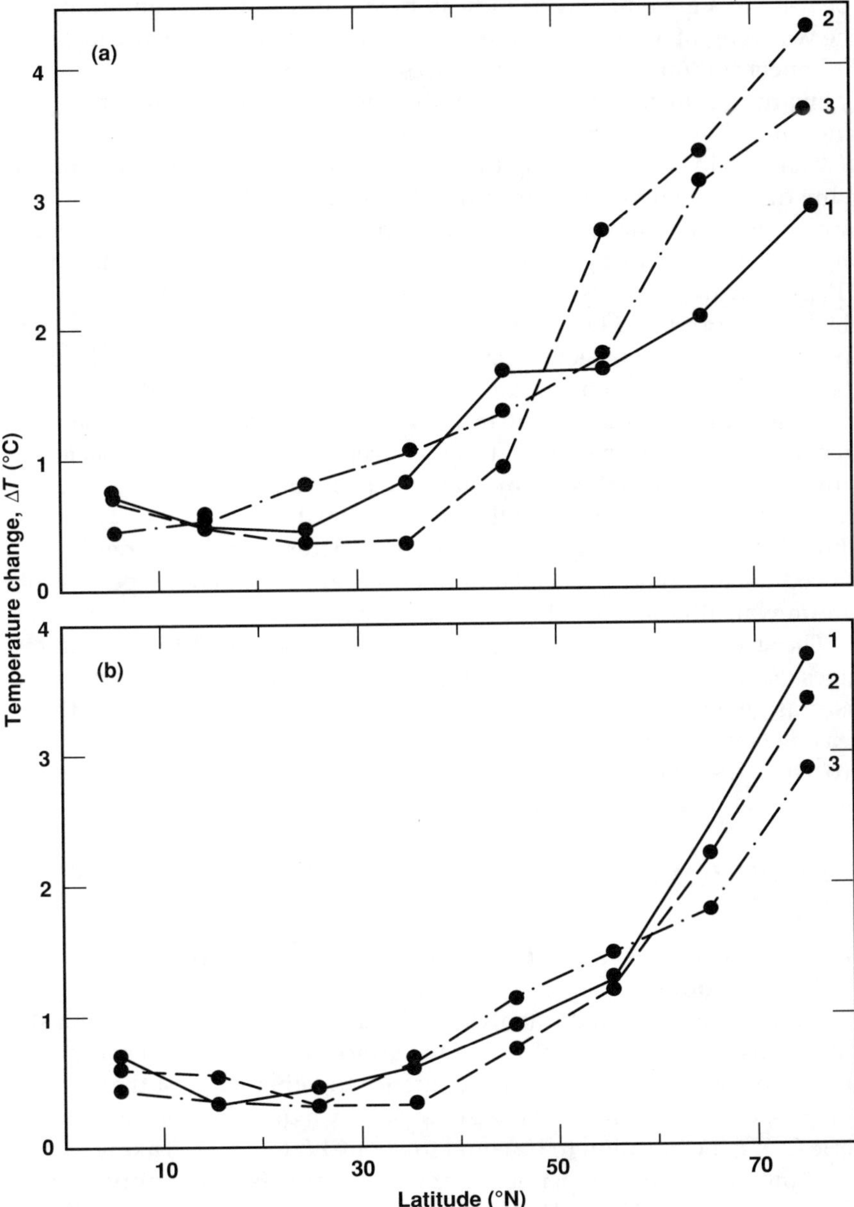


**Figure 6.3.** Relative changes in surface air temperature for (a) winter and (b) summer versus latitude for the Northern Hemisphere, normalized to a mean hemispheric temperature increase of 1°C (based on empirical data). The periods considered are (1) the Holocene optimum, (2) the Eemian interglacial, and (3) the Pliocene optimum.

can be used in a general way as indicators of future climatic conditions if (1) the warming of at least one of the periods occurred due to an increase in the concentrations of greenhouse gases, or (2) the factors that caused global warming during at least one of the past epochs were similar to the forcing caused by the increase in greenhouse gas concentrations.

Changes in climate during the Tertiary likely resulted from changes in many different factors, of which the changing concentration of $CO_2$ in the atmosphere may have been one of the most important (see Chapter 2). Changes in climate during the Pleistocene (including during the Holocene and Eemian warm periods) were largely related to changes in two factors: (1) the Earth's orbit and (2) changing levels of $CO_2$ and $CH_4$ in the atmosphere. The unmistakable periodicities of about 20,000, 40,000, and 100,000 years in Pleistocene ocean and land records for the past million years are a clear indication of the influence of orbital changes. Similarly, the records of changes in $CO_2$ and $CH_4$ concentrations in the Vostok ice core over the last 160,000 years indicate the probable importance of greenhouse gases in influencing the Earth's global temperature. These combined effects, acting in ways not yet fully determined, apparently produced the alternating glacial and interglacial cycles and the periods of optimum warmth (Barnola et al., 1987).

The strong influence of $CO_2$ in both Tertiary and Pleistocene climate changes gives some support to the hypothesis that these intervals may be used as general scenarios for future changes in climate caused by the increasing concentrations of $CO_2$ and other greenhouse gases. This approach is described in more detail in the next section, where analog and empirical methods are used together.

## 6.5. EMPIRICAL ESTIMATES OF FUTURE CHANGES IN CLIMATE

### 6.5.1. Future Changes in the Temperature and Precipitation Distributions

Model calculations, the analysis of instrumental observations for the last 100 years, geological estimates of the atmospheric chemical composition in the past, and paleoclimatic evidence together suggest that the mean global air temperature increase caused by a doubling of the atmospheric $CO_2$ concentration to be equal to about 1.5 to 4.5°C. Taking into account the logarithmic dependence of the mean surface air temperature on the atmospheric $CO_2$ content (including the equivalent effect of other trace gases), this approach suggests that the mean global air temperature will increase by about 1°C by 2000, by about 2°C by 2025, and by about 3 to 4°C by the middle of the twenty-first century (Budyko and Izrael, 1987). If this is the case, these increases will be similar to the ones obtained

---

[1] See Chapter 2 for a discussion of the spatial pattern, homogeneity, and reliability of reconstructions of past temperature patterns.

from paleoclimatic reconstructions for the Holocene, Eemian, and Pliocene optima.

In order to use these past climates as indicators of future climatic conditions, the dependence of the latitudinal pattern of the anomalies must be related to the rise in mean global air temperature. If the anomalies associated with different levels of warming are found to be directly proportional to the mean global air temperature changes, the application of the empirical approach is made easier because the data from different periods can be used to cross-check the approach. If the relationship is not linear, the possibility for such testing is limited.

Within the limitations described in Chapter 2, comparison of the normalized, latitudinal anomaly patterns shows good agreement for air temperature changes for the three warm epochs over the range 0° to 80°N (Budyko et al., 1989). Comparison of the reconstructions of the winter and summer air temperature increase for the three warm epochs with the air temperature change since the nineteenth century also shows that the latitudinal anomaly patterns are quite similar. All three reconstructions reveal a distinct latitudinal variability of air temperature anomalies: The higher the latitude, the greater the anomaly in both summer and winter. Winter reconstructions show greater anomalies in the regions where climate is more continental (Budyko and Izrael, 1987; Budyko et al., 1989).

The latitudinal similarity of the reconstructions of air temperature anomalies for the three warm periods suggests that the distributions of the temperature anomalies may be approximately proportional to the increase in the mean global air temperature and, consequently, that regional distributions of the surface air temperature anomalies may also be similar (Budyko and Izrael, 1987). The similarities of the reconstructions also suggest that the empirical approach can be useful for estimating the spatial distribution of temperature anomalies in the future.

Taking advantage of the similarity in patterns, the three winter reconstructions and three summer reconstructions can be averaged to obtain averaged warming patterns for each season (Budyko and Izrael, 1987; Budyko et al., 1989). In both works, reconstructions were developed for the mean summer and winter temperature changes in the Northern Hemisphere, normalized to a mean global temperature increase of 1°C (see Fig. 6.4).

Applying the approach to estimating the influence of annual precipitation over continental regions is more complex. The hypothesis that precipitation change is not directly proportional to the level of warming has been suggested in many studies (e.g., Budyko, 1980; Drozdov, 1981). This suggestion has been based on empirical analyses which show that temperature differences between the equator and the poles decrease as the globe warms. As a consequence, there are changes in the atmospheric pressure field and general circulation that noticeably affect the precipitation regime such that the subtropical, high-pressure

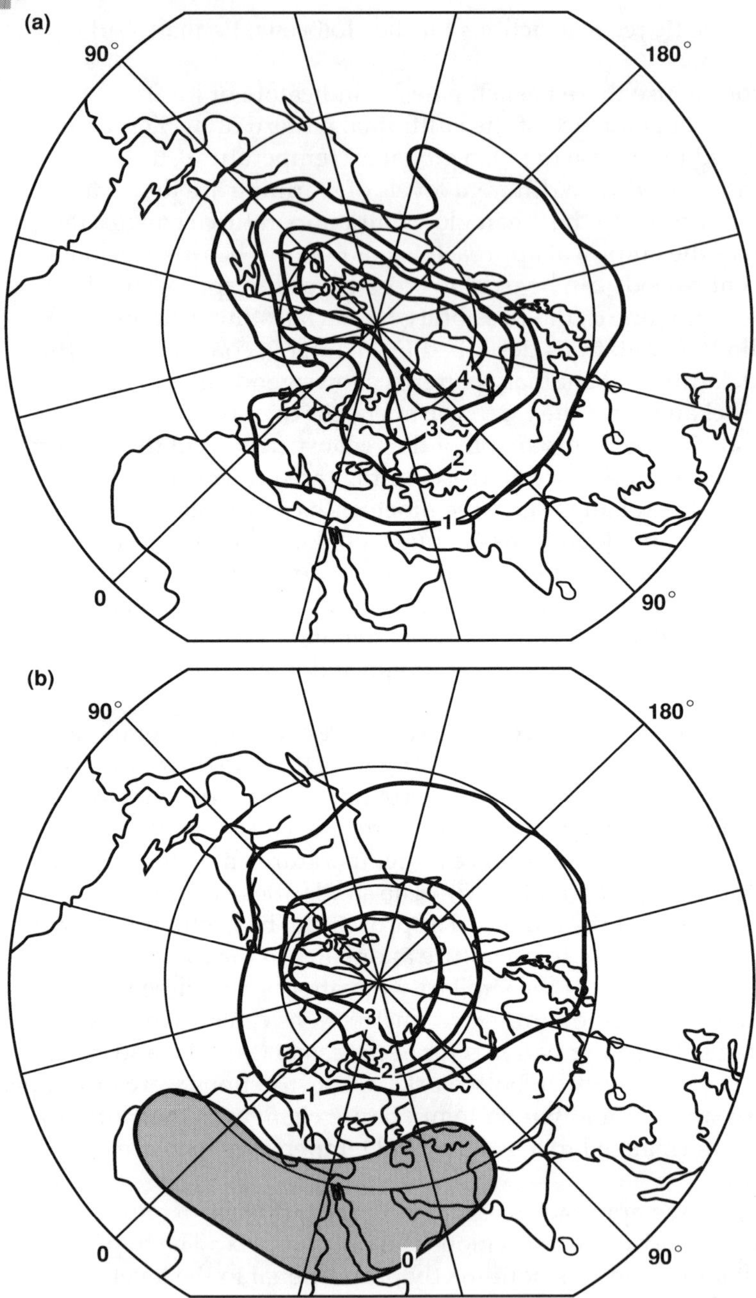


**Figure 6.4.** Maps of relative changes in surface air temperature (°C) for the Northern Hemisphere averaged over three past geological warm periods for (a) winter and (b) summer. Shading indicates areas of decreased temperature.

zone in the Northern Hemisphere would extend to higher latitudes as the world begins to warm (Budyko and Izrael, 1987). This would lead to more frequent droughts in some midlatitude continental regions. As warming intensified further, there would be a noticeable increase in the absolute humidity, which, in turn, would lead to a significant increase in precipitation. This would make moisture conditions on the continents more homogeneous, which is one of the characteristic features of the warm epochs of the past. These suppositions are not contradicted by the paleoclimatic reconstructions of annual precipitation changes during the Holocene, Eemian, and Pliocene optima (Budyko and Izrael, 1987; Budyko et al., 1989).

### 6.5.2. The Reliability of Estimates of Future Climate

The limits in accuracy of the empirical estimates of future climate change have been discussed by Budyko and Izrael (1987) and other Soviet authors. These studies indicate that the accuracy of these estimates is significantly limited by factors outside the realm of atmospheric studies. The most important factor is the accuracy of projections of economic activity and the consequent influence on the future atmospheric chemical composition. In particular, estimates of $CO_2$ emissions from fossil-fuel consumption have a major influence on the global energy balance.

Because of the long time necessary to change energy systems, estimates of future $CO_2$ emissions will likely be quite close for the next 10 to 20 years. Forecasts of further changes in the concentrations of greenhouse gases become less accurate farther into the future, with the forecasts for the second half of the twenty-first century having quite low reliability.

In the early twenty-first century, the projected increase in the global average temperature due to the increasing concentrations of greenhouse gases will likely exceed the natural variations experienced over the past 100 years. However, farther into the next century, other factors could limit the accuracy of estimates of the long-term temperature changes.

Budyko (1972) made the first quantitative forecast that can be compared to observed variations in the mean air temperature of the Northern Hemisphere. Such a comparison shows that the observed anomaly in mean air temperature differs from the predicted value by about 15%. The very good agreement between observations and the forecast indicates a comparatively modest influence of the unpredictable natural factors on the temperature over the last 10 years.

Such accuracy, however, cannot be achieved in the forecasts of regional changes of temperature and precipitation. The correlation coefficients presented in Tables 6.3 and 6.4 provide an indication of the consistency of the Northern Hemisphere latitudinal and regional temperature anomaly patterns from the three different warm periods. Although the changes were induced by somewhat different factors, correlation coefficients close to 1.0 for the latitudinal comparison suggest that north-south patterns are similar (see Fig. 6.3). The correlation coefficients in Table 6.4 for regional patterns are noticeably lower than the latitudinal correlation coefficients in Table 6.3. The lowest correlation coefficients, which seem to indicate differences in the temperature distribution for the Holocene optimum winter, may be a result of the poor sensitivity of the available methods for estimating paleotemperatures during winter in the middle and high latitudes. Note that correlation coefficients are positive in all cases and that, except for the Holocene optimum winter, these coefficients are high enough to suggest a similarity among past warm epochs sufficient to consider these periods as useful for projecting possible future climatic conditions (Budyko and Izrael, 1987).

Tables 6.5 and 6.6 display similar statistics for the results of model simulations of the temperature anomaly for a $CO_2$ doubling (see also Fig. 6.5). The results of models and paleoclimates compare well for winter temperature anomalies, but for summer the agreement is much less satisfactory (see Fig. 6.6). Although not shown, the patterns of spatial distribution of precipitation change for different regions of the Earth

**Table 6.3.** Correlation coefficients comparing Northern Hemisphere latitudinal patterns of relative temperature anomalies for the Mid-Holocene (1), Eemian (2), and Pliocene (3) optimum warm epochs.

| Warm | Winter | | | Summer | | |
|---|---|---|---|---|---|---|
| epoch | 1 | 2 | 3 | 1 | 2 | 3 |
| 1 | 1.0 | — | — | 1.0 | — | — |
| 2 | 0.92 | 1.0 | — | 0.99 | 1.0 | — |
| 3 | 0.92 | 0.95 | 1.0 | 0.97 | 0.97 | 1.0 |

**Table 6.4.** Correlation coefficients comparing the patterns of relative temperature and precipitation for 12 dispersed continental regions in the Northern Hemisphere for the Mid-Holocene (1), Eemian (2), and Pliocene (3) optimum warm epochs.

| Warm | Winter temperature | | | Summer temperature | | | Annual precipitation | | |
|---|---|---|---|---|---|---|---|---|---|
| epoch | 1 | 2 | 3 | 1 | 2 | 3 | 1 | 2 | 3 |
| 1 | 1.0 | — | — | 1.0 | — | — | 1.0 | — | — |
| 2 | 0.5 | 1.0 | — | 0.6 | 1.0 | — | —[a] | 1.0 | — |
| 3 | 0.3 | 0.7 | 1.0 | 0.9 | 0.8 | 1.0 | 0.7 | —[a] | 1.0 |

[a] The near lack of reconstructed precipitation data for the Eemian over North America prevents a representative calculation of the correlation coefficient.

calculated by the set of climate models are not in good agreement (Grotch, 1988).

It should be noted, however, that the results in Table 6.3 and 6.4 are not directly comparable to the results in Tables 6.5 and 6.6. Although the correlation coefficients among the warm epochs are limited by the accuracy of paleoclimatic reconstructions, the correlation coefficients among models may also be unrepresentative because the comparison does not reveal errors common to all of the models. These comparisons led to the conclusion in *Anthropogenic Climatic Changes* (Budyko and Izrael, 1987) that the analog method provides more reliable information about future climate than do presently available climate models.

**Table 6.5.** Correlation coefficients comparing Northern Hemisphere latitudinal patterns of relative temperature anomalies from four different general circulation models for a doubling of $CO_2$.[a]

| | Winter temperature | | | | Summer temperature | | | |
|---|---|---|---|---|---|---|---|---|
| Model | 1 | 2 | 3 | 4 | 1 | 2 | 3 | 4 |
| 1 | 1.0 | — | — | — | 1.0 | — | — | — |
| 2 | 0.98 | 1.0 | — | — | 0.10 | 1.0 | — | — |
| 3 | 0.80 | 0.90 | 1.0 | — | 0.50 | 0.80 | 1.0 | — |
| 4 | 0.97 | 0.90 | 0.70 | 1.0 | 0.50 | 0.60 | 0.90 | 1.0 |

[a] Model results are from (1) Manabe and Wetherald (1987), (2) Hansen et al. (1984), (3) Washington and Meehl (1984), and (4) Schlesinger and Mitchell (1987).

**Table 6.6.** Correlation coefficients comparing the patterns of relative temperature for 12 dispersed continental regions in the Northern Hemisphere for results from the four different general circulation models for a doubling of $CO_2$ used in Table 6.5.

| | Winter temperature | | | | Summer temperature | | | |
|---|---|---|---|---|---|---|---|---|
| Model | 1 | 2 | 3 | 4 | 1 | 2 | 3 | 4 |
| 1 | 1.0 | — | — | — | 1.0 | — | — | — |
| 2 | 0.8 | 1.0 | — | — | −0.1 | 1.0 | — | — |
| 3 | 0.6 | 0.5 | 1.0 | — | 0.0 | 0.0 | 1.0 | — |
| 4 | 0.8 | 0.8 | 0.8 | 1.0 | −0.2 | 0.4 | 0.3 | 1.0 |

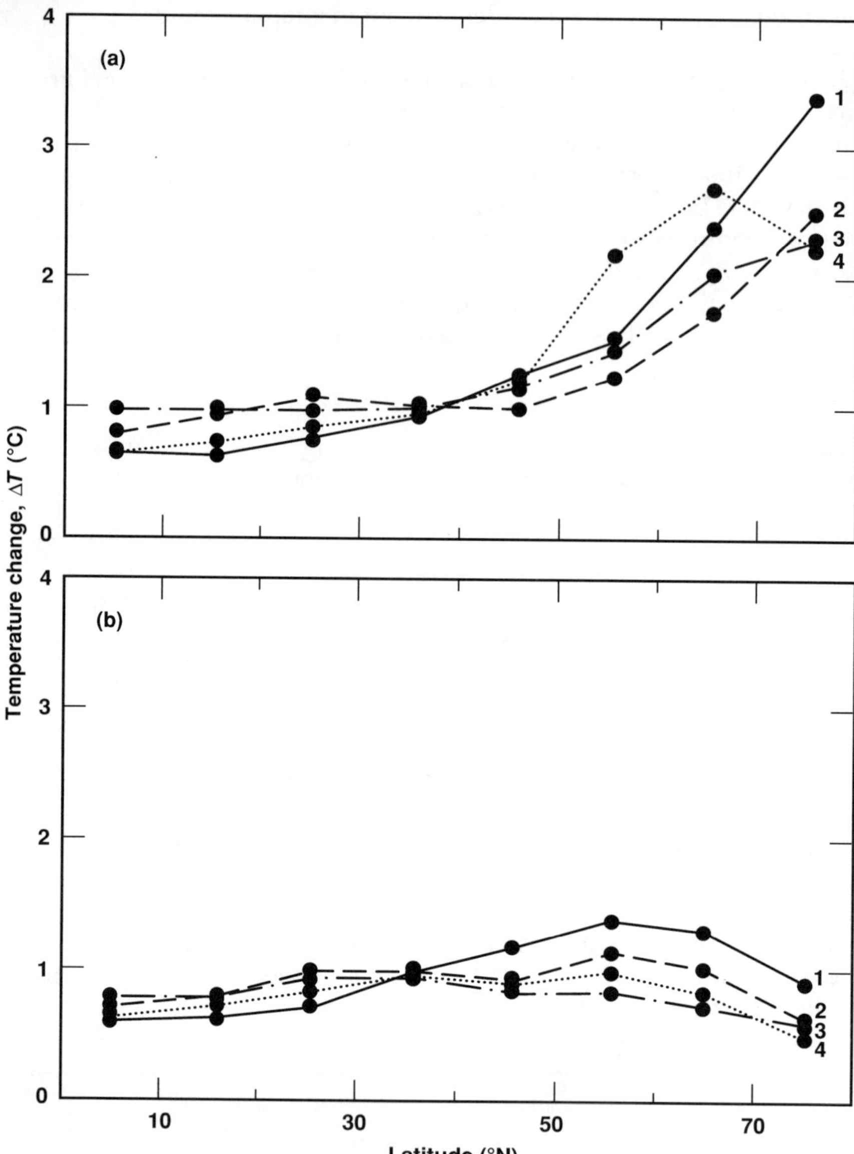


**Figure 6.5.** Relative surface air temperature changes versus latitude for the Northern Hemisphere for a $CO_2$ doubling for (a) winter and (b) summer. Temperature changes have been normalized to 1°C. Results are presented for the climate models of (1) Manabe and Wetherald (1987); (2) Schlesinger and Mitchell (1987); (3) Hansen et al. (1984); and (4) Washington and Meehl (1984).

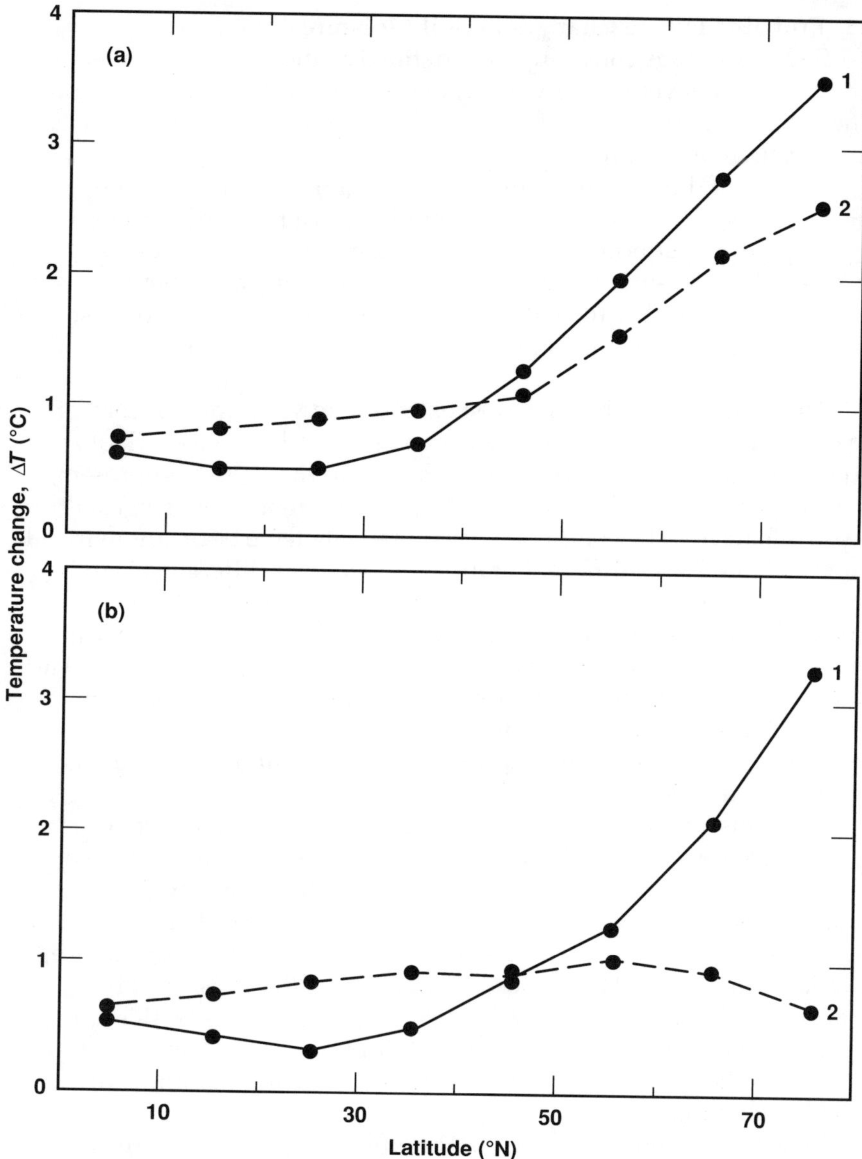


**Figure 6.6.** Relative surface air temperature increases with latitude normalized for an increase of mean hemispheric temperature of 1°C for empirical data (curve 1) and averaged for four climate models (curve 2) for (a) winter, and (b) summer.

### 6.5.3. Probable Future Changes in Soil Moisture

Soil moisture is conventionally defined as the amount of moisture in the active soil layer (usually the uppermost 1 m). Over many regions of the Earth, soil moisture is the major factor limiting crop productivity and ecosystem development.

Several of the climate models project significant summertime desiccation, especially of continental interiors, as a result of changes in the precipitation patterns and increases in the evaporation rates (see Fig. 5.14). Because such a drying could lead to significant adverse impacts, it is interesting to evaluate this finding in the context of past periods of climatic warming and the record of changes over the past century.

The Soviet Union has the most extensive measurement record of active soil-layer moisture, with measurements at a large network of agrometeorological stations extending back into the 1930s.[2] At present, there are about 3000 stations conducting observations. The mean soil moisture characteristics of the observational data have been summarized for areas planted with different crops averaged over different administrative and economic regions of the country (e.g., Zhukov, 1986). Analyses of these data have been presented in a number of monographs (Verigo and Rasumova, 1973; Meshcherskaya et al., 1982; Kelchevskaya, 1983). Unfortunately, soil moisture in agricultural fields depends on the type of crop and on the agricultural technique. As a result, these measurements are not always representative of the undisturbed areas adjoining these fields.

Soil moisture measurements in areas with natural plant cover (usually meadows) have also been carried out at existing hydrometeorological stations in the Soviet Union since 1967. The measurements are conducted every 10 days by a method that measures the change in soil weight during drying. The results are expressed in terms of the amount of available moisture. The amount of moisture in the soils, but not available to plants, is determined experimentally at each site; this value for unavailable moisture is held constant through the year and is subtracted from the measured values of total moisture to determine the available moisture.

In order to compare results for different soil types, data are often expressed in terms of relative soil moisture, which is the ratio of soil moisture content to its field capacity when both values are expressed in terms of the amount of available moisture. Figure 6.7 presents maps of the relative soil moisture content of the 1-m soil layer for summer and winter based on multiyear measurements from 50 of the most representative stations (Vinnikov and Yeserkepova, 1989). In northern regions, the relative moisture of the soil active layer is more than 100%.

---

[2] Regular network measurements of soil moisture have generally not been established outside the Soviet Union. Remote sensing methods (including satellite observation) are being developed but are not yet widely used, and data from them are scanty and not yet reliable.

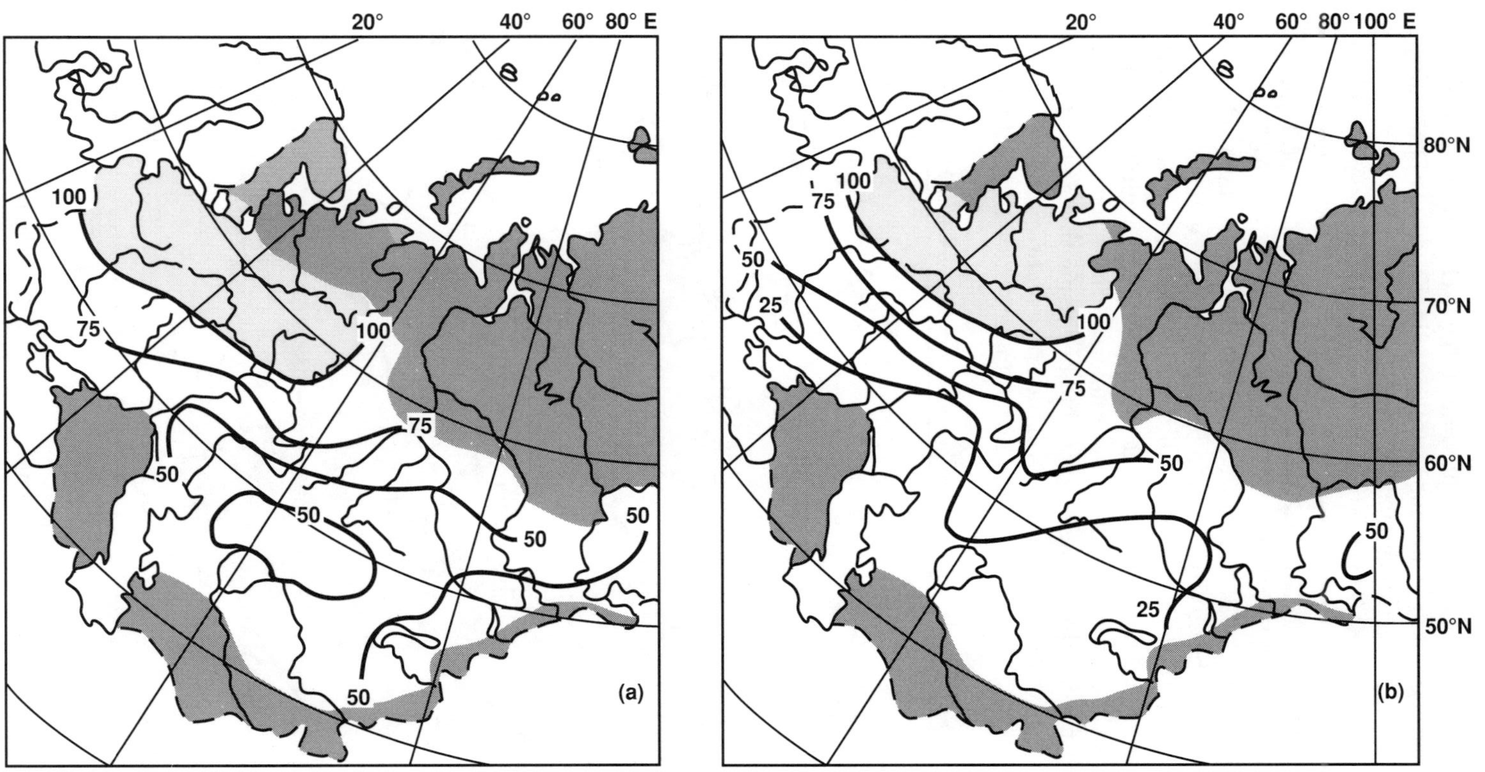


**Figure 6.7.** Relative wetness of a 1-m soil layer (%) for (a) winter and (b) summer. The averages for each station are generated from measurement periods of up to 30 years ending in 1985. Light-shaded areas have excess soil moisture. Measurements are not available in dark-shaded areas.

This situation occurs when the groundwater table is close to the surface, so that the observed value of soil moisture exceeds the value of the field capacity and approaches full capacity.

An alternative method for estimating the moisture content of the active soil layer is to calculate it from the components of the water balance derived from the observed mean values of major meteorological variables. The method, developed by Budyko (1971), is often used for this purpose (Korzun, 1974; Zubenok, 1976). Its main advantage is its consistency with the basic laws (heat- and water-balance equations), which ensures that the results will be physically reasonable. The method uses two parameterizations: one for calculating evapotranspiration and a second for calculating runoff. Both parameterizations have been experi mentally substantiated, and the results obtained using this method generally agree with modern data on soil moisture and runoff in many regions where groundwater level does not influence the active soil layer.

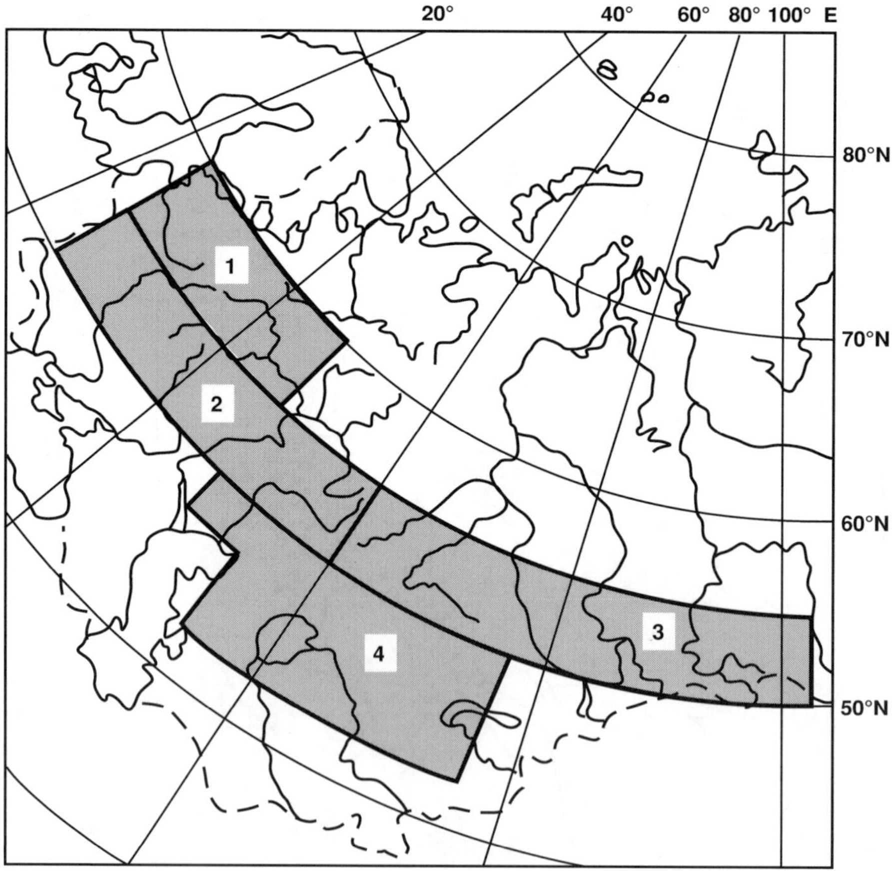


**Figure 6.8.** Four regions in the Soviet Union for which soil moisture trends have been evaluated.

Figure 6.8 shows the four regions in the Soviet Union for which soil-moisture trends have been estimated, and Fig. 6.9 shows empirical estimates of the linear trend in soil moisture for the period 1972 to 1985 for each of the four regions. This period is interesting, both because of the relative completeness of the data set and because Northern Hemisphere mean air temperature was rising at a rate of about 0.2°C/decade over this period. At the same time, precipitation increased almost everywhere in these regions, even though the temperature trends in each region were quite variable through the seasons.

In the three regions in the latitudinal belt 50° to 60°N, the rate of mean annual soil moisture content increase was 1.5 to 2 cm/10 yr, with the positive trend occurring for all months of the year. The increase in the level of the Caspian Sea starting in 1978 also suggests that the trend estimates are reliable. In the fourth region, closest to the Caspian Sea, which was more southern, the soil moisture trend was almost absent. Thus, over the 1972 to 1985 period when local temperature trends were

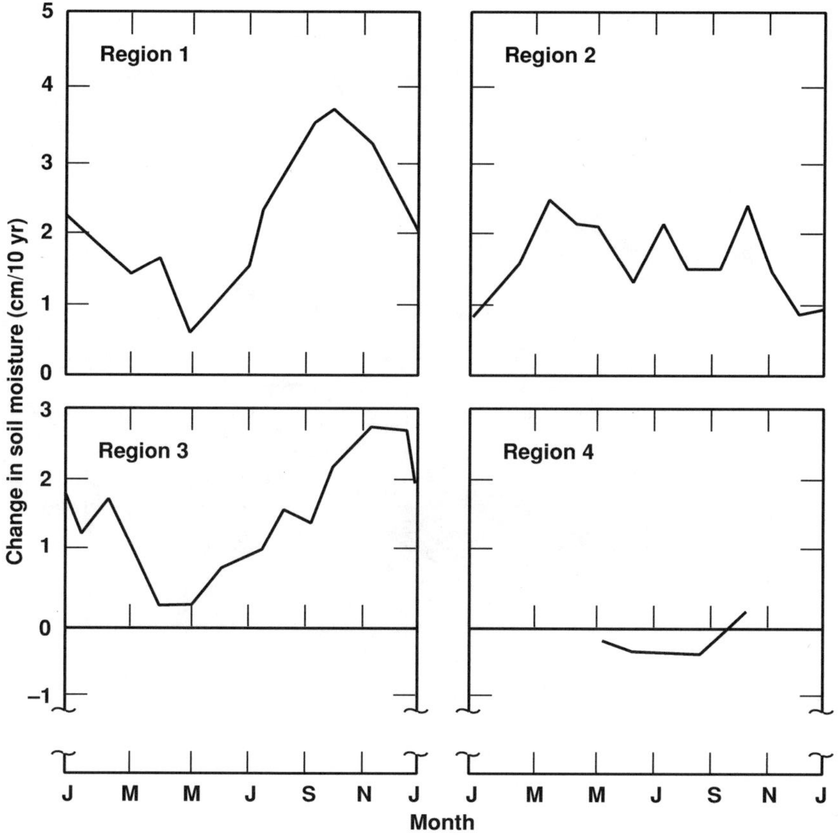


**Figure 6.9.** Observed trends in soil moisture for the period from 1972 to 1985 for the four regions in the Soviet Union shown in Fig. 6.8.

variable, soil moisture in these regions increased at the same time as a relatively rapid increase in mean temperature occurred in the Northern Hemisphere. These data suggest that the increases in mean annual precipitation and soil moisture content, which some climate models suggest will occur in some continental regions as a result of global warming, may not be accompanied by decreasing soil moisture in summer, which some models suggest will also occur. To better understand the potential for soil moisture changes, there is a need for the direct intercomparison of the calculation methods used for soil moisture in different models and of the model results and the observations of soil moisture.

Empirical estimates of potential changes in soil moisture can also be developed from climatic reconstructions for the past warm epochs (Budyko and Izrael, 1987). Vinnikov and Lemeshko (1987) and Vinnikov et al. (1989) used Budyko's method to derive soil moisture estimates for the Holocene climatic optimum (5000 to 6000 years ago) and for the Eemian interglacial. As shown in Fig. 6.10, the summer soil moisture during the

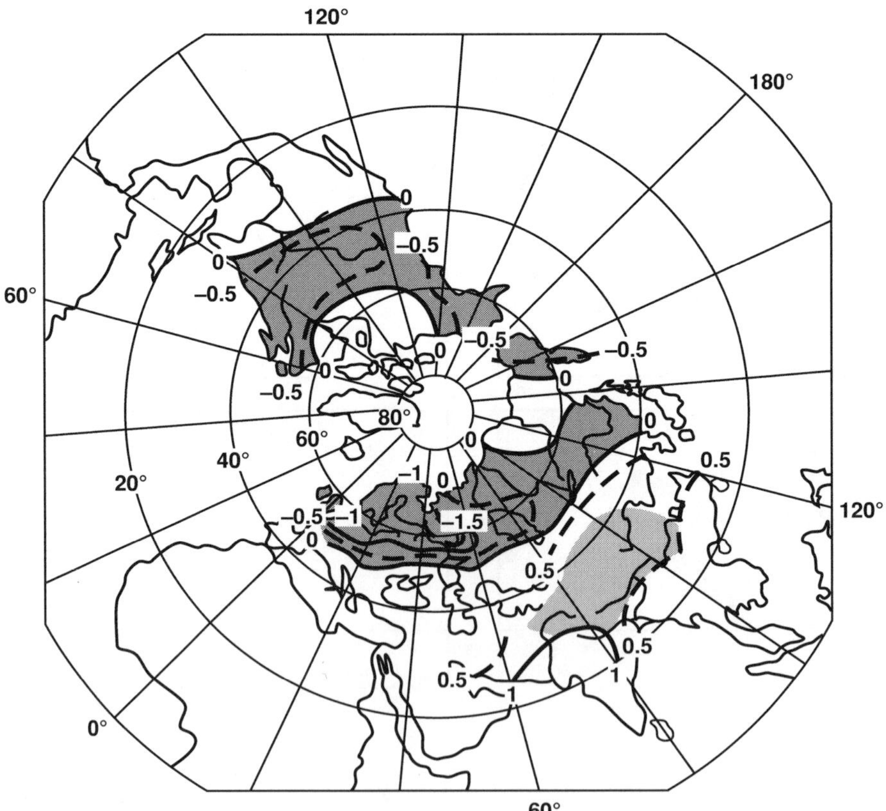


**Figure 6.10.** Estimates of change in summer moisture content (cm) of the 1-m thick active soil layer for the Holocene optimum as compared to present-day values. Dark hatching indicates areas where soil moisture decreased. No data are available in mountain areas shown by light hatching.

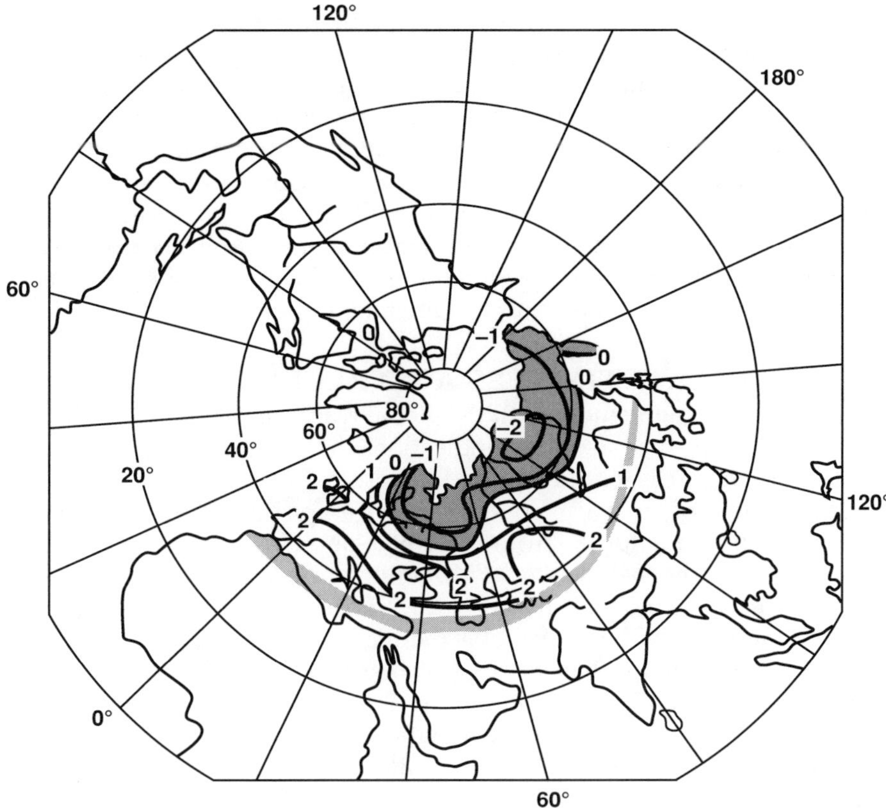


**Figure 6.11.** Estimates of change in moisture content (cm) of the 1-m thick active soil layer for the Eemian interglacial as compared to present-day values. Dark hatch areas show decreases in soil moisture, and light hatching indicates the southern boundary of the reconstruction.

Holocene climatic optimum is estimated to have been somewhat less than at present in a number of midlatitude regions; the southern boundary of the belt with decreased soil moisture in North America was located at about 40°N and in Eurasia at about 50°N. The largest decrease in soil moisture occurred in northern Europe and eastern North America; the 1- to 1.5-cm decrease in water amount in a 1-m soil layer amounts to about a 10% drop compared to the present. Increases in soil moisture are most evident in the arid zones of Central Asia; estimates for these regions, however, are preliminary.

For the last interglacial, estimates are available only over Eurasia north of 40°N. As indicated in Fig. 6.11, the soil moisture content decreases in summer poleward of about 55° to 60°N. However, this decrease is small and does not cover the entire continent. Of most importance is the 2- to 2.5-cm increase in moisture content (as much as 10 to 30% of the modern normal) in the forest, forest-steppe, and steppe zones of Eurasia.

Thus, data from past warm epochs were characterized by some decrease in soil moisture in wet regions and an increase in regions now having arid climates. Summer desiccation over the continents did not take place in these past warm epochs. The scale of the changes for these warm epochs is also comparable to the scale of changes in the soil moisture that are evident in measurements made over the Soviet Union in recent decades. Analyses of paleoclimatic reconstructions of soil moisture for the Holocene optimum and the Eemian interglacial do not show large changes in soil moisture that could be considered unfavorable for the biosphere in the extratropical part of the Northern Hemisphere continents.

## 6.6. EMPIRICAL ESTIMATES AND RECENT TRENDS

In addition to agreement with the global warming trend discussed in Section 6.3.2, there are also some indications that the regional pattern of recent climate change is tending toward patterns similar to those of the Holocene, the first of the warm epochs that the empirical method suggests may serve as an analog for the future. Budyko and Groisman (1989) compared data on the distribution of air temperature and precipitation anomalies over the extratropical part of the Northern Hemisphere for the first half of the 1980s with the available evidence on those climate changes in the Holocene optimum that are suggested as an analog of climatic conditions for the year 2000. Although the 5-year time interval is, in general, insufficient for reliable determination of regional features of the developing global climate changes, the similarities found between distributions of winter temperature anomalies and annual precipitation anomalies during the 1980s warming and similar changes for the Holocene (and expected by about the year 2000) deserve further study.

These relationships suggest augmenting the previously described empirical method for projecting future changes in the regional climate by extrapolating the trends of anomalies occurring during warming periods into the future. The reliability and testing of this approach should increase with lengthening of the time interval over which the trends in the anomalies of meteorological elements are determined. For example, over the extratropical latitudes of eastern continents of the Northern Hemisphere, Budyko and Groisman (1989) found good agreement between precipitation anomalies for the period 1981 to 1985 and the changes expected in the year 2000 based on the Mid-Holocene reconstruction. The correlation coefficient for the indicated anomalies was 0.87, which implies the presence of a very close relationship between the two precipitation distributions obtained by these independent empirical methods.

As the warming develops further over the next 5 or 10 years, it is suggested in *Anthropogenic Climatic Changes* (Budyko and Izrael, 1987) that this method for estimating future changes in global mean air temperature and the spatial distributions of these changes should become more applicable. The availability of these two independent empirical approaches for projecting regional climate changes should considerably increase the utility of these techniques. Combining results from these empirical approaches with results from climate models should further increase the reliability of the results, while at the same time helping to improve both methods.

## 6.7. LESSONS AND OPPORTUNITIES FOR THE FUTURE

Past climates can provide important insights into the general characteristics of the future climate, even without an exact correspondence of any past period with the future.

(1) *What have we learned from the empirical approach about the sensitivity of climate to the increasing $CO_2$ concentration?*

The geological evidence, which provides estimates of past atmospheric composition and reconstructions of past climatic conditions, suggests that each doubling of the atmospheric $CO_2$ concentration has been associated with an increase in global average surface temperatures of about 3°C, the relationship being approximately logarithmic in the $CO_2$ concentration. There are also indications that changes in the concentration of $CH_4$ may have contributed to these changes in climate.

(2) *What have we learned from past climates about the possible future changes in climate?*

The estimates of climate sensitivity suggest that temperature will warm significantly in the next century. During past warm periods, the warming has been greatest in middle and high latitudes and nearly nonexistent at low latitudes, with some regions that are now hot and arid being cooler as a result of increased precipitation. During the three best-documented warm epochs, the normalized latitudinal patterns that have been reconstructed are quite similar. There is also some similarity among the regional climate patterns for each of the three periods.

(3) *What have we learned from past climates about the association of temperature and precipitation changes?*

For many regions, vegetation patterns from past warm periods indicate that increased soil moisture and precipitation have accompanied increases in temperature. Although evidence of arid regions may not be as well preserved as evidence of less-arid conditions, the association of warmth with moisture for paleoclimates seems both extensive and persistent.

(4) *What have we learned from the empirical approach about the initiation of greenhouse warming?*

The warming over the past century has been generally consistent with estimates from the empirical approach, depending on how account is taken of the thermal inertia of the ocean. The warming pattern has also been generally consistent with the patterns evident in past warm epochs.

## *Prospects for Future Climate*

The empirical approach has provided a number of important insights. Further research could significantly enhance the understanding necessary to accurately project future climate changes.

(1) Intensified analyses of paleoclimatic conditions are needed to better understand the relative contributions of various causes of climate change.

(2) Analyses should be extended to lower latitudes and to the Southern Hemisphere as paleoclimatic reconstructions become available.

(3) Regional climatic reconstructions should be further analyzed to refine indications that warm and moist conditions occurred simultaneously in midlatitude continental interiors.

(4) Empirical methods should be developed to provide estimates of climatic variability during warm epochs of the past.

# *CHAPTER 7. IMPACTS OF CLIMATE CHANGE ON WATER RESOURCES AND AGRICULTURE*

## 7.1. INTRODUCTION

The changes in climate projected to result from increasing concentrations of greenhouse gases will lead to impacts on important resources, including agriculture, fresh water, natural ecosystems, and coastal developments. A growing body of climate impact research already suggests that important effects will be felt in all countries, sometimes in severe and dramatic ways. This chapter focuses on the potential impacts of changes in climate on water resources and agriculture in the United States and the Soviet Union, although many other impacts will also occur. These other effects include a rising sea level (see Chapter 3) that will threaten coastal regions and natural ecosystems; altered productivity of ocean and freshwater fisheries as a result of changes in temperatures, ocean currents, and nutrient flows; worsened urban air quality if rising temperatures increase the formation of low-level ozone; forest migration and diebacks, increased pest outbreaks, and greater frequency of fires; and more frequent and more intense storms. Better understanding of potential impacts, and the consequences of the relatively rapid rate at which they may occur, requires intensified efforts.

Quantitative estimates of changes in major long-term climatic variables such as air temperature, precipitation, and evapotranspiration are needed in order to project resource availability and quality. With such projections of climate change, simulation models of watersheds and agricultural production could be used to evaluate societal impacts of changes in climate and the performance and design of such systems under different climatic conditions. In the absence of accurate and detailed projections that incorporate the complexities of regional meteorology and the time-dependent response of the climate, self-consistent scenarios of plausible future conditions must be used to investigate the potential impacts on societal resources. Scenarios representing the possible ranges of future climate change can be derived in several ways, including developing purely hypothetical estimates for use in sensitivity studies (e.g., 2, 3, or 4°C), generating plausible ranges from the results of general circulation models (GCMs), and relying on paleoclimatic reconstructions of past climates or more recent periods that may be similar to future conditions. The complexities of going beyond climate scenarios to incorporate consideration of potential economic, social, and technological factors have been left to other forums.

## 7.2. IMPACTS OF CLIMATE CHANGE ON WATER RESOURCES

Global climate changes caused by increased concentrations of atmospheric trace gases will have important and diverse implications for fresh water resources. Although we cannot see clearly the direction or magnitude of many important changes, research in the past few years has highlighted regional vulnerabilities and has improved understanding of appropriate methods for answering hydrologic and water-resource questions.

Changes in climatic conditions will greatly complicate the design, operation, and management of water systems. History has repeatedly

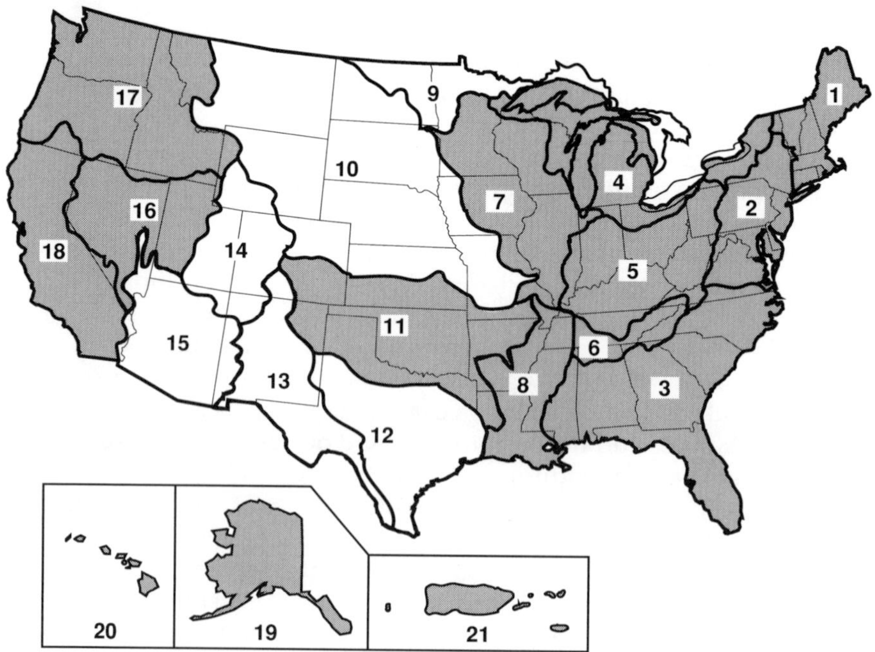


**Figure 7.1.** Flooding and drought risks. River basins vulnerable to flooding and drought are shaded. In these basins, the ratio of reservoir volume to annual supply is less than 0.60 (from Gleick, 1990).

shown that water resources are vulnerable to both natural and human-induced disruptions, including flooding and drought, overpumping of groundwater, and contamination problems. Figures 7.1 to 7.4 highlight those water-resources assessment areas of the United States that are vulnerable to floods and droughts (Fig. 7.1), water supply or demand changes (Fig. 7.2), disruptions of hydroelectricity generation (Fig. 7.3), and excessive groundwater pumping (Fig. 7.4) (Gleick, 1990). These characteristics have led to the development of an extensive and expensive system of aqueducts, pumping stations, hydroelectric facilities, and flood-control reservoirs to handle the temporal and spatial discrepancies between water demand and supply.

1. New England
2. Mid-Atlantic
3. South Atlantic Gulf
4. Great Lakes
5. Ohio
6. Tennessee
7. Upper Mississippi
8. Lower Mississippi
9. Souris-Red-Rainy
10. Missouri
11. Arkansas-White-Red
12. Texas-Gulf
13. Rio Grande
14. Upper Colorado
15. Lower Colorado
16. Great Basin
17. Pacific Northwest
18. California
19. Alaska
20. Hawaii
21. Caribbean

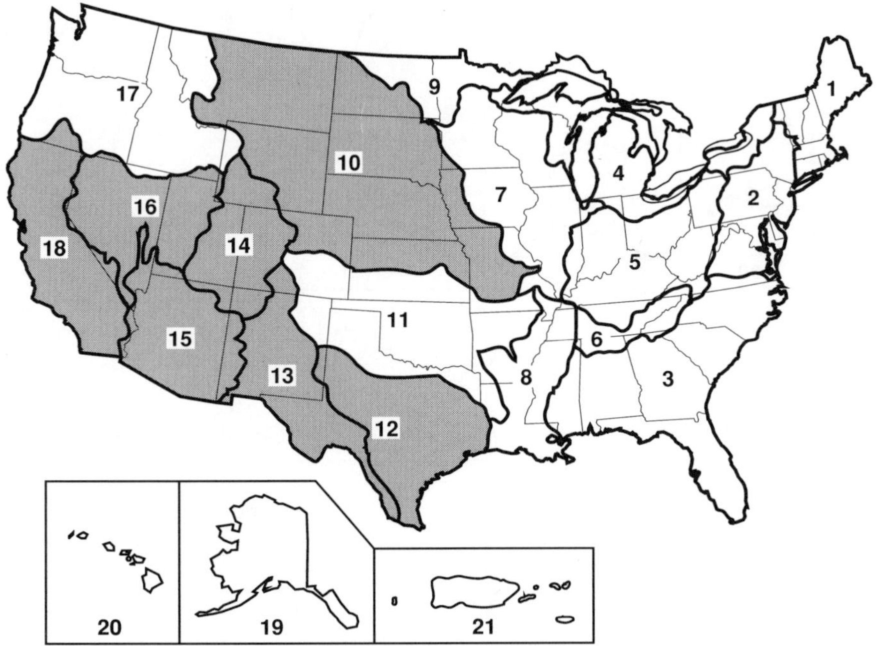


**Figure 7.2.** Supply risks. River basins where annual demand is large relative to annual supply are shaded. Climate-induced changes in either demand or supply could cause water-resource problems (from Gleick, 1990).

1. New England
2. Mid-Atlantic
3. South Atlantic Gulf
4. Great Lakes
5. Ohio
6. Tennessee
7. Upper Mississippi
8. Lower Mississippi
9. Souris-Red-Rainy
10. Missouri
11. Arkansas-White-Red
12. Texas-Gulf
13. Rio Grande
14. Upper Colorado
15. Lower Colorado
16. Great Basin
17. Pacific Northwest
18. California
19. Alaska
20. Hawaii
21. Caribbean

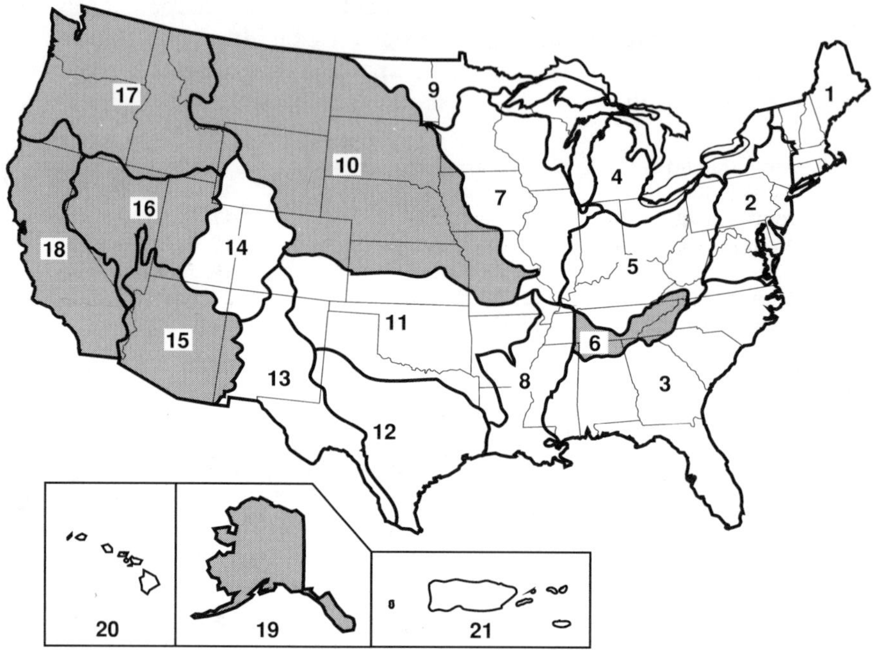


**Figure 7.3.** Hydroelectric power production risks. Regions where hydroelectric production is greater than 25% of total annual electricity production are shaded (from Gleick, 1990).

1. New England
2. Mid-Atlantic
3. South Atlantic Gulf
4. Great Lakes
5. Ohio
6. Tennessee
7. Upper Mississippi
8. Lower Mississippi
9. Souris-Red-Rainy
10. Missouri
11. Arkansas-White-Red
12. Texas-Gulf
13. Rio Grande
14. Upper Colorado
15. Lower Colorado
16. Great Basin
17. Pacific Northwest
18. California
19. Alaska
20. Hawaii
21. Caribbean

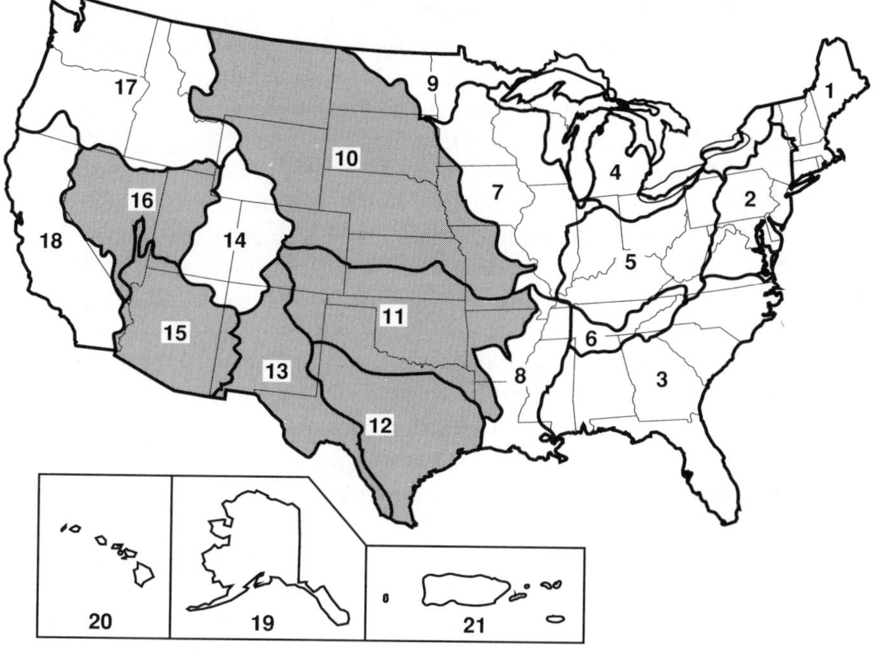


**Figure 7.4.** Groundwater overdraft risks. Regions where groundwater overdraft is more than 25% of total groundwater withdrawals are shaded (from Gleick, 1990).

### 7.2.1. Methodological Approaches

In the last 10 years, there has been a dramatic increase in the number of studies of the hydrologic implications of anthropogenic climate change (see Gleick, 1989). The studies have used a wide variety of methodological approaches, climate scenarios, and data. Studies can be classified into four groups:

(1) Analysis of past, long-term variations of runoff and meteorological variables. This approach includes the use of regression analyses such as used by Stockton and Boggess (1979) and Revelle and Waggoner (1983) for hydrologic basins in the United States, and the study of basin response to past climatic anomalies, such as the work of Schwarz (1977) and the collected studies in Glantz (1988).

(2) Application of long-term water balance methods. Glantz and Wigley (1987) and Vinnikov et al. (1989) have used water balance methods to study water basins in the US and USSR, respectively. This technique is relatively simple and requires few basic data (such as annual runoff, precipitation, and air temperature); however, use of this method leads to general results that must be regarded with caution.

(3) Application of general circulation models. The results of large-scale GCMs have been used to evaluate regional hydrologic consequences of climate changes in the Northern Hemisphere. In particular, changes in runoff, soil moisture, and evaporation have been evaluated (US EPA, 1984; Sanderson and Wong, 1987; Singh, 1987). Great care must be taken, however, in using GCM hydrologic output directly because of the poor spatial resolution of GCM output and the very simplified hydrologic parameterizations included in these models.

(4) Application of deterministic hydrologic models. Physically based models of hydrologic processes offer considerable advantages for the study of the impacts of climate change. Although many different types of models exist, more detailed regional information can be obtained with better temporal resolution than is available from some of the above-mentioned techniques. This approach has been used by many researchers (Nemec and Schaake, 1982; Cohen, 1986; Gleick, 1986a, 1986b, 1987a, 1987b; Mather and Feddema, 1986; Flaschka et al., 1987; Bultot et al., 1988; and Shiklomanov, 1989).

Considerable attention has been given to using the wide range of deterministic, physically based hydrologic models. Many deterministic models have been developed to analyze different types of hydrologic phenomena. The models vary in their ability to reproduce small- and large-scale features of watersheds: narrowly focused models study short-time periods and incorporate site-specific characteristics, whereas general models are capable of representing water balances in a large region. Each type of model has strengths and limitations, depending on model design, data requirements, and the objectives of the analyst.

### 7.2.2. Review of Recent Research

Numerous studies of the hydrologic impacts of climate change have recently been completed (for summaries, see Klemes, 1985; Beran, 1986, 1987; Dooge, 1986; Changnon, 1987; WMO, 1987; Shiklomanov, 1988; Gleick, 1989). Table 7.1 lists some of the more important recent studies on the effects of climate changes in climate on surface runoff.

**Table 7.1.** Studies of hydrologic impacts using hypothetical scenarios of changes in temperature (*T*) and precipitation (*P*).

| Study | Region | Change in *T*(°C) | Change in *P*(%) | Percent change in runoff (%) Annual | Summer | Winter |
|---|---|---|---|---|---|---|
| Stockton and Boggess (1979) | Western US | +2 | –10 | –40 to –70 | | |
| Nemec and Schaake (1982) | Arid basin | +1 | +10 | +50 | | |
| | | +1 | –10 | –50 | | |
| | Humid basin | +1 | +10 | +25 | | |
| | | +1 | –10 | –25 | | |
| Revelle and Waggoner (1983) | Colorado | +2 | +10 | –18 | | |
| | | +2 | –10 | –40 | | |
| Gleick (1986a, 1987b) | Sacramento | +2 | –10 | –18 | –32 | – 9 |
| | | +2 | +10 | +12 | –12 | +25 |
| | | +4 | –10 | –21 | –68 | +14 |
| | | +4 | +10 | +7 | –56 | +54 |
| Flaschka et al. (1987) | Great Basin | +2 | –10 | –17 to –28 | | |
| | | +2 | –25 | –33 to –51 | | |
| Shiklomanov (1989) | Forest-steppe | +1.8[a] +0.5[b] | +4 | –10 | +8 | +76 |
| | Steppe | +1.5[a] +0.5[b] | +12 | +49 | +300[c] | +730[c] |

[a] Winter.
[b] Summer.
[c] The large percentage increases in this basin are small absolute increases.

Although different techniques and scenarios were used for many of these studies, they all show that runoff is very sensitive to even slight changes in temperature and precipitation. For example, an increase in surface air temperature of 1 to 2°C and a decrease in precipitation of 10% may lead to decreases in annual runoff of 40 to 70% in arid and semi-arid regions.

Although additional work is needed to evaluate more fully the effects of changes in storm frequencies and intensities, changes in vegetation, and other hydrologic complications, the hydrologic effects described above would have severe societal ramifications.

### 7.2.3. Assessment of Possible Changes in USSR River Runoff

The effects of anthropogenic climate change on the water resources of the USSR rivers were first evaluated at the State Hydrological Institute by Budyko et al. (1978). Those assessments were based on the use of paleoclimatic reconstructions as scenarios of future conditions. In recent years, scenarios have been developed for annual runoff for Soviet rivers and portions of the Northern Hemisphere for the years 1990, 2000, and 2020 (Budyko and Izrael, 1987; Shiklomanov, 1988; Vinnikov et al., 1989). Figure 7.5 shows isolines of runoff changes for the USSR for a global warming of 1°C (for approximately the year 2000). These assessments based on paleoreconstructions indicate that increases and decreases of runoff of 5 to 20% are possible by the end of the century, with significant ramifications for long-term water planning and protection of the environment.

More detailed regional and temporal information is needed, however, if decisions are to be made concerning runoff management, water transfers, and ecological protection. Using physically based water balance models with a 10-day time step, Shiklomanov (1989) computed runoff changes for two sub-basins (the Sosna and Chir) in the Don River Basin, and for the much larger Volga River Basin. Scenarios of climate change were based on the paleoclimatic reconstructions discussed in Chapters 2 and 6. In this study, winter temperatures were assumed to increase by 1.5 to 1.8°C and summer temperatures by 0.5°C. Precipitation was assumed to increase by between 3 and 12%. The results indicated that these changes in climatic characteristics could lead to a significant transformation of the seasonal distribution of stream flow due to changes in snowfall and snowmelt. In all three river basins studied, winter runoff was projected to increase (see Table 7.2). Summer runoff was projected to decrease in the forest-steppe and southern forest zones and to increase in the more southern steppe zone of the Chir Basin. Annual runoff in the Volga and Sosna Basins was projected to decrease about 10%, whereas annual runoff in the smaller Chir Basin was projected to increase by nearly 50%.

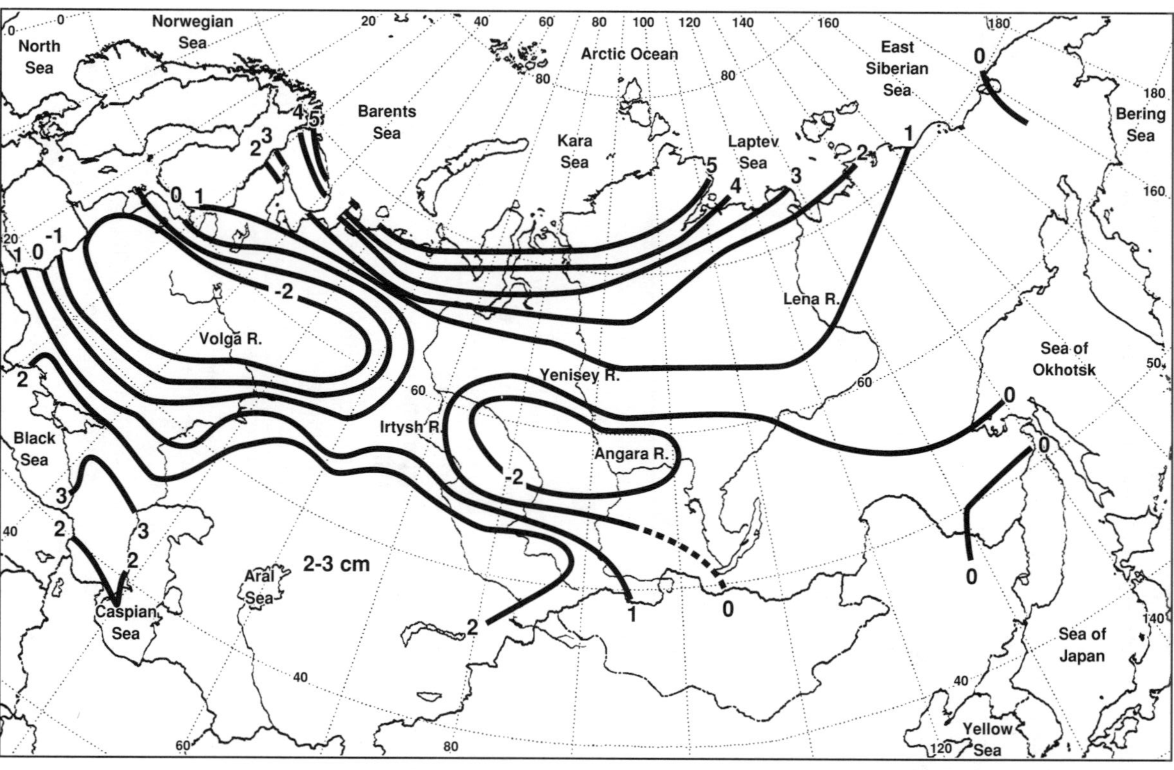


**Figure 7.5.** Change of annual river runoff of the USSR (in cm) for a global warming of 1°C.

For more significant changes in climate, the warmer paleoclimatic reconstructions suggest that annual precipitation over the whole of the USSR could rise, leading to more uniform increases in runoff, even for basins for which the moderate warming reconstruction suggests that runoff may decrease. The reliability of these estimates is not high, however, because of uncertainties about long-term precipitation estimates.

**Table 7.2.** Probable changes in seasonal river runoff in the USSR for a Holocene-like warming of about 1°C [from Shiklomanov (1989)].

| River | Drainage area ($10^3$ km$^2$) | Vegetation zone | Runoff[a] (mm/yr) Annual | Winter | Spring | Summer & autumn |
|---|---|---|---|---|---|---|
| Volga at Volgograd | 1360 | Forest/ steppe | 187/165 | 22/35 | 107/94 | 58/36 |
| Sosna at Elets | 16.3 | Forest/ steppe | 144/129 | 30/53 | 102/63 | 12/13 |
| Chir at Oblivskoje | 8.47 | Steppe | 47/70 | 3/22 | 43/45 | 1/3 |

[a] Numerator indicates mean long-term natural conditions; denominator indicates computed values in the case of global warming of 1°C.

Shiklomanov (1989) also applied similar techniques to evaluate the impacts of global warming on water levels in the Caspian Sea, the largest closed water body in the world. The future level of the Caspian Sea depends on natural inflows (80% come from the Volga River Basin), on precipitation onto and evaporation from the water surface, and on human activities. A detailed water balance model of the basin has been developed and used to evaluate two future scenarios: (1) stationary climate to 2020, and (2) anthropogenic changes of climate.

In the first case, empirical data on water balance components from 1880 were used, and projections of future human activities were incorporated. Inflows to the Caspian Sea, even without changes in climate, are projected to decrease from the long-term average by 40 km$^3$/yr (13%) by 1989, 55 km$^3$/yr (18%) by 2000, 60 km$^3$/yr (20%) by 2010, and 65 km$^3$/yr (22%) by 2020. In the second case, anthropogenic changes in climate projected for 2000 and 2020 were taken into account. Variations in inflow to the Caspian Sea were estimated with two independent approaches: using the runoff estimates from Fig. 7.5, and using the detailed water balance model described above. If the climate changes as the reconstruction suggests, the results indicate that water levels are likely to decrease by the end of the century, followed by eventual increases due to precipitation increases in the Volga watershed (see Fig. 7.6). Although these results are preliminary, they show the need to account for future changes of climate in planning water use in the Volga River Basin.

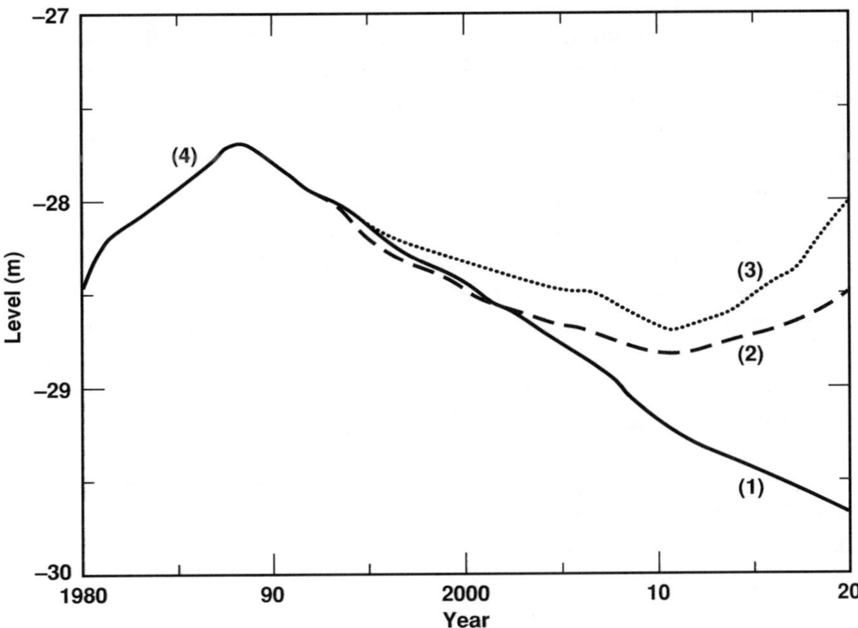


**Figure 7.6.** Levels of the Caspian Sea relative to an absolute value. Curves correspond to the following scenarios: (1) Stationary climate with human impacts; (2) anthropogenic changes of climate (model) and human impacts; and (3) anthropogenic changes of climate (reconstructed map for Mid-Holocene) and human impacts. Curve (4) shows observed variations in level for years before 1988.

### 7.2.4. Assessment of Possible Changes in North American Hydrology

In one of the earliest studies of climate and water, Stockton and Boggess (1979) used the empirical relationships derived by Langbein et al. (1949) to predict changes in runoff for the 18 water-resource regions of the coterminous United States. Using only hypothetical annual-average changes, they concluded that warmer and drier shifts in climate would be the most problematic for water availability. Revelle and Waggoner (1983) also used the empirical relationships of Langbein to evaluate hypothetical climate-change scenarios for the Colorado River Basin. They concluded that annual runoff in this region is especially sensitive to changes in temperature.

More recently, a number of studies have offered new insights into hydrologic vulnerabilities to greenhouse warming. These studies involve more accurate and comprehensive regional water-balance models and incorporate climate scenarios developed from general circulation models (Cohen, 1986; Gleick, 1986b, 1987a, 1987b; Flaschka et al., 1987; US EPA, 1988). Table 7.1 lists those studies that used hypothetical or general circulation model scenarios of climate change as input to regional hydrologic models.

Cohen (1986) evaluated the implications of GCM temperature and precipitation scenarios for water levels in the Great Lakes. In this study, the net basin water supply of the Great Lakes was predicted to decline in response to a wide range of climate scenarios, although the overall results were sensitive to changes in wind speed and other assumptions about variables that might alter lake evaporation rates.

In a study of the Sacramento Basin in California, Gleick (1986a, 1986b, 1987a, 1987b) identified hydrologic impacts that were consistent over a wide range of both hypothetical and GCM-generated climate-change scenarios. These included large decreases in summer soil-moisture levels, decreases in summer runoff volumes, major shifts in the timing of average-monthly runoff throughout the year, and large increases in winter runoff volumes. When calculated using eight different GCM temperature and precipitation scenarios that included both increases and decreases in average precipitation, summer runoff decreased by between 30 and 68%; winter runoff increased 16 to 81%; and summer soil moisture decreased 14 to 36% (Fig. 7.7). The principal mechanism that would drive these hydrologic effects would be a dramatic change in snowfall and snowmelt conditions. Due to higher temperatures, a greater fraction of precipitation would fall as rain rather than as snow, reducing the total annual snowpack. In addition, the snowpack would begin to melt earlier and faster in the spring, leading to less spring and summer snowmelt runoff and to decreases in summer soil moisture. This effect has now been noted in other regions [Shiklamanov, 1987; Bultot et al., 1988 (for Belgium); US EPA, 1988; Martinec and Rango, 1989] and has been identified in GCM results (Manabe and Wetherald, 1986; Wilson and Mitchell, 1987).

Two other results are particularly noteworthy. First, annual runoff in humid basins appears to be more sensitive to changes in precipitation than to changes in temperature (Flaschka et al., 1987; Gleick, 1987b; Karl and Riebsame, 1989; Shiklomanov, 1989), an effect described theoretically by Wigley and Jones (1985). Second, in watersheds with a seasonal snowfall and snowmelt pattern, the monthly distribution of runoff and soil moisture is more sensitive to temperature than to precipitation (Gleick, 1987b; Bultot et al., 1988; US EPA, 1988; Shiklomanov, 1989). In these watersheds, higher temperatures reduce the ratio of snow to rain during the winter, hasten the onset of spring snowmelt, and increase the rate of snowmelt runoff, as described above.

Recently, the American Association for the Advancement of Science (AAAS) convened a panel to evaluate the implications of climate change and variability for water resources in the United States (AAAS, 1990). Under the auspices of the AAAS, the entire scope of the problem was studied, from basic physical, climatological, and hydrologic issues to questions of political and economic allocation of water resources under new climatic conditions.

Despite this work, some fundamental questions have yet to be answered about how greenhouse warming will alter regional precipitation patterns and how water availability and quality will be affected.

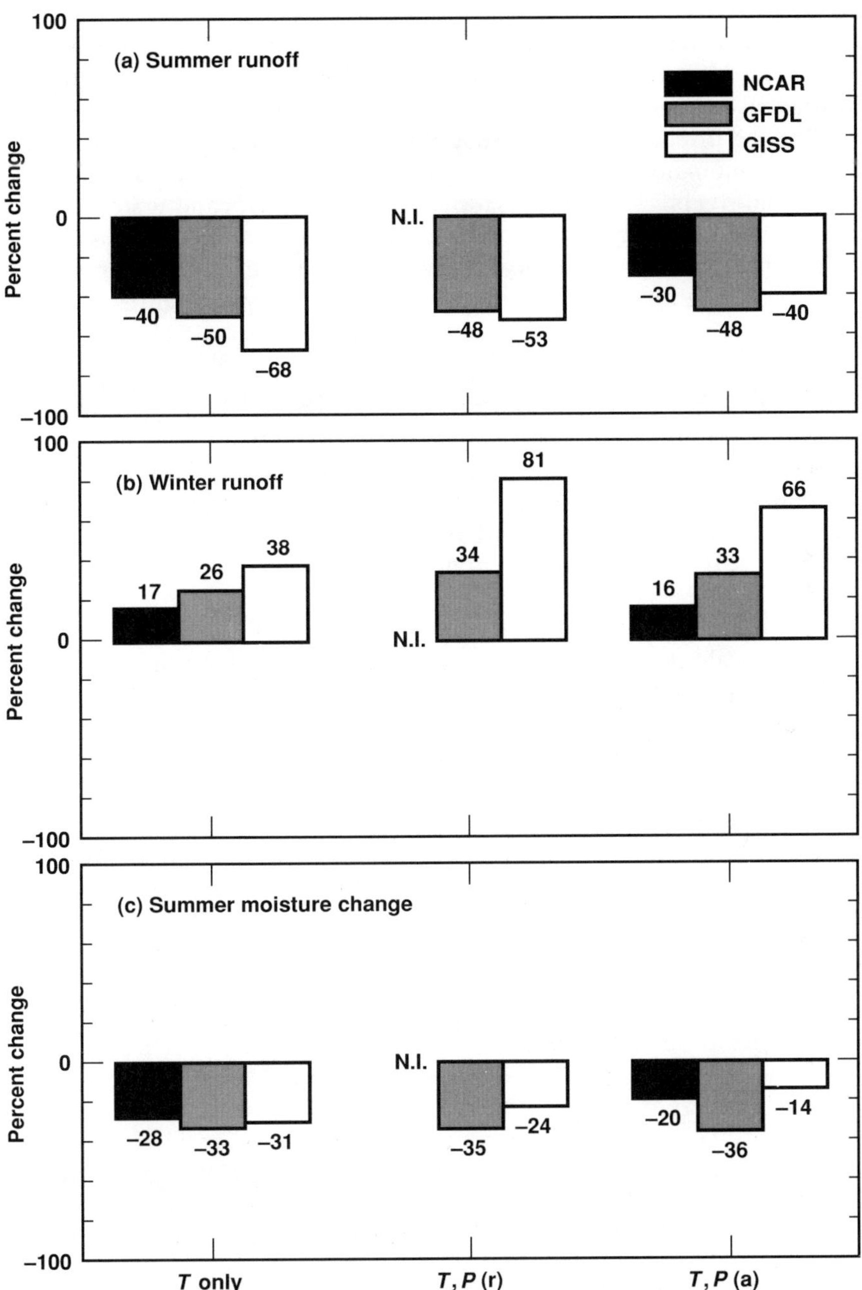


**Figure 7.7.** Climate-induced hydrologic changes in the Sacramento Basin, California, as derived from results of three different GCMs assuming only temperature changes (*T* only), relative (%) precipitation change [*T*, *P*(r)] and temperature and absolute precipitation change [*T*,*P*(a)]; from Gleick (1987b). N.I. indicates no information was available. The GCMs used here are those of the National Center for Atmospheric Research (NCAR), the Geophysical Fluid Dynamics Laboratory (GFDL), and the Goddard Institute for Space Studies (GISS). Summer is June, July, August. Winter is December, January, February.

Few watersheds have been studied in detail using appropriate models or methods. Little work has been done on the interactions among climate, vegetation responses, and water resources. The role of management in mitigating the impacts on water resources has been inadequately assessed. Questions remain about the implications of climate change for shared international rivers. The role of international water laws and water treaties in resolving climate-induced disputes is also uncertain. These issues are of great importance to society if critical impacts are to be identified and if concerted efforts are to be made to reduce the consequences of climate change.

## 7.3. CRITICAL WATER RESOURCES ISSUES

Research into the impacts of climate change on water resources has identified the following issues of special concern: water quantity and water management problems, water quality, soil moisture, the frequency and severity of extreme events, and hydroelectric generation. Joint US/USSR research in these areas should be encouraged.

### 7.3.1. Water Management and Planning

Large-scale changes in global climate that cause a temporal and spatial redistribution of precipitation and runoff would inevitably require changes in plans for the development, distribution, and management of major water users. Two major changes may be expected: (1) Changes in the availability of water resources and (2) changes in the demand for water resources from agriculture, industry, and population.

Global warming would cause the water supply situation to improve in some regions and to worsen in others. At present, only general conclusions can be drawn concerning prospects of water supply and trends in water project development for particular regions because of remaining uncertainties and the low accuracy of climatic projections. The most research has been done for North America and the Soviet Union; for other regions, such studies are limited.

Many parts of the United States have existing problems with water availability, as shown in Fig. 7.2. Changes in climate that reduce either the overall quantity of water or the timing of water availability will have important effects on agriculture, industrial and commercial activities, and urban development. On the other hand, changes in climate that increase overall water availability either could be beneficial or could increase the risk of flooding.

Both large-scale general circulation models and the paleoclimatic record indicate that the midlatitude, midcontinental region of the United States will undergo drying during summer months due to higher temperatures and evaporation and shifts in precipitation patterns (Manabe and Wetherald, 1986; Budyko and Izrael, 1987; Mitchell and Warrilow, 1987). More detailed regional hydrologic models have shown that runoff in major semi-arid river basins, such as the Colorado River Basin, is particularly sensitive to changes in temperature and precipitation (see Gleick, 1988a).

Many groundwater reservoirs in the western US are already subjected to overuse. For example, the large fossil-water reservoir of the Ogallala Aquifer has been oversubscribed for many years. This groundwater reservoir underlies approximately 600,000 $km^2$ in the Great Plains of Texas, New Mexico, Oklahoma, Kansas, Colorado, and Nebraska and serves agricultural, industrial, and domestic purposes. Because of the ease of access to this groundwater, and because of the lack of other major surface-water supplies, these states have become dependent on it as a primary source of supply. Yet, its location corresponds closely with the region identified as vulnerable to soil-moisture drying due to climate changes. Wilhite (1988) and Glantz and Ausubel (1988) discuss the implications of climate change for this region and the possibility of using the present problems of the Ogallala as an example of the difficulties to be addressed in considering appropriate policy responses to future changes in climate.

Large-scale water-transfer projects are often proposed in regions with water shortages. For several reasons, new large transfers are increasingly unlikely, even in the event of climate-induced shortages:

(1) Unfavorable anthropogenic changes of climate may cover large areas, including the basins planned for water diversions;
(2) Uncertainties concerning the nature of regional changes in climate will complicate both the planning and the economic evaluation of such projects; and
(3) Such projects often have large-scale, adverse environmental impacts.

In general, Soviet water resources are poorly distributed in space compared to the demands of population, industry, and agriculture. More than 80% of river runoff is concentrated in the northern areas of the European USSR, Siberia, and the far eastern USSR with its limited economic infrastructure. Meanwhile, densely populated regions in the southern European USSR, the Caucasus, Kazakhstan, and central Asia are characterized by extremely limited water resources and growing water consumption and population (Shiklomanov and Markova, 1987).

To improve water availability in the southern regions of the USSR, large-scale water transfer projects were planned during the 1970s. These projects were to transfer water from the northern rivers in the European USSR to the Volga Basin, for example, from the Ob River to Kazakhstan and central Asia and from the Danube to the lower reaches of the Dnieper. Considering the possible climate-induced changes in runoff shown in Fig. 7.5 and Table 7.2, water transfers appear to be increasingly less attractive and less economically necessary because increases in water availability may occur in these regions. However, these conclusions are very preliminary, given the limitations in our understanding of the applicability of the paleoreconstructions.

If the paleoreconstructions are correct, potential reductions in annual runoff of 8 to 12% in the Volga Basin would suggest a need for additional water in that basin. However, two considerations suggest that water transfers from the northern rivers in this basin may be unnecessary: (1) Present high water levels in the Caspian Sea (Fig. 7.6), and (2) suggestions from paleoclimatic reconstructions (but not model simulations) that the further warming in the early twenty-first century may eventually lead to significant increases in both runoff in the Volga Basin and total water availability in the southern areas. The latter point can be seen from Table 7.3, which uses runoff suggested by the paleoclimatic reconstructions to show data on streamflow distribution in the Volga Basin over seasons, mean monthly natural distribution, and runoff estimates, including the effects of man's activity and regulation by a system of reservoirs. These data are quite approximate, but they suggest that if the predicted global warming occurs, river runoff in midlatitudes in the Northern Hemisphere may be subject to changes similar to those produced by the intensive control of a system of reservoirs.

**Table 7.3.** Streamflow ($km^3$) distribution in the Volga for various seasons.[a]

| Runoff characteristic | Winter | Spring | Summer/ autumn | Annual |
|---|---|---|---|---|
| Mean monthly natural runoff | 31 | 144 | 79 | 254 |
| Present runoff, regulated | 65 | 92 | 77 | 234 |
| Runoff for Holocene-like conditions (without control) | 47 | 128 | 49 | 224 |
| Runoff for Eemian-like conditions | 103 | 101 | 61 | 265 |

[a] Source: Shiklomanov (1989).

### 7.3.2. Surface Water Quality

Despite the importance of the problem, very little work has been done on the implications of climate changes for surface water quality. Numerous river basins in the US and the USSR are already vulnerable to water-quality problems during times of the year when flow is low or during dry years. The Colorado River, for example, has severe salinity problems in certain reaches due to agricultural runoff, which led in the 1960s to an international dispute between the US and Mexico. In the USSR, similar severe salinity problems occur in the Amu Darya and other rivers of the Aral Sea Basin.

Other water-quality problems will arise in rivers where the flushing of pollutants is a function of water supply, in lakes where higher temperatures may lead to accelerated eutrophication and changes in fish species, and in coastal areas where rising sea level may combine with changes in freshwater flows to raise the salinity of water supplies (Jacoby, 1989). Higher temperatures in subalpine lakes in California, for example, could increase annual primary production by as much as 87%, leading to degraded water quality and changed species compositions (Byron et al., 1988).

### 7.3.3. Soil Moisture

Soil moisture is critical for agriculture and for ecosystems. Yet, present soil-moisture parameterizations in climate models are oversimplified, and reconstructions of past soil-moisture conditions are limited by missing data and incomplete regional information. Even today, consistent data on actual soil moisture conditions are not available in most parts of the world.

Some of the regional results described above support recent suggestions that, during summer, midlatitude soil-moisture reductions may occur in many regions of the world; others suggest a slight moistening. The principal physical mechanisms tending toward a drying include the decrease in snow as a fraction of total winter precipitation, an earlier and faster disappearance of winter snowpack due to higher average temperatures, and a more severe evapotranspiration demand during the warmer summer months: Such factors are both physically plausible and consistent with the hydrologic mechanisms that lead to regional summertime soil-moisture drying in some of the GCMs. Other, countervailing hydrometeorological features, such as cloud cover and evapotranspiration feedbacks, can tend to counteract these influences. In particular, if the increases in precipitation identified by the paleoclimatic reconstructions and some model simulations exceed the increased evapotranspiration, soil moisture in these regions could increase (see Chapters 5 and 6).

### 7.3.4. Frequency of Extreme Events

Current water problems in many parts of the US and USSR are caused, in part, by periodic extreme events such as coastal storms, droughts, and high-intensity precipitation events that cause flooding. At present, we are unable to predict with certainty that the frequency or intensity of storm events will change, although, for example, in Northern California there has been an increase in the variability of natural runoff in the last decade, as shown in Fig. 7.8.

Other evidence suggests that seasonal changes in water availability may increase the probability of severe flooding in mountainous regions. This risk arises from a change in snowfall and snowmelt conditions due to the expected rise in average temperatures. Higher temperatures mean that the proportion of rain to snow in mountainous regions will increase, leading to greater prompt runoff and a higher risk of winter flooding.

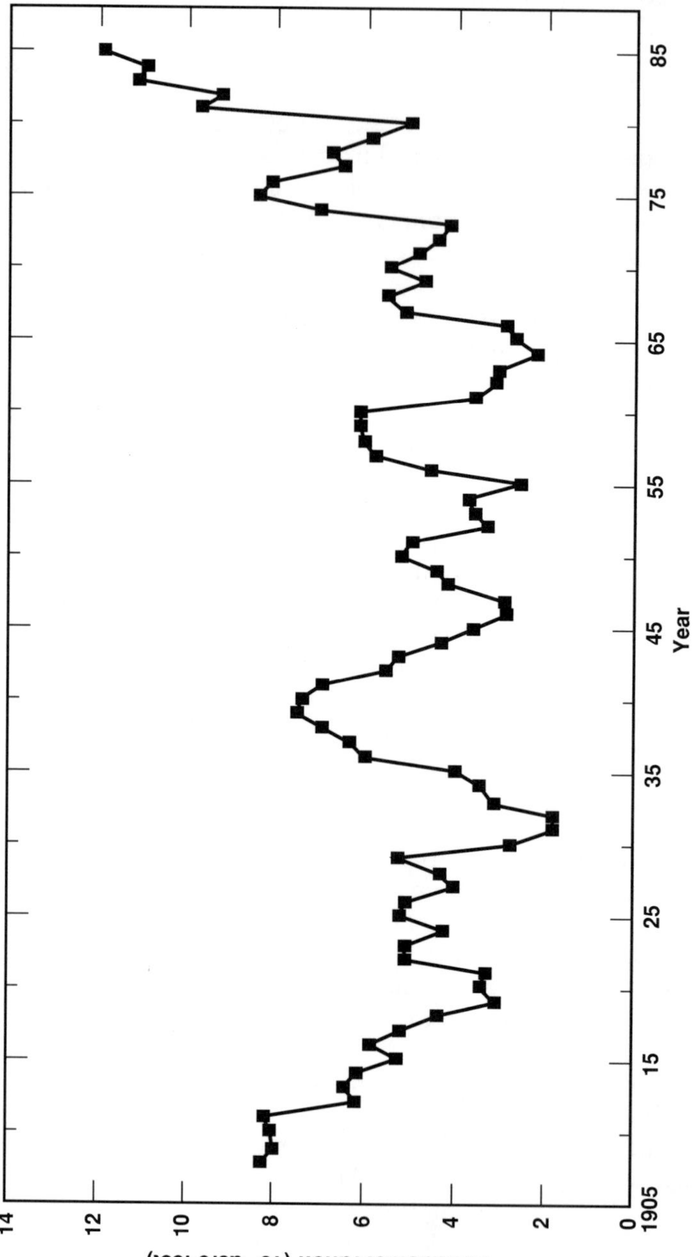


**Figure 7.8.** Historical variability of runoff in the Sacramento Basin, California. The plot shows the 5-year moving average of the standard deviation of runoff (from Gleick, 1989).

### 7.3.5. Generation of Hydroelectricity

Many regions are highly dependent upon hydroelectricity for a large fraction of their total electricity supply. Figure 7.3 shows such areas for the US. Even under present climatic conditions, droughts can substantially reduce hydroelectric generation, leading to increases in the use of other energy resources, usually fossil fuels. For example, during a normal water year, California's hydroelectric output provides approximately 20% of the state's total electrical energy supply. In 1977, the second year of a severe drought, hydroelectric output dropped to only 7% of California's total annual energy production. Initially, some of the deficit was made up by importing surplus electricity from the states in the Pacific Northwest, but these imports dropped to one-third their normal level when the Pacific Northwest began to suffer from similar drought effects. The remaining electricity deficit was made up by burning additional fossil fuels at an estimated direct cost of half a billion dollars (Gleick, 1988b).

Similar effects have been seen in the USSR. From 1977 to 1983, low river runoff in southern Siberia decreased power production at hydroelectric power stations. This caused serious consequences for total power supplies, economic activities, and the population in that vast region. As precipitation patterns begin to shift, the operation and management of existing reservoirs may have to be altered, and the design of new systems may be greatly complicated; however, little analysis in these areas has been completed (e.g., see Miller, 1990).

## 7.4. IMPACTS OF CLIMATE CHANGE ON AGRICULTURE

Despite highly developed technology, there remains heavy reliance on agriculture—the production of human food from soil, water, plant, and animal resources. This dependence on agriculture makes projecting the agricultural consequences of the predicted global climate change a crucial task. This section examines how global climate change may affect agriculture in the US and the USSR, two of the largest and most important food-producing countries in the world.

A brief comparison of agriculture in the US and the USSR reveals great similarities and great differences. Table 7.4 provides some agricultural statistics compiled for 1983 by the Food and Agriculture Organization (FAO, 1985) of the United Nations. Although the USSR has more arable land and land in permanent crops, and produces more root crops, pulses (non-oil crop legumes), and milk than the US, the US produces more cereals and meat. US exports have considerably higher value than USSR exports, but the two countries spend about the same amount to import agricultural products.

The US and the USSR both have rich soil resources, especially in the interior grasslands. The chernozems (rich, organic soils) of the steppes of the USSR and the similar mollisols of the former North American tallgrass prairies form the basis for large cereal-production regions. However, the continentality of these interior regions far from oceanic moisture sources also leads to drought vulnerability, as well as to extreme variations in seasonal temperatures. Severe drought, accompanied by warming, has been recurrent in both regions, most notably in the 1930s when dry years induced agricultural disaster in both the USSR and the US. Persistent drought in that decade in the US contributed to about 200,000 farm bankruptcies or involuntary transfers and the migration of more than 300,000 people out of the southern Great Plains. Even today, with advances in agricultural technology and irrigation techniques, crop production in both countries remains sensitive to interannual weather variations.

**Table 7.4.** Agricultural statistics for 1983 in the US and the USSR.[a]

| Characteristic | Descriptor | Unit | US | USSR |
|---|---|---|---|---|
| Land use | Total land | 1000 ha | 916,660 | 2,227,200 |
| | Arable land and permanent crops | 1000 ha | 189,915 | 232,390 |
| | Irrigated lands | 1000 ha | 19,831 | 19,146 |
| Population | Total | Thousands | 234,500 | 272,540 |
| | Agricultural | Thousands | 4,410 | 39,613 |
| Labor Force | Total | Thousands | 107,288 | 136,261 |
| | Agricultural | Thousands | 2,018 | 19,805 |
| Production | Cereals | 1000 Mt | 207,875 | 186,898 |
| | Root crops | 1000 Mt | 15,696 | 83,060 |
| | Pulses | 1000 Mt | 925 | 8,205 |
| | Oil crops (oil equiv.) | 1000 Mt | 9,660 | 3,261 |
| | Meat | 1000 Mt | 25,302 | 16,450 |
| | Milk | 1000 Mt | 63,488 | 96,415 |
| Means of production | Agricultural tractors in use | Thousands | 4,550 | 2,697 |
| | Fertilizer consumption | 1000 Mt | 19,852 | 22,948 |
| External Trade (exports) | Agricultural products | $1,000,000 | 37,537 | 2,386 |
| External Trade (imports) | Agricultural products | $1,000,000 | 17,680 | 18,560 |

[a] Source: FAO (1985).

Both countries are major producers of wheat, with the USSR producing more spring wheat than winter wheat, while in the US more winter wheat is grown. Major crops in the USSR (in addition to wheat) are oats, barley, sunflower, and sugarbeet; in the US, maize (known in the US as corn) and soybeans are the other major crops. The nations' crop productions reflect their geography: the USSR spans higher latitudes (important agricultural regions extend from 45°N to well poleward of 60°N), and growth of short-season crops such as barley, sunflowers, and spring wheat is dictated by temperature and growing-season limitations. The US is bounded by Canada on the north at the 49°N parallel, but agricultural production extends south to 25°N, which allows more growth of crops of subtropical origin, such as maize, and crops tolerant of hot weather, such as soybeans. These latitudinal differences of the two countries may be important in projecting the shifts in crop geography that may accompany global climate change. For example, the USSR has northern-latitude regions into which agricultural regions may expand, soil resources permitting, whereas subtropical agricultural crops or grasslands may become more prevalent in the US.

Change in crop geography is only one aspect of the impact of global climate change on agriculture. In order to gain a broad understanding of the potential agricultural future, in light of predicted climate change, many factors must be considered. Biophysical processes are temperature and moisture dependent; crops, insects, weeds, livestock, and diseases will all respond according to their individual dependences. Carbon dioxide ($CO_2$) itself is a necessary building block of photosynthesis. As climatic factors change, a cascade of effects will occur throughout the agricultural system, as human decisions involving infrastructure, transportation, markets, and trade respond. Ultimately, impacts of climate change on agriculture may reverberate throughout the international food economy and thus through the global society as a whole.

### 7.4.1. Review of Recent Studies of Agricultural Impacts

Numerous methods have been used to study the impacts of climate change on agriculture. These include controlled experiments, historical studies, spatial analysis, estimation of potential production, dynamic modeling of crop growth, risk analysis, and integrated modeling of agricultural and economic systems. The choice of approach and method depends on the sphere of analysis and the research questions posed.

Contradictory conclusions sometimes result because of differences in scenarios. For example, paleoclimatic reconstructions and some GCM results suggest that higher levels of precipitation will lead to increased crop productivity, whereas other GCM simulations that suggest significant and dramatic drying of midcontinental regions would lead to the opposite conclusion. These different climatic scenarios lead to different impacts on agriculture—one favorable and the other unfavorable.

Resolving these differences with more research is critical to understanding climatic impacts on resources and society, and such research is an important subject for US and Soviet collaboration.

Controlled experiments have been conducted in phytotrons, greenhouses, and field chambers to determine the direct physiological consequences of specified climatic regimes. Much research in the US has focused on the direct effects of $CO_2$ on crops. Acock and Allen (1985) and Cure (1985) have reviewed these experimental studies: For crops growing in experimental settings with doubled $CO_2$ (660 ppmv) and current climate conditions, there is an average increase in yields of about 30% and increases in water-use efficiency. More recent experimental work has attempted to characterize the physiological effects of both increased $CO_2$ and higher temperatures on crop growth (see, for example, Baker et al., 1989), but results have been inconsistent.

Historical studies of periods in which warming and/or drying have occurred offer insight into possible responses of an agricultural system to future warming. Work by Menzhulin et al. (1983), Menzhulin and Nickolaev (1987), and others show that Soviet grain yields decreased in the 1930s by 25 to 30% (Fig. 7.9), about the same as the US yield decreases (Warrick, 1984). Decreases in Soviet cereal yields during the 1960s are partly attributable to the development of virgin land in the west-central USSR. This Soviet research shows that the period 1975 to 1988 was also unfavorable for agriculture in the USSR due to increased temperature and decreased precipitation (Table 7.5). The changes in the aridity index in the US and USSR shown in Fig. 7.10 indicate that aridity changes in the past few years are not quite as severe as those that occurred in the 1930s. The mean deviation of the aridity index[1] over this 13-year period shows that year-to-year variability has increased recently in much of the US and the USSR (Fig. 7.11).

Spatial analysis consists of identifying critical environmental limits of specific crops, applying climate change scenarios, and calculating the resulting spatial shifts in crop regions. This agroclimatic method provides an approximation of possible changes in crop areas from a biological perspective, but does not address potential changes in either yield or production. Rosenzweig (1985) specified environmental requirements for North American wheat-growing regions and projected that wheat could still be grown in most parts of the US under a doubled $CO_2$ climate. Under such conditions, however, fall-sown spring wheat would replace hard winter wheat where warmer winter temperatures prohibit vernalization (Fig. 7.12).

Another research method is the estimation of potential production from climatic variables or indices such as length of growing season, precipitation, evapotranspiration, solar radiation, and temperature.

---

[1] The aridity index is defined as $s_L = \Delta T_i/\sigma_T - \Delta P_i/\sigma_P$ where $\Delta T_i$ and $\Delta P_i$ are the anomalies of temperature and precipitation, and $\sigma_T$ and $\sigma_P$ are their standard deviations.

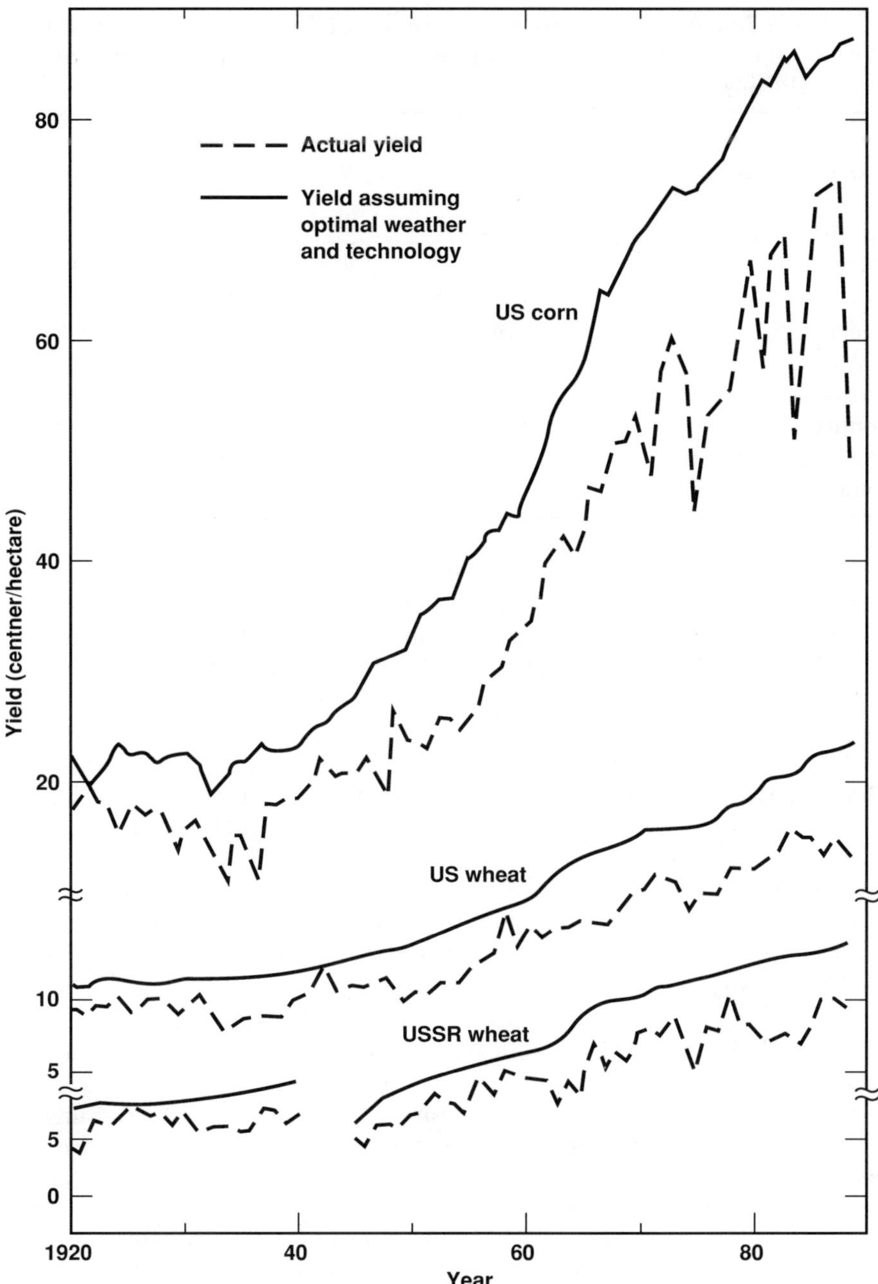


**Figure 7.9.** Trends in major cereal crops in the US and USSR. Dashed lines show actual yields, and solid lines show the potential yield obtained by assuming optimal weather conditions and up-to-date technology (from Menzhulin and Nickolaev, 1987). One centner equals 50 kg; one hectare equals 10,000 square meters (or 2.47 acres).

**Table 7.5.** Aridity index changes in the USSR during selected warm intervals of the present century. A more positive index indicates more arid conditions.

| Economic regions | 1914 to 1926 | 1927 to 1939 | 1962 to 1974 | 1975 to 1988 |
|---|---|---|---|---|
| North Caucasus | –0.35 | –0.25 | +0.23 | +0.60 |
| West Siberia | –0.23 | +0.30 | –0.02 | +0.38 |
| Donetsk-Dnieper | –0.58 | –0.19 | –0.01 | +0.42 |
| Ural | –0.20 | +0.36 | –0.37 | +0.44 |
| Volga | –0.59 | +0.03 | –0.03 | +0.01 |
| Kazakhstan | –0.18 | +0.15 | +0.14 | +0.39 |
| South of Ukraine | –0.42 | +0.03 | +0.03 | +0.23 |
| Southwest of Ukraine | –0.81 | +0.06 | +0.12 | +0.05 |
| Central Chernozem | –0.51 | –0.12 | +0.19 | +0.15 |

This technique was used in the FAO Agro-ecological Zone Project (Kassam, 1977). One problem with this approach is that high-temperature effects on crop biomass and crop yields are often opposite. That is, annual net primary productivity of perennial vegetation should increase with warmer temperatures; however, many agricultural crops are annuals that respond to higher temperatures by hastening their rate of development and by decreasing yields. The potential production method also tends to smooth the high degree of variability in production attributable to other factors that affect crop growth, such as management and economics. Dudek (1989), using the potential production method to estimate the climatic effects of doubled-$CO_2$ change scenarios and direct $CO_2$ effects on agricultural crops in California, found that the statewide average yields of most crops, except cotton and some vegetables, would decrease slightly (less than 10%) for a doubling of $CO_2$ (Fig. 7.13).

Dynamic-process models of crop growth simulate the response of agricultural plants to climate, soil, and management factors. These models may be used with climate change scenarios either to explore consequences to yields and stages of development or to determine thresholds of crop-growth sensitivity to changing climate variables. They are also useful for testing possible adaptations to climate change, such as altered planting dates, irrigation scheduling, or crop variety.

Peart et al. (1989), Ritchie et al. (1989), and Rosenzweig (1989) used the CERES and SOYGRO crop models to project the effects on yields in the US Southeast, Great Lakes, and Great Plains regions of $CO_2$ -induced climate changes and the direct effects of increased $CO_2$ on crop growth. In northern latitudes, where warmer conditions lengthened the frost-free growing season, higher temperatures increased yields. In the humid Great Lakes and corn belt regions, the changes in temperature used in their scenario hastened maturity and decreased yields. In the southern

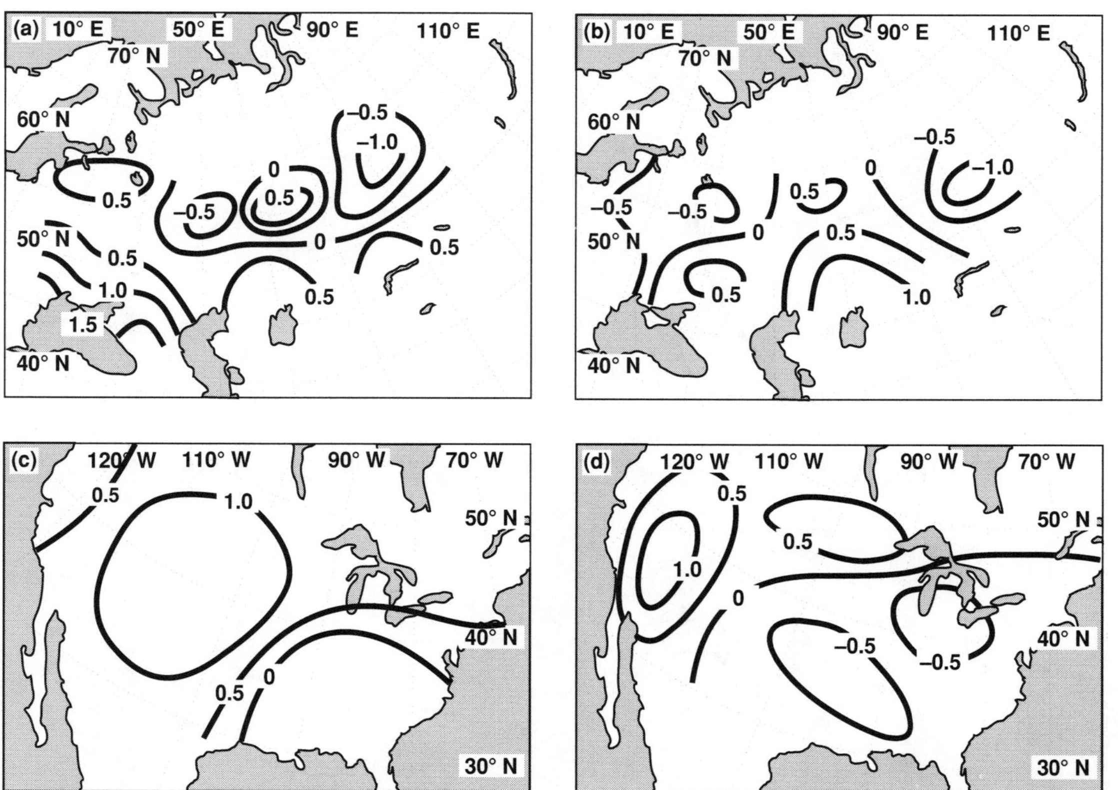


**Figure 7.10.** Aridity index changes from the beginning to the end of time periods (a) 1914 to 1939, (b) 1962 to 1988 in the USSR, (c) 1914 to 1939, and (d) 1962 to 1988 in the US. Positive numbers indicate areas where aridity has increased.

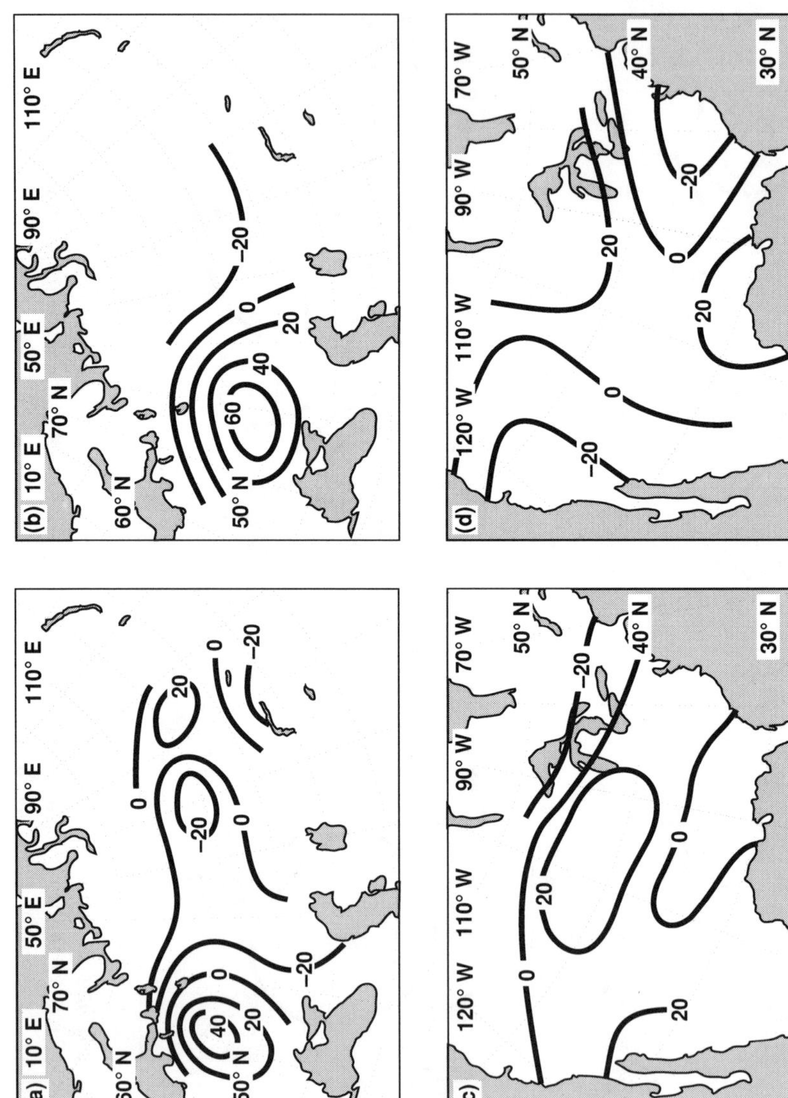


**Figure 7.11.** Agrometeorological variability index changes (in %) from the beginning to the end of time periods (a) 1914 to 1939, (b) 1962 to 1988 in the USSR, (c) 1914 to 1939, and (d) 1962 to 1988 in the US.

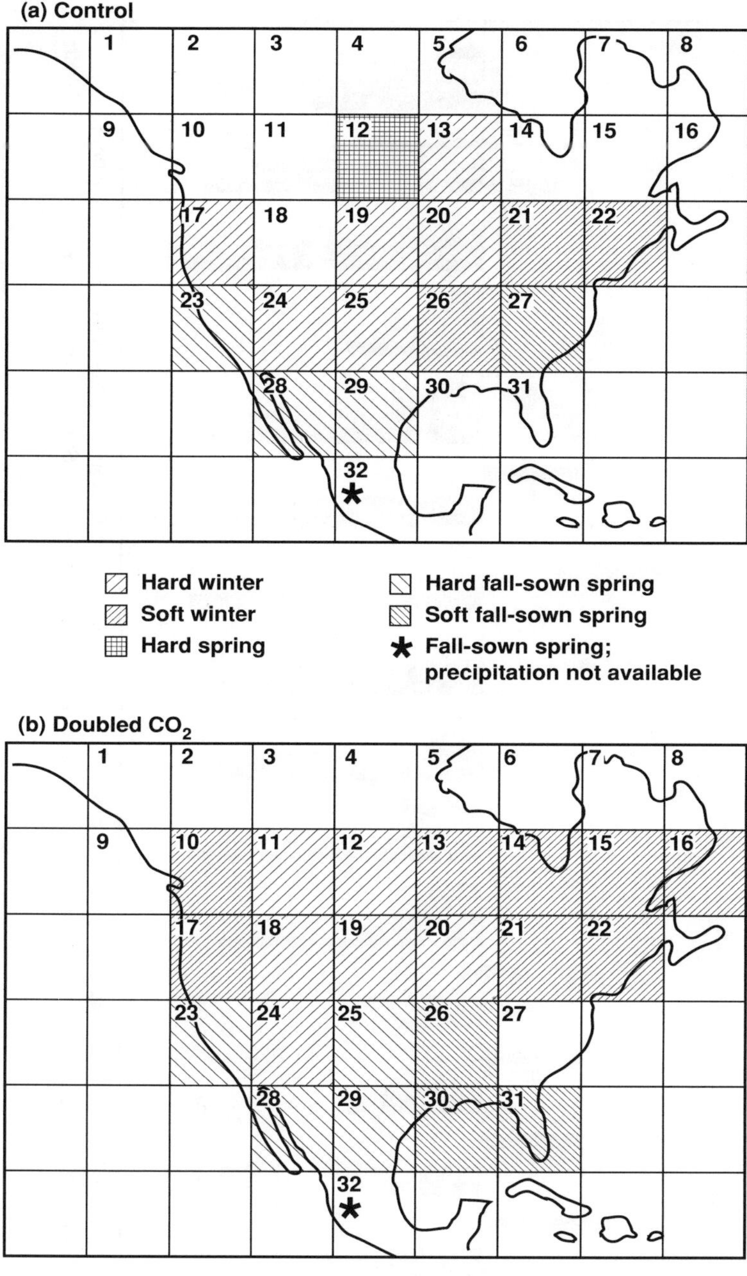


**Figure 7.12.** North American wheat regions assuming climatic conditions drawn from (a) the GISS GCM control run and (b) doubled $CO_2$ runs (Rosenzweig, 1985).

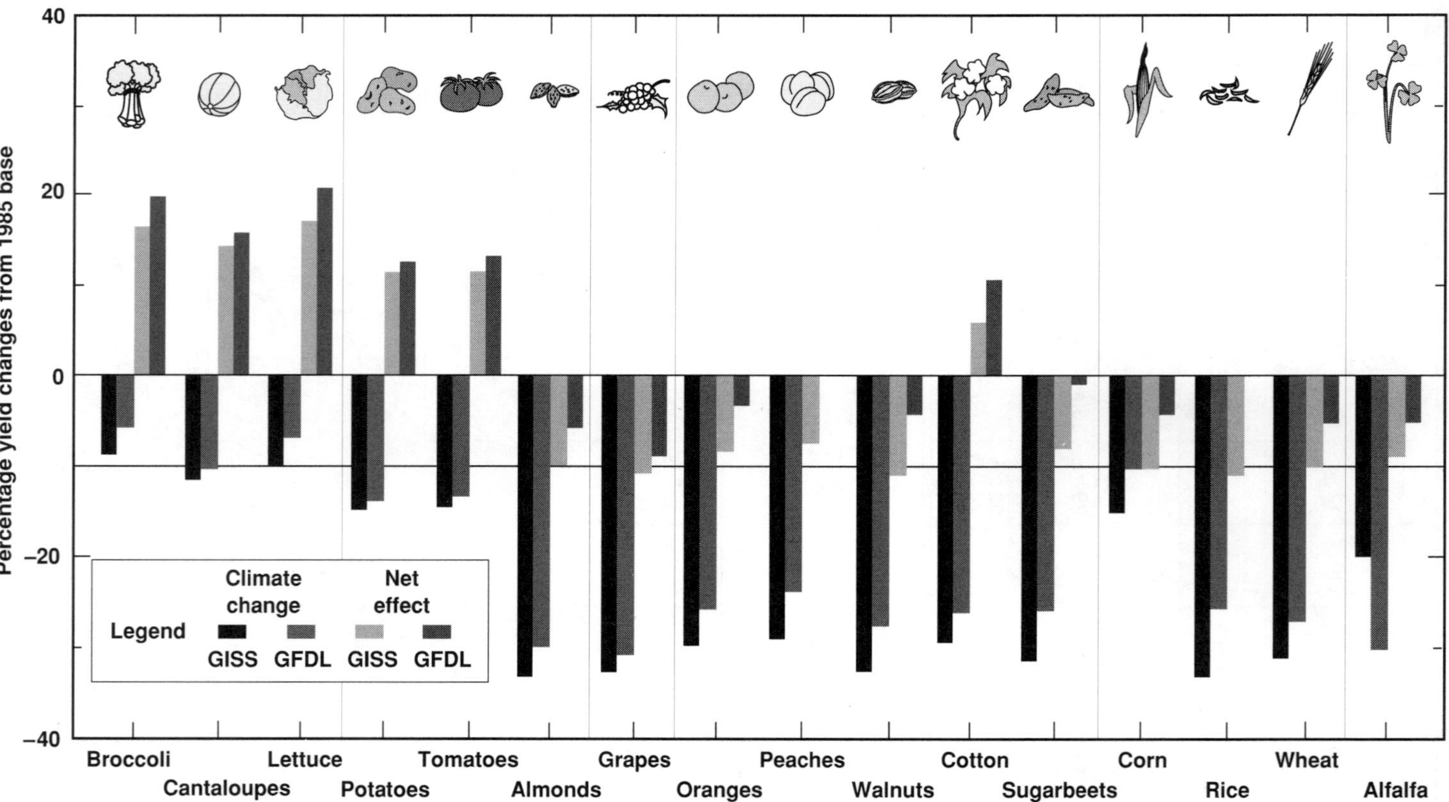


**Figure 7.13.** Average yield changes for various crops grown in California for four climate-change scenarios (Dudek et al., 1989).

plains, irrigated yields were higher and less variable than rain-fed yields, but they still declined when measured against the base-case irrigated yields. In the Southeast, average yields decreased significantly when climate change was modeled without the beneficial direct effects of increased $CO_2$, but declined less when the direct effects of $CO_2$ were included. These results extrapolated to other US regions are shown in Fig. 7.14.

In the USSR, Sirotenko and Boiko (1980) used a soil-plant-atmosphere simulation model to estimate the effects of varying $CO_2$ concentrations on agricultural productivity. Pavlova and Sirotenko (1985) have also evaluated the use of dynamic models for estimating the effects of climate change on crops.

Risk analyses seek to determine changes in the probabilities of either extreme climatic events, such as drought or high temperatures, or the ensuing changes in probabilities of low or high yields. Mearns et al. (1984) found that the relationships between changes in mean temperature and corresponding changes in the probabilities of extreme-temperature events are nonlinear and that relatively small changes in mean temperature could result in relatively large changes in event probabilities. For example, in Des Moines, Iowa, the likelihood of a five-day maximum temperature occurrence of at least 35°C is about three times greater for a relatively small 1.7°C increase in mean temperature than for current climatic conditions.

Coupled agricultural and economic modeling studies combine results from the above methods with economic modeling. The scales of these studies range from individual farms to regional agricultural economies, the national agricultural sector, and finally to global food trade. The US Environmental Protection Agency (EPA) has recently completed a set of studies that integrated climate change, direct $CO_2$, and economic effects on US agriculture, both regionally and nationally (Smith and Tirpak, 1989). The EPA studies concluded that climate change would result in northward shifts in cultivated land, with associated regional economic changes in both southern and northern parts of the country. Crop irrigation requirements were projected to increase, as well as infestations of agricultural pests and diseases, leading to continuing trends of environmental hazards.

An international study of both agronomic and economic impacts of climate change was conducted by the International Institute of Applied Systems Analysis (IIASA) and the United Nations Environment Program (UNEP) (Parry et al., 1988). As part of that effort, Pitovranov et al. (1988) studied the effects of climatic variations on agriculture in the northern regions of the USSR. The results showed small increases in yields of spring wheat and winter rye in more northern regions with climate warming, and lower yields of barley and potatoes in the central region of the European USSR. Cropped areas of winter wheat and corn that

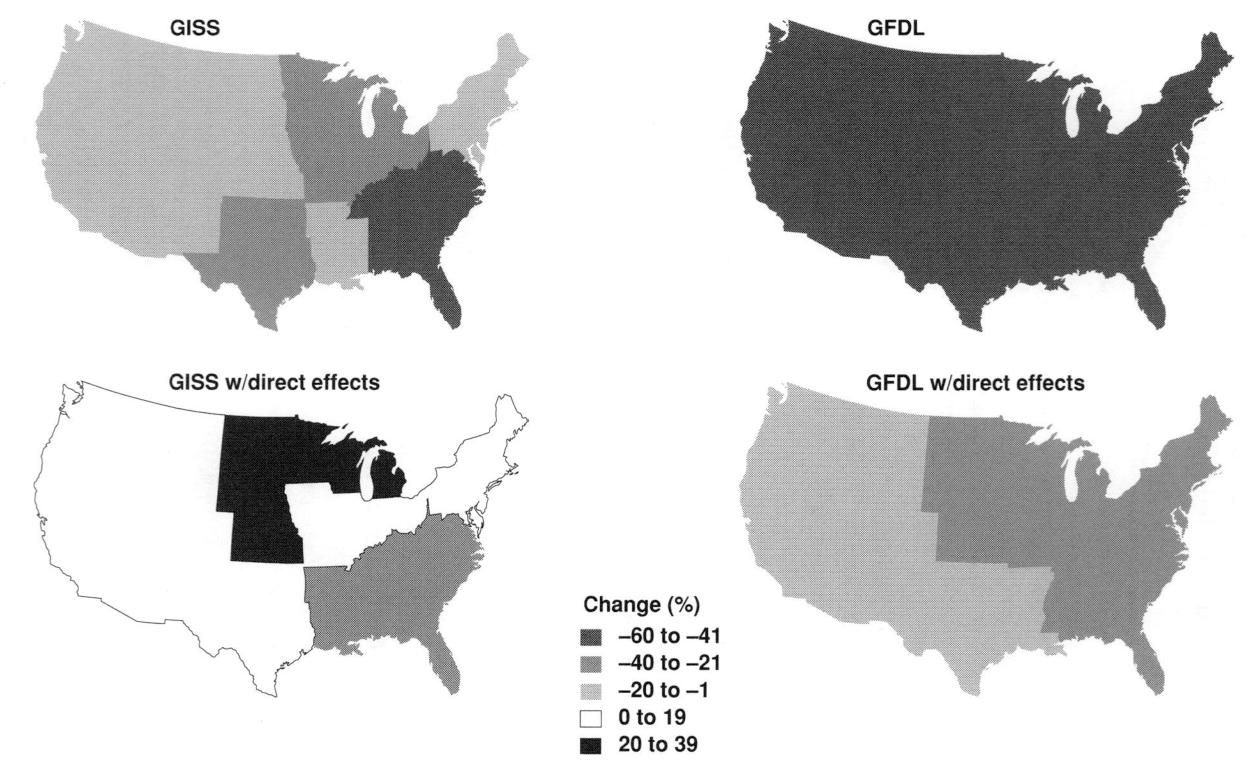


**Figure 7.14.** Projected changes in US crop yields estimated by the CERES and SOYGRO models, obtained by assuming the climate is modified as estimated by the GISS and GFDL climate models in response to a $CO_2$ doubling, with and without including the beneficial direct effects of $CO_2$ doubling.

respond favorably to temperature increases were projected to grow at the expense of areas of barley and potatoes, which are better adapted to moister, cooler conditions.

### 7.4.2. Direct Effects of $CO_2$

The study of agricultural impacts of trace-gas-induced climate change is complicated by the fact that increasing the atmospheric $CO_2$ concentration has other effects on crop plants besides affecting their climatic regime. Because of their beneficial physiological nature, these are known as "direct" or "fertilizing" effects. Specifically, most plants growing in a $CO_2$ concentration higher than the current ambient level exhibit increased rates of net photosynthesis (i.e., total photosynthesis minus respiration). This increase is manifested in higher leaf area, dry-matter production, and yield for many crops (Acock and Allen, 1985; Cure, 1985).

Carbon dioxide enrichment also tends to close plant stomates, which are small openings in leaf surfaces through which $CO_2$ is absorbed and water vapor is released. By so doing, $CO_2$ reduces transpiration per unit leaf area while enhancing photosynthesis. Thus, it often improves water-use efficiency, defined as the ratio between crop biomass accumulation or yield and the amount of water used in evapotranspiration (Mauney et al., 1978).

Increasing $CO_2$ concentrations have been demonstrated to decrease photorespiration in so-called C3 plants (important crops with this pathway are wheat, rice, potatoes, cotton, and soybeans). In the process of photorespiration, a considerable fraction of the $CO_2$ initially fixed into carbohydrates is reoxidized back to $CO_2$ ; with increased $CO_2$ , more $CO_2$ is fixed as carbohydrate (Laysk, 1977). In C4 plants (e.g., maize, sugar cane, and sorghum), $CO_2$ is first trapped in the mesophyll cells and then concentrated in the bundle sheath cells that carry on photosynthesis. These plants, while more efficient photosynthetically under current $CO_2$ levels than C3 plants, are less responsive to $CO_2$ enrichment.

The beneficial direct effects of $CO_2$ have also been shown to occur at less than optimal levels of light, water, and mineral-nutrition levels. The percent increase in yields with high $CO_2$ is often relatively higher for water-stressed plants than for nonstressed plants (Gifford, 1979), but if the water deficit is severe and long-lasting, water stress-induced inhibition of photosynthesis may dominate the photosynthetic enhancement of high $CO_2$ . Nitrogen-limited plants have also responded positively to high $CO_2$ (Wong, 1979).

Although these direct effects of high $CO_2$ have been observed under controlled experimental conditions, their magnitude and significance in the open field are still uncertain. Although beneficial effects seem fairly unequivocal, there is still uncertainty as to their importance relative to large-scale climatic effects. One reason for this uncertainty is that greenhouse and field-chamber environments are not identical to field conditions, being much smaller and less variable and having different, less windy, evaporative regimes. Results of the first open-air field experiments on cotton organized by the US Department of Energy are eagerly awaited. Experimental research has also concentrated mostly on the process of photosynthesis and on a limited number of crops, especially soybeans and cotton, with some work on maize and wheat. Data on $CO_2$ effects on other processes, such as respiration, or on other crops important to the US and the USSR, such as sugarbeets and sunflowers, are notably lacking.

Another reason for uncertainty is that physiological feedback mechanisms may limit the extent to which direct $CO_2$ effects are realized. Crop plants have sometimes been observed to acclimate eventually to higher $CO_2$ levels; their photosynthetic rates decline to levels similar to or only slightly higher than those observed at present atmospheric concentrations. This may be caused either by a decrease in the rubisco enzyme content in the chloroplasts (Keerberg and Vijl, 1982) or by an accumulation of starch in leaves that tends to inhibit further carbohydrate production. A limited sink (i.e., any growing, storing, or metabolizing tissue) for the assimilates, such as small size or quantity of the kernel or grain, may also serve to dampen the effects of $CO_2$ on photosynthetic rates.

Most experiments have tested a step change of $CO_2$ over only one annual crop-life cycle. This approach cannot be used to determine the long-term evolutionary response to higher $CO_2$ levels, which may tend toward less efficient photosynthesis. Studies of the interactive effects of elevated $CO_2$ concentration and higher temperatures, under different water and nutritional regimes, are also few in number, and results have been inconsistent. For example, enhanced growth of cotton with combined high $CO_2$ and high temperatures is reported by Kimball et al. (1986), but growth and development responses of soybeans leveled off at higher temperatures in elevated $CO_2$ experiments by Baker et al. (1989).

Most biological processes respond to temperature with curves moving from minimal rates at lower temperatures through optimal rates in some median temperature range, then descending to a minimal rate with higher temperatures. Because regional differences in temperature are predicted with climate change, the relative magnitudes of combined $CO_2$ and temperature effects will likely be different in different places. In cooler places, higher temperatures will shift biological process rates toward optimum, and beneficial effects are likely to ensue. In locations where increased temperatures move beyond optima, negative consequences may dominate. Controlled experiments and studies with dynamic-process crop models are needed to determine if these relationships are likely to change with higher $CO_2$ concentrations.

## 7.5. CRITICAL ISSUES IN AGRICULTURAL IMPACT ASSESSMENT

Despite the progress that has been made in understanding the agricultural consequences of climate change, many questions remain.

This section presents several critical issues that are under study or that need to be studied to evaluate more fully the effects of climate change on global and regional food availability.

### 7.5.1. Variability

Predictions of climate change have usually been expressed as changes in the mean values of temperature and precipitation. If changes in daily and interannual variability occur, however, important agricultural consequences are likely to ensue. One aspect of these variability changes is the importance of extreme events—such as droughts, heat waves, and storms—to agricultural crops and livestock. Runs of extreme climatic variables (e.g., drought, long periods of high humidity, or prolonged heat spells during the time of grain filling) can decrease crop productivity. For rain-fed crops, the frequency, intensity, and duration of extreme climatic events can be more consequential to crop yields than changes in mean values.

Unfortunately, GCMs have not yet been validated to project these types of changes with confidence. Initial studies, however, show that interannual and diurnal variability and annual amplitude of temperature are likely to decrease, whereas interannual and diurnal variability of precipitation may increase (Rind et al., 1989; Mearns et al., 1990). Characterizing potential changes in climate and yield variability is a crucial research topic that demands more attention in future studies.

### 7.5.2. Drought

The anticipated global warming has led to concern about future water availability because, in the historical record, warmer summers have often been correlated with reduced precipitation. In addition, hydrothermal conditions in most agricultural regions of the USSR are, in general, less favorable for crop production than in the US. Studies addressing the question of future droughts have shown the potential for large decreases in runoff in the western US (Revelle and Waggoner, 1983; Gleick, 1988a) and summer drying in midcontinental regions of both the US and USSR (Manabe and Wetherald, 1987). Because precipitation changes are not consistent in GCM doubled-$CO_2$ experiments, the use of soil moisture as a drought predictor has produced mixed indications of future climatic conditions (see, for example, Kellogg and Zhao, 1988).

Using a new drought index based on the difference between potential evaporation (the atmospheric demand for water) and precipitation (the atmospheric supply for water), drawn from results of the Goddard Institute for Space Studies (GISS) GCM, predictions for transient climate change show increased drought frequency for current emissions growth rates (Rind et al., 1990). Using these GCM scenarios, droughts, which are induced by increasing temperature rather than reduced precipitation, begin to increase in frequency in the 1990s. Severe droughts over all land areas, which occur only 5% of the time in the current climate, are projected to occur about 50% of the time by the 2050s (Fig. 7.15).

Droughts of this frequency would have severe consequences on agriculture. These results are in direct contrast to the increases in precipitation (and reduced drought frequency) for some regions estimated by the Soviet paleoclimatic studies described above. The GCM-based study does show lower drought severity and precipitation increases at higher latitudes, where temperature is lower. Determining the likelihood of future drought is a critical question for regional, national, and international agricultural studies.

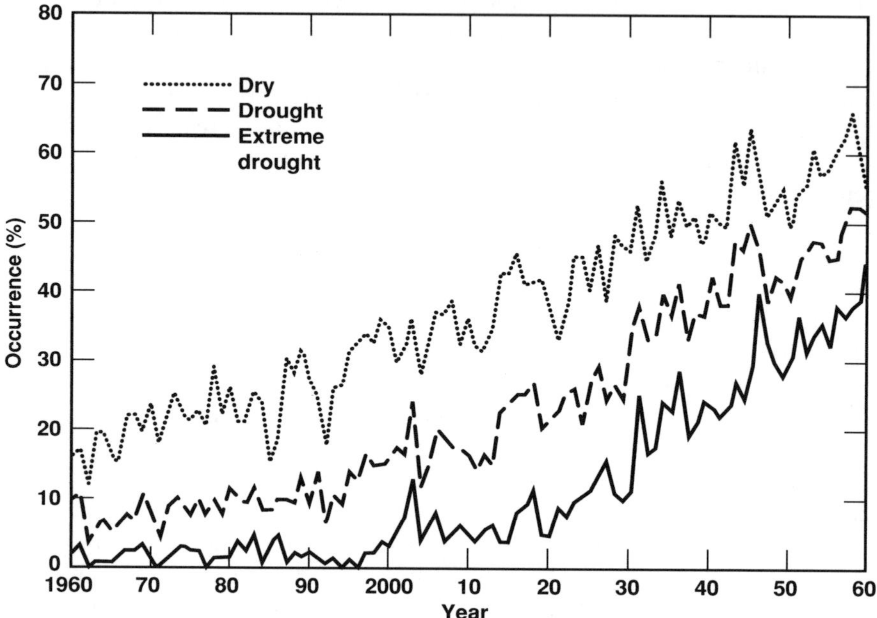


**Figure 7.15.** Drought occurrence as a function of time during the period June to August as generated from the GISS GCM Supply Demand Drought Index (SDDI). Results are from the transient run of the model and are averaged over all grid boxes that are at least 90% land (except Antarctica) (from Rind et al., 1990). In the control run, dry conditions have a frequency of occurrence of 16%, drought conditions 5%, and extreme drought 1%.

### 7.5.3. Technology

Next to climate, technology is the most critical factor affecting crop yields. Yield increases in the past 50 years have been driven by such technical advances as mechanization, chemical fertilizer application, development of hybrid varieties, and chemical herbicides and pesticides. Menzhulin et al. (1983) and Menzhulin and Nickolaev (1987) have derived technology trends for US corn and wheat yields and for USSR wheat yields for the period 1920 to 1985 and have compared these optimal yields to the actual yields.

A major issue in projecting the impacts of future climate change is that of estimating future technological improvements. The climate changes predicted for increasing trace gases may not be realized until the latter part of the next century. An agricultural impact analysis of this sort ideally should include a projection of prevailing technological conditions. However, this requires numerous assumptions about population growth, rates of technical progress, and resource availability. One approach is to conduct sensitivity tests with alternative technology scenarios to explore a range of projections with differing dynamic adjustments, as was done by Adams et al. (1990) (Table 7.6). Another approach is to construct a moving baseline that accounts for technological improvements.

**Table 7.6.** Economic effects of GISS climate change with increased technology, in billions of (1982) dollars.[a]

| Analysis | Total economic surplus[b] | Change in surplus from 1981 to 1983 base[c] | Change in surplus due to climate change[c] |
|---|---|---|---|
| Base[d] | 94.577 | — | — |
| GISS climate change[e] | 88.724 | –5.853 | –5.853 |
| Technological change to 2060 | 126.987 | +32.410 | — |
| Technological change to 2060 with climate change | 124.854 | +30.277 | –2.133 |

[a] Source: Adams et al. (1990), modified.
[b] Income equivalent associated with production and consumption of agricultural products at base period prices.
[c] Changes reflect income lost or gained by producers and consumers due to technological improvements and/or climate change.
[d] The base case reflects economic, agronomic, and environmental conditions for 1981 to 1983. Yields and water availability are specified at actual 1981 to 1983 levels.
[e] Analysis is based on estimates of the change in crop yields from the CERES and SOYGRO models for corn, wheat, and soybeans plus a yield adjustment for all other crops in the model (cotton, barley, rise, sorghum, oats, and hay) equal to the average change of corn, wheat, and soybeans for each region. It also includes changes in irrigation water demand (by crop and region) as well as changes in regional ground- and surface-water supplies for irrigation. See Adams et al. (1990) for details of models and methods.

### 7.5.4. Global Food Trade and Vulnerable Regions

Previous studies of the impacts on agriculture of climate change have, for the most part, dealt with isolated regions or nations and have ignored the interconnected nature of the world food system. Understanding the potential impacts in other parts of the world is necessary to characterize regional or national outcomes. Both long- and short-term changes in climatic conditions in different countries and regions interactively affect global food production, demand, prices, and trade. These factors, in turn, depend on changes in agricultural production potential, crop and livestock systems, and the actual adjustment mechanisms adopted by different countries. Studies that include all these interactive biophysical and economic mechanisms are needed to give a full picture of the potential impacts of climate change on global agriculture.

Furthermore, changes in global agriculture cannot be accurately characterized without analyzing regions that are vulnerable to food deficits. Food production in many countries and regions is currently subject to climatic stresses. Will these countries continue to be vulnerable in the future? Will other regions become vulnerable to increased hunger? What may happen to world food stocks? Will there be net annual surpluses or deficits? These are important questions to answer with future impact studies.

### 7.5.5. Environmental Quality

Natural environments may also suffer because of agricultural responses to climate change. For example, increased demand for water for irrigation, greater need for pesticides, the potential for increased soil erosion, and loss of wildlife habitats may all adversely affect natural habitats. Shifts in regional production patterns imply changes in ground- and surface-water pollution, and chemical pesticide usage may change to control both crop and livestock pests. Loss of biological diversity may diminish the germ-plasm base on which the adaptive capacity of agricultural crop breeding programs depend. Soils resources may be exposed to increased wind and water erosion, if agricultural regions expand into areas previously protected by natural land cover.

Agriculture-related environmental crises are occurring in the present, and may be exacerbated by changes in climate in the future. In the US, selenium contamination from irrigated regions in California has caused the death of migratory birds in the Kesterson Reservoir. In the USSR, removal of irrigation water from the major source rivers of the Aral Sea has caused its surface area and volume to shrink dramatically. Salts from the emptied basin are picked up by wind and deposited on surrounding cropland. Both of these environmental disasters are related to agriculture; with climate change, the potential for increased demand for irrigation due to higher temperatures may increase the occurrences of such situations.

The current predictions of global warming caused by increasing trace gases, potential increases in ultraviolet radiation from stratospheric ozone depletion, acid precipitation from industrial effluents, and increas-ing air pollution all raise the probability that environmental quality will continue to be threatened. These stresses are likely to occur simultaneously, and studies are needed to address their interactive effects on agriculture.

It is also important to remember that agriculture is not passive in its relation to the greenhouse effect, but is itself an active contributor. Clearing of forested land for agriculture often involves burning of vegetation, converting fixed carbohydrates to atmospheric $CO_2$ and other gases (e.g., CO, $NO_x$, etc.). Agricultural activities release other radiatively and chemically active trace gases as well. Flooded rice paddies emit methane as a product of anaerobic decomposition of organic matter, and ruminants (cattle, sheep, goats, etc.) release methane as a product of their digestive processes. Soils change a portion of the nitrogenous fertilizer applied to them to nitrous oxide, another greenhouse gas.

### 7.5.6. Adaptation

Examples of farm-level adaptations to climate variability are changes in planting and harvest dates, tillage and rotation practices, substitution of crop varieties or species more appropriate to the changing climate regime, increased fertilizer and pesticide applications, and improved irrigation and drainage systems. Improvements in agricultural technology will ease such adjustments.

Agricultural research centers should encourage the development of effective approaches for preparing for climate change, such as the continued breeding of heat- and drought-resistant crops. Research should be directed to determining what are the heat- and drought-tolerance limits of both presently grown and alternative crops and varieties. The adaptation of crop architecture and physiology to extreme environmental conditions may be possible with modern genetic techniques.

At the regional level, agricultural development and water- and land-planning institutions should take future climate change into account, with attention to maintaining flexibility to the increased hazard of flooding from sea-level rise in low-lying coastal agricultural areas, and to drought planning for areas likely to suffer increasing aridity. Nationally, food-security issues and food-support needs of vulnerable regions should be addressed in light of potential climate change. Efforts such as these will ease adjustment to the considerable warming that is predicted to occur, even if increases in trace-gas emissions were to stop immediately.

## 7.6. LESSONS AND OPPORTUNITIES FOR THE FUTURE

The increasing availability of possible scenarios of future climate has permitted increased investigation of the sensitivity of hydrological and agricultural systems to climate change, thereby providing important insights into the future.

(1) *What have we learned about how future changes and impacts will be different than past responses to short-term climate anomalies?*

One of the greatest difficulties facing society will be identifying a long-term climate change, as opposed to short-term climatic variability. The appropriate responses to these two conditions are often different.

Unless these two effects can be separated, inappropriate management decisions could be made. Thus, a short-term water or food shortage might be mitigated by looking for surplus supplies from a neighboring region or from the world market. A long-term shift or shortage, however, would require changes in demand, pricing, management options, and perhaps in the international infrastructure itself.

(2) *What have we learned about what options may be best suited to responding to climate change?*

Resource managers have learned a great deal from experiences with past climatic extremes and variability. Indeed, these kinds of events have already led to the development of a wide range of options for responding and adapting to climatic variability and climate change. For short-term variability, such as extreme drought, options might include changing existing water-allocation agreements, using crop storage, rationing, altering cropping types and patterns, temporary protection of coastal developments, and so on. For long-term changes in climate, however, the options might be very different. The design and construction of new physical structures such as dams, aqueducts, and levees may be needed. New crop varieties may need to be developed. Developments may have to be excluded from the margins of the ocean, or flood-insurance policies may have to change.

(3) *What have we learned about the difficulties of responding to climate change?*

Many of the problems that arise from climatic variability have their roots both in structural limitations, such as dams or aqueducts being improperly sized or located or developments placed too close to vulnerable coastlines, and in institutional constraints, such as inflexible operating rules and markets. For example, it is now understood that to try to completely eliminate damages from floods and droughts through physical means is economically and environmentally unrealistic. As a result, management measures such as zoning are receiving increasing attention. Addressing the problem of future climate change will require innovative use of these measures to enhance the overall reliability and flexibility of our systems.

(4) *What have we learned about whether there will be winners and losers from future climate change?*

Considering the potential for different societal impacts of climate change in different locations, the question necessarily arises of "winners and losers," as defined in conventional economic or political terms. New paradigms of social and environmental thinking may be required that eliminate the need for analyzing the effects of global climate change in terms of a narrow competition focusing on how some regions or nations may win and lose. The complex international connections that exist among the nations of the world make it extremely likely that adverse impacts will be felt widely: flooding in Bangladesh will have consequences that reverberate throughout southern Asia; the potential for yield increases in one region of the world may be offset by decreases in other regions and by changes in the global food markets; changes in water availability in northern Africa will affect dozens of countries. These interconnections and consideration of intergenerational equity emphasize the growing need for global environmental cooperation rather than global economic competition.

Many critical questions about climate change and societal impacts remain unanswered, as does the question of appropriate action on the larger global warming issue. The challenge is to devise appropriate responses to a problem that is uncertain in many of its dimensions, but that may have enormous consequences.

A sustained effort of cooperative research on the impacts of climate change on society and on adaptive mechanisms appropriate to the needs and vulnerabilities of each region is crucially important.

(1) The detail of scenarios must be improved, both in terms of spatial and temporal resolution and in terms of the set of variables made available.

(2) Consideration of potential impacts on hydrologic resources must be extended and coupled to consideration of potential response measures.

(3) The potential for adaptation of agricultural crops must be further investigated to determine the full range of soil, climate, and other controlling factors.

# CHAPTER 8. PROSPECTS FOR FUTURE CLIMATE

The different climates of the past and model simulations of future climates convincingly indicate that the continuing emissions of greenhouse gases will lead to significant global warming and to changes in precipitation and other climatic variables. The projected changes in atmospheric composition and, consequently, in climatic conditions will be unique and more rapid than at any time in the past. The developing understanding of the chemical cycles controlling atmospheric composition and of the processes and behavior controlling the climate system can provide significant guidance about how the future climate will change.

This chapter first summarizes the many scientific advances described in the preceding chapters that can help us better understand and describe the climate system and the resulting agricultural and hydrological impacts of these changes in climate. The chapter then draws from this understanding to outline the prospects for future climate.

## 8.1. FINDINGS FROM STUDIES OF ANCIENT CLIMATES

Climate has varied significantly over the Earth's history. There have been glacial periods when the global average temperature may have been about 5°C lower than at present, and there have been substantially warmer periods when the global average temperature may have been as much as 5°C to even 10°C higher than at present. Although individual locations may experience such departures from average conditions on a daily, monthly, and sometimes seasonal basis, these past changes in the global average climate have caused dramatic changes in the distribution and character of plant and animal life.

Many factors caused past climates to differ from the present climate. Geological evidence from the past 10 million years suggests that the most important of these factors included changes in carbon dioxide ($CO_2$) concentration, in continental extent and location, and in the heights of mountain ranges. During the colder periods of the last million years, orbitally induced changes in the distribution of solar radiation over the year and over latitude have paced changes in the extent of continental ice sheets such that cycling is evident at frequencies of about 20,000, 40,000, and 100,000 years. Geological evidence and correlations between the $CO_2$ concentrations and temperature changes as seen in polar ice cores suggest that large changes in atmospheric composition have been associated with both the glacial/interglacial climate change and with the much warmer climates of earlier epochs. Higher concentrations of greenhouse gases were generally associated with warmer epochs and lower concentrations with colder epochs, but there are periods when changes in ice volume do not closely coincide with concentration changes even though the correlation is very high.

Climates in polar regions have been much more variable than climates in lower latitudes, probably primarily because of the albedo and thermal changes associated with changes in sea ice cover. Substantial geological evidence indicates that when the temperature was higher, there was much more warming in polar latitudes than in low latitudes; when the temperature was lower, there was much greater cooling in middle and high latitudes than in low latitudes. These changes are also found in model simulations of past climates.

Changes in soil moisture and precipitation have been quite complex. Evidence does not indicate that there have been simple zonal shifts in precipitation zones. Rather, there are indications that changes have varied regionally, being dependent on general circulation patterns and on the orographic and geographic features of each particular region.

## 8.2. FINDINGS FROM STUDIES OF CLIMATES OF THE RECENT PAST

The decadal-average global temperature over the past several thousand years has been quite stable, varying from the long-term mean by only about 1°C. Thus, civilization has developed its important and complex agricultural and hydrologic systems under conditions that are generally dependable and repetitive. The past level of natural variability—resulting from both variations in various forcing factors (e.g., volcanic eruptions, solar variability) and from random fluctuations inherent in the climate system—is relatively small compared with the projections of possible future climatic warming through the twenty-first century. The coming climatic conditions are thus significantly different from those to which societies have become acclimated.

The climate of particular regions can vary significantly from season to season and year to year, even seemingly persisting in anomalously warm or cool conditions for up to a few decades. This seemingly chaotic behavior of climate, however, has not led to excursions in annual-average global temperature of more than several tenths of a degree. For temperature variations that are prolonged or are greater than about 1°C, the global climate appears to be much more determined by definable and measurable factors than by random or chaotic behavior.

Factors other than changes in atmospheric composition have not caused large or prolonged changes in global average temperature over historical times. Neither major volcanic eruptions nor multi-decade periods with higher or lower volcanic activity have been associated with increases or decreases in global average temperature of more than a few tenths of a degree. Changes in solar output apparently have been small compared with the projected radiative perturbations from changing atmospheric composition, and the associated global average temperature changes have generally been less than about 0.1°C.

Analyses of the instrumental temperature record extending back about 100 years indicate an average global warming of about 0.4 to 0.5°C, with an uncertainty of about 0.1°C. The warming is evident even after various efforts to account for biases in the measurement processes. The warming has been somewhat more regular in the Southern Hemisphere than in the Northern Hemisphere. The warming may be less obvious in the Northern Hemisphere because (1) there are greater possibilities for factors such as volcanic eruptions, ocean circulation changes, and continentality to induce enhanced variability, and (2) the thermal inertia is reduced in the Northern Hemisphere due to the lower fraction of ocean area compared to the Southern Hemisphere, allowing natural variations to be larger than in the Southern Hemisphere. This natural variability could either have added to or diminished the apparent greenhouse warming over the past century by a few tenths of a degree. It is unlikely, however, that natural variations have been more important than the temperature increase caused by the greenhouse effect.

## 8.3. LESSONS FROM ANALYSES OF ATMOSPHERIC COMPOSITION

Pioneering ice core measurements have shown that the present atmospheric concentration of $CO_2$ has risen to the highest level in 160,000 years. Measurements at an extensive network of stations show an 11% increase in $CO_2$ concentration from about 315 ppmv to about 351 ppmv over the last 30 years. Carbon dioxide concentrations further back in time varied from less than about 200 ppmv during the maximum cooling of the last two glacial cycles up to about 280 ppmv during the warm interglacial periods. The latter include the several-thousand-year period prior to the nineteenth century expansion of industry and mechanized agriculture.

Emissions of fossil-fuel $CO_2$ and reductions in the amount of soil carbon and vegetation have altered the distribution of carbon in the atmosphere-ocean-biosphere system. Calculations based on compilations of fossil-fuel use and depletion of $^{14}C$ indicate that about 200 PgC has been released over the past 150 years; biospheric releases are estimated to have totaled about 60 to 90 PgC over a similar period, although this is quite uncertain. Over the 150-year period, these releases have been about equally divided between the atmosphere and the land and ocean reservoirs. The mechanisms governing uptake by the land (probably in the soils) are poorly understood; the relatively slow uptake by the oceans results from the slow mixing of carbon downward into the cold, deep ocean from the relatively stable, upper mixed layer of the ocean.

The concentrations of trace gases having radiative properties similar to $CO_2$ have also increased significantly. The concentration of methane ($CH_4$) has doubled since 1800, and the concentrations of man-made chlorofluorocarbons have reached levels that significantly affect infrared radiation fluxes. The concentrations of these other trace gases have started to influence the atmospheric concentrations of other atmospheric constituents. Ozone, in particular, absorbs solar ultraviolet radiation, thereby warming the stratosphere, and acts like $CO_2$ to trap infrared radiation, thereby warming the troposphere.

Emissions of the various man-made gases have been increasing and are continuing to increase, and they will accelerate the changes in atmospheric composition. The rate of emission of $CO_2$ is increasing at 1 to 2% per year. If coal becomes the major fuel for raising the global standard of living and expanding industrial development, the rate of increase in $CO_2$ emissions seems likely to increase still more. The rate of emission of $CH_4$ appears tied primarily to the rate of increase of agricultural production (which is increasing to meet the needs of the increasing world population) and to the increasing use of fossil fuels. The rate of chlorofluorocarbon production has been increasing rapidly, but is to be slowed and stopped in order to limit further damage to the stratospheric ozone layer.

## 8.4. SIMULATION OF THE GLOBAL CLIMATE

Comprehensive numerical models based on the fundamental laws of physics have been developed to simulate the global climate. These models are capable of simulating with reasonable fidelity many features of the global climate, including latitudinal belts of temperature and precipitation from tropical to subtropical to temperate to polar regions, the general midlatitude locations of the major storm-generating pressure centers, and the seasonal evolution through winter and summer. Because even the fastest supercomputers are not fast enough, the horizontal resolution of the models available for climate studies is such that areas of up to about 250,000 $km^2$ (roughly the size of Italy or New Zealand) must be treated as having uniform climatic conditions. This necessary limitation, which also smooths mountain ridges, coupled with as yet incomplete treatments of clouds, hydrology, and oceans, limits the ability of the models to reproduce the many important details of the local and regional climate that can often control agriculture, water resources, and other societal activities.

Nevertheless, climate models can help explain many aspects of past climates. When forced by specified changes in $CO_2$ concentration, orbital parameters, and glacial conditions, simulations of the climate changes at 3000-year intervals since the last glacial maximum about 18,000 years ago reproduce many of the climatic features deduced from geological evidence, indicating both that large-scale features in the models are reproduced reasonably accurately and that there is increasing understanding of the causes of past changes in climate. Model results suggest that substantial enhancement of the concentrations of greenhouse gases may have had to be a critical contributing factor to the very warm climates of the past. Changes in oceanic circulation, shifting of the continental plates, and orographic uplift and descent may also have been important factors.

The sensitivity of the equilibrium climate to changes in atmospheric composition and to other related forcing factors has been explored in a range of model studies. Doubling of the atmospheric concentration of $CO_2$ is projected to lead to an increase in global average atmospheric temperature of about 1.5 to 4.5°C, once a new equilibrium climatic state is reached. Studies using more than a dozen climate models from many countries indicate that subtle, but differing, treatments of cloud cover and optical characteristics are the primary cause of the disagreements among model-based estimates of climatic sensitivity. More detailed comparisons of the responses of climate models indicate general agreement that the warming will be greater than the global average in high latitudes during the winter as the cold season contracts, and that the warming will be moderated somewhat in low latitudes. There is little certainty, however, concerning the regional patterns of the responses of temperature, precipitation, and soil moisture to the increasing $CO_2$ and greenhouse gas concentrations, with some models suggesting midcontinental summer drying and others suggesting no change or even moistening. Assessments of model simulations of potential changes in the frequencies and intensities of inter-annual variations and extreme events are particularly limited, despite the potential importance of determining such changes. These limitations also apply to calculations estimating sensitivity to the non-$CO_2$ trace gases, with the additional difficulty that atmospheric chemistry would need to be treated interactively.

Because of dissatisfaction with climatic sensitivity simulations for a doubled $CO_2$ concentration, new climate models are being used to study the more realistic situation of slowly increasing concentrations of $CO_2$ and, in some cases, other radiatively active trace gases. Such calculations require proper treatment of the coupling of the upper ocean to the large heat capacity and cold waters of the deep ocean in order to account for the delay (or lag) in the warming that will be induced as the system waits for the deep ocean to warm. Very large time constants, and hence substantial computer resources, are needed to treat the global ocean interactively. Models including allowance for changes in ocean currents have so far made only limited progress, but significant advances are starting to be made. If substantial emissions of greenhouse gases continue, the rate of increase in the global average temperature should be considerably more in the next century than the rate of about 0.05°C/decade experienced over the past century.

## 8.5. EMPIRICAL ANALYSES OF THE CLIMATE RECORD

Reconstructions of past climatic conditions suggest an apparent coupling to past changes in the $CO_2$ concentration. The climatic sensitivity to doubling of the $CO_2$ concentration suggested by empirical analyses of these paleoclimates is about 3°C, which is near the center of the range of model estimates. Although the relationships between changes in past atmospheric composition and distributions of climate must be further refined, the apparent logarithmic relationship is in general agreement with theoretical estimates.

Estimates of future climatic patterns deduced from relationships apparent during past warm climates suggest a global warming over the preindustrial baseline of about 0.5°C in the 1980s, increasing thereafter at a rate of as much as about 0.75°C every 25 to 50 years, depending on the rate of increase in the concentrations of greenhouse gases and the lag effect of the oceans.

Warmer climates have often been accompanied by wetter conditions. Changes in precipitation and soil moisture can be exceedingly complex, indicating that changes in regional meteorological conditions must be carefully interpreted. However, it is very interesting that the midcontinental desiccation found in some model results for a $CO_2$ doubling is found in some, but not all, reconstructions of past warm periods. Better understanding of the relative role of past changes in $CO_2$ concentration and of changes in orbital forcing could allow more effective use of the reconstructed warm climates as indicators of future conditions related to continuing increases in the $CO_2$ and trace gas concentrations.

## 8.6. PROJECTIONS OF FUTURE CLIMATE

The results of these various approaches—analysis of paleoclimates, theoretical simulation of future climate, and development of empirical relationships between past and future climates—all suggest that the increasing emissions of greenhouse gases will accelerate the ongoing changes in atmospheric composition and the climatic warming that appears to be starting to exceed the bounds of past natural fluctuations. Over the next century, the projected growth in greenhouse gas emissions will, based on current numerical models and empirical analyses, raise the global average temperature by a few degrees Celsius above pre-industrial conditions. This warming might be intermittent, the hesitations being caused by various natural and random components during the next century; overall, these components may have a rather modest effect. There will likely be amplification of the warming in high latitudes and over midlatitude continental regions, especially in winter.

Empirical estimates of the influence on warming of increased concentrations of trace gases in the atmosphere indicate that the mean air temperature early in the twenty-first century will have increased to about 1°C above the pre-industrial level. Temperature increases would then continue to rise to about twice that amount by the middle of the century compared with the pre-industrial epoch, according to this method. For later in the century, the probable error in estimating global temperature change will undoubtedly grow. Nevertheless, for the period up to 2025 to 2050, the various estimates of increasing mean temperature generally agree in sign and magnitude, and the errors of these estimates are likely to be less than 50% of the value of the expected temperature change. Estimates of future changes in precipitation, however, are much less certain.

## 8.7. PROSPECTS FOR IMPACTS ON WATER RESOURCES AND AGRICULTURE

Changes in precipitation and evaporation will have important effects on the quantity of water available, on water quality, on soil moisture, and on hydroelectric generation. They may also affect the frequency of extreme events. Spatial redistribution of water resources as the climate changes could exacerbate existing water-resource limitations, especially in important midcontinental agricultural regions, depending on the actual changes in climate. Changes in seasonal water availability may complicate water quality problems, especially where salinity is now a problem. A rising snowline may lead to increased potential for winter and spring floods in mountainous regions.

Agriculture is critically dependent on warmth and soil moisture occurring in the regions where soils are rich and suitable. There remains considerable uncertainty about how the suggested increases in precipitation suggested for different regions and some seasons and the increasing $CO_2$ concentration will combine with the rising temperatures to affect soil moisture during the growing season and plant productivity. The potential for increased variability, which may bring increased drought frequency and/or intensity, is suggested for some scenarios of changes in climatic conditions.

## 8.8. RESEARCH DIRECTIONS FOR THE FUTURE

The US-USSR collaboration has broken new ground in understanding past and current trends in climate. Comparison of the empirical and modeling methods for estimating future climatic conditions has identified major new research questions that must be addressed. Ongoing US-USSR collaborative efforts are needed to address the limitations and weaknesses in approaches.

Major scientific issues that have been identified in this report include:

(1) *Extension of the areal coverage and temporal span of accurate paleoclimatic reconstructions.*

This will require coordinated sampling programs, intercomparison of methods, exchanges of data, and collaborative development of anomaly maps.

(2) *Assembly and analysis of comprehensive data sets for temperature, precipitation, and other climatic variables.*

These efforts must be intensified to allow clearer identification of greenhouse-gas-induced trends and removal of bias and contributions by non-greenhouse gases.

(3) *Gathering and exchange of data sets derived from surface, oceanic, atmospheric, cryospheric, satellite, chemical, and paleoclimatic measurements.*

Accelerated efforts will allow more rapid investigation of the processes controlling global change and the rates and magnitude of its occurrence.

(4) *Determination of the sources, sinks, and interactions of $CO_2$ and other greenhouse gases.*

The coupling of atmospheric chemistry and the climate system must be more carefully investigated, documented, and analyzed so that the rates of future change in atmospheric composition and the roles of climate system feedbacks can be better estimated, whether or not actions are taken to limit the release of the greenhouse gases.

(5) *Development, testing, and verification of coupled atmosphere, ocean, ice, land, and biosphere models.*

This effort must be intensified, with initial focus on improving the representations of individual processes and components, including especially clouds, hydrology and soil moisture, chemistry, and oceanic circulation and heat uptake.

(6) *Relationships among climatic patterns occurring during past warm periods.*

More thorough analysis can ensure proper attribution of the changes among $CO_2$, $CH_4$, orography, orbital elements, and other factors affecting the global and regional climate.

(7) *Comparison of the results of modeling and empirical approaches.*

Special attention should be given to cases for which the magnitude of global change in the models matches that which is determined from reconstructions of past global change.

(8) *Development of improved climate scenarios.*

Scenarios with more detailed spatial resolution and describing a wider set of climatic variables must be developed for use in impact studies.

(9) *Detailed analyses of hydrological, agricultural, and other environmental impacts.*

These analyses are needed to better understand societal sensitivity to future climate change.

## 8.9. WORKING GROUP VIII AND FUTURE CLIMATE

For more than 15 years, Working Group VIII under the US-USSR bilateral agreement on protection of the environment has fostered collaborative research on climate. Significant improvements in scientific understanding have been made, helping to transform the issue of anthropogenic climate change from a little-known concern to one now receiving wide international recognition. This strong record of environmental cooperation is one that can serve as a model for the ongoing global efforts, providing the scientific groundwork for soundly based development of appropriate policies. By this focus on important scientific issues, a bridge has been built between our nations. This bridge can and must be broadened in the future.

# REFERENCES [1]

## CHAPTER 1

Budyko, M. I., and M. C. MacCracken, 1987, "US/USSR meeting of experts on causes of recent climate change," *Bulletin of the American Meteorological Society* **68**, 237–243.

Porter, S., and H. Wright (Eds.), 1984, *Late Quaternary Environments of the United States*, University of Minnesota Press, Minneapolis, Minnesota, *Vol. 1*, 408 pp., *Vol. 2*, 277 *Vol. 2*, 277 pp.

US-USSR Meeting Report, 1982, "Climatic effects of increased atmospheric carbon dioxide," *Proceedings of the Soviet-American Meeting on Studying Climatic Effects of Increased Atmospheric Carbon Dioxide*, 15–20 June, 1981, Gidrometeoizdat, Leningrad, 56 pp. (R).

Velichko, A. A., H. E. Wright, and C. W. Barnosky (Eds.), 1984, *Late Quaternary Environments of the Soviet Union*, University of Minnesota Press, Minneapolis, Minnesota (H. E. Wright and C. W. Barnosky, Eds. of English translation), 327 pp.

## CHAPTER 2

Ager, T., 1983, "Holocene vegetational history of Alaska," in *Late-Quaternary Environments of the United States* (H. E. Wright, Jr., Ed.), *Vol. 2, The Holocene*, University of Minnesota Press, Minneapolis, Minnesota, pp. 128–141.

Andrews, J. T., P. T. Davis, W. N. Mode, H. Nicols, and S. K. Short, 1981, "Relative departures in July temperatures in northern Canada for the past 6000 years," *Nature* **289**, 164–167.

Arrhenius, S., 1896, "On the influence of the carbonic acid in the air upon the temperature of ground," *Philosophical Magazine* **41**, 237–275.

Arrhenius, S., 1908, *Worlds in the Making*, Harper, New York.

Axelrod, D. I., 1984, "An interpretation of Cretaceous and Tertiary biota in polar regions," *Palaeogeography, Palaeoclimatology, Palaeoecology* **45**, 106–147.

Baker, R. G., 1983, "Holocene vegetational history of the western United States," in *Late-Quaternary Environments of the United States* (H. E. Wright, Jr., Ed.), *Vol. 2, The Holocene*, University of Minnesota Press, Minneapolis, Minnesota, pp. 109–127.

Bandy, O. C., R. E. Casey, and R. C. Wright, 1976, "Late Neogene planktonic zonation, magnetic reversals, and radiometric dates: Antarctic to tropics," *Antarctic Oceanology* **1**, 1–26.

Barash, M. S., 1985, "Reconstruction of the Quaternary ocean paleotemperatures by using planktonic foraminifera," in *Methods of Paleoclimatic Reconstructions*, Nauka, Moscow, pp. 134–142(R).

[1] Throughout this list of references, (R) indicates that the text of the referenced document is in Russian.

Barnola, J. M., D. Raynaud, Y. S. Korotkevich, and C. Lorius, 1987, "Vostok ice core provides 160,000-year record of atmospheric $CO_2$" *Nature* **329**, 408–414.

Barron, E. J., 1987, "Cretaceous paleogeography," *Palaeogeography, Palaeoclimatology, Plaeoecology, Special Issue* **59**, 1–3.

Bartlein, P. J., T. E. Webb III, and E. Fleri, 1984, "Holocene climatic changes in the northern Midwest: Pollen derived estimates," *Quaternary Research* **22**, 361–374.

Berner, R. A., A. C. Lasaga, and R. M. Garrels, 1983, "The carbonate-silicate geochemical cycle and its effect on atmospheric carbon dioxide over the past 100 million years," *American Journal of Science* **283**, 641–683.

Bezusko, L. G., and Klimanov, V. A., 1988, "The climatic conditions of the Ukraine during the Holocene and Late Glacial," in *Paleoclimate of the European part of the USSR during the Holocene*, Nauka, Moscow, pp. 125–135 (R).

Borzenkova, I. I., 1981, "On the global temperature trend in the Cenozoic," *Meteorologiya i Gidrologiya* **12**, 25–35 (R).

Borzenkova, I. I., 1987, "Features of land moistening in the Northern Hemisphere during the geological past," *Meteorologiya i Gidrologiya* **10**, 53–61 (R).

Borzenkova, I. I., and V. A. Zubakov, 1984, "Climatic optimum of the Holocene as a model of the global climate at the beginning of the 21st Century," *Meteorologiya i Gidrologiya* **8**, 69–77 (R).

Borzenkova, I. I., and V. A. Zubakov, 1985, "The changing climate during the Late Miocene and Pliocene," in *Proceedings of the State Hydrological Institute* **339**, pp. 93–118 (R).

Bowen, R., 1966, *Paleotemperature Analysis*, Elsevier, Amsterdam-London-New York, 207 pp.

Bradley, R. S., 1985, *Quaternary Paleoclimatology, Methods of Paleoclimatic Reconstruction*, Allen and Unwin, Boston–London, 472 pp.

Budyko, M. I., 1974, *Climatic Changes*, Gidrometeoizdat, Leningrad, 280 pp. (R) (English translation: American Geophysical Union, 1977, 261 pp.).

Budyko, M. I., 1984, *Evolution of the Biosphere*, Gidrometeoizdat, Leningrad, 487 pp. (R) (English translation: D. Reidel, Dordrecht, The Netherlands, 1986).

Budyko, M. I., and A. B. Ronov, 1979, "The evolution of the atmosphere in the Phanerozoic," *Geochemistry* **5**, 654–653 (R).

Budyko, M. I., A. B. Ronov, and A. L. Yanshin, 1985, *The History of the Earth's Atmosphere*, Gidrometeoizdat, Leningrad, 209 pp. (R) (English translation: Springer-Verlag, 1987, 139 pp.).

Clark, D. L., 1979, "Arctic ocean ice cover and its Late Cenozoic history," *Bulletin of the Geological Society of America* **82**, 3313–3324.

CLIMAP Project Members, 1984, "The last interglacial ocean," *Quaternary Research* **21**, 123–224.

COHMAP Members, 1988, "Climatic changes of the last 18,000 years: Observations and model simulations," *Science* **241**, 1043–1052.

Crowley, T. J., 1983, "The geologic record of climatic change," *Reviews of Geophysics and Space Physics* **21**, 828–877.

Crowley, T. J., 1989, "Paleoclimate perspectives on a greenhouse warming," in *Climate and Geosciences: A Challenge for Science and Society in the 21st Century* (A. L. Berger, J. -C. Duplessy, and S. H. Schneider, Eds.) D. Reidel, Dordrecht, The Netherlands (in press).

Davis, M., 1983, "Holocene vegetational history of the eastern United States," in *Late Quaternary Environments of the United States, Vol. 2.*, University of Minnesota Press, Minneapolis, pp. 166–181.

Delcourt, P. A., and H. R. Delcourt, 1983, "Late Quaternary vegetational dynamics and community stability reconsidered," *Quaternary Research* **17**, 265–271.

Duan Wanti, Pu Qiuyyi, and Xihao, 1980, "Climatic variations in China during the Quaternary," *GeoJournal* **4**, 515–524.

Emiliani, C., 1966, "Isotopic paleotemperature," *Science* **154**, 851–857.

Frakes, L. A., 1979, *Climates through Geologic Time*, Elsevier, Amsterdam, 310 pp.

Gallimore, R. G., and J. E. Kutzbach, 1989, "Effects of soil moisture on the sensitivity of a climate model to Earth orbital forcing at 9000 yr BP," *Climatic Change* **14**, 175–205.

Gaven, C., C. Millaire-Marcell, and N. Petit-Maire, 1981, "Pleistocene Lacustrine episode in south-eastern Libya," *Nature* **290**, 131–133.

Genthon, C., J. M. Barnola, D. Raynaud, C. Lorius, J. Jouzel, N. I. Barkov, Y. S. Korotkevich, and V. M. Kotlyakov, 1987, "Vostok ice core: Climatic response to $CO_2$ and orbital forcing changes over the last climatic cycle," *Nature* **329**, 414–418.

Golbert, A. V., 1987, *The Fundamentals of Regional Paleoclimatology*, Nedra, Moscow, 221 pp. (R).

Grichuk, V. P., 1961, "The fossil floras as the paleontological basis of Quaternary stratigraphy," in *Relief and Stratigraphy of the Quaternary Deposits of Northwestern Russian Plain*, AN USSR, Moscow, pp. 25–71 (R).

Grichuk, V. P., 1969, "Reconstruction of some climatic elements in the Northern Hemisphere during the Atlantic period of the Holocene," in *Holocene*, Nauka, Moscow.

Grichuk, V. P., 1982, "Paleoecological reconstructions based on paleobotanic data," in *Paleogeography of Europe for the Last 100,000 Years*, (I. P. Guerassimov and A. A. Velichko, Eds.) Nauka, Moscow, pp. 120–121.

Grichuk, V. P., 1985, "The reconstruction of climatic indicators by floras and the estimate of its accuracy," in *Methods of Paleoclimatic Reconstructions*, Nauka, Moscow, pp. 20–23 (R).

Gurtovaya, Ye. Ye., 1987, "The climatic changes of the Russian plain and Siberia during the Mikulino (Kazantsevo) Interglacial," *Izvestiya AN SSR* **2**, 54–62.

Hays, J. D., J. Imbrie, and N. J. Shackleton, 1976, "Variations in the Earth's orbit: Pacemaker of the ice ages," *Science* **195**, 1121–1131.

Hecht, A. D. (Ed.), 1985, *Paleoclimate Analysis and Modelling*, John Wiley & Sons, New York, 445 pp.

Herman, Y., and D. M. Hopkins, 1980, "Arctic oceanic climate in Late Cenozoic time," *Science* **409**, 557–562.

Imbrie, J., and K. P. Imbrie, 1979, *Ice Ages: Solving the Mystery*, Enslow, Hillside, New Jersey, 224 pp.

Jessen, K., and V. Milthers, 1928, "Stratigraphical and paleontological studies of interglacial freshwater deposits in Jutland and northwest Germany," *Danm. Geol. Unders.* **48**, 379.

Jouzel, J., C. Lorius, J. R. Petit, C. Genthon, N. I. Barkov, V. M Motlyakov, and V. N. Petrov, 1987, "Vostok ice core: A continuous isotope temperature record over the last climatic cycle (160,000 years)," *Nature* **329**, 403–408.

Jouzel, J., G. Raisbeck, J. P. Benoist, F. Yiou, C. Lorius, D. Raynaud, J. R. Petit, N. I. Barkov, Y. S. Korotkevitch, and V. M. Kotlyakov, 1988, "The Antarctic climate over the late glacial period," *Quaternary Research* **31**, 135–150.

Khotinskiy, N. A., 1977, *The Holocene of the northern Eurasia*, Nauka, Moscow, 198 pp. (R).

Khotinskiy, N. A., and S. S. Savina, 1985, "Paleoclimatic patterns of the USSR in boreal, Atlantic and subboreal periods of the Holocene," *Izvestiya, AN USSR, Geography Series* **4**, 18–34 (R).

Klimanov, V. A., 1981, "The connection of subfossil pollen spectra with present-day climatic conditions," *Izvestiya, AN SSSR, Geography Series* **5**, 101–114.

Klimanov, V. A., 1982, "Climate of eastern Europe during the climatic optimum of the Holocene (from palynological data)," in *Development of the Environment of the USSR in the Late Pleistocene and Holocene*, Nauka, Moscow, pp. 251–258 (R).

Krasilov, V. A., 1987, *The Cretaceous*, Nauka, Moscow, 239 pp. (R).

Kulkova, I. A., 1987, "The climatic optimum of Eocene in the northern Asia," in *Earth's Climates in the Geological Past*, Nauka, Moscow, pp. 84–89 (R).

Kutzbach, J. E., 1983, "Modeling of Holocene climates," in *Late Quaternary Environments of the United States, Vol. 2* (S. Porter and H. Wright, Eds.), University of Minnesota Press, Minneapolis, Minnesota, pp. 271–277.

Kutzbach, J. E., and R. G. Gallimore, 1988, "Sensitivity of a coupled atmosphere/mixed-layer ocean model to changes in orbital forcing at 9000 yr BP," *Journal of Geophysical Research*, **93**, 803–821.

Kutzbach, J. E., and P. J. Guetter, 1986, "The influence of changing orbital parameters and surface boundary conditions on climate simulations for the past 18,000 years," *Journal of the Atmospheric Sciences* **43**, 1726–1759.

Kutzbach, J. E., P. J. Guetter, W. F. Ruddiman, and W. L. Prell, 1989, "The sensitivity of climate to Late Cenozoic uplift in southern Asia and the American West: Numerical experiments," *Journal of Geophysical Research* **94**, 18393–18407.

Kutzbach, J. E., and F. A. Street-Perrott, 1985, "Milankovitch forcing of fluctuations in the level of tropical lakes from 18 to 0 kyr BP," *Nature* **317**, 130–134.

Leroi-Gourhan, A., 1968, "Denominations des oscillations wurmiennes," *Bulletin de l'Association Francaise pour l'etude du Quaternaire* **4**.

Liberman, A. A., M. V. Muratova, and I. A. Suetova, 1985, "Using linear interpolation for the paleoclimatic reconstruction," in *The Methods of the Paleoclimatic Reconstruction*, Nauka, Moscow, pp. 48–52.

Lorius, C., J. Jouzel, C. Ritz, L. Merlivat, N. I. Barkov, Y. Korotkevich, and M. V. Kotlyakov, 1985, "A 150,000 year climatic record from Antarctic ice," *Nature* **316**, 591–596.

Lowenstam, G. A., 1968, "Paleotemperatures of the Permian and Cretaceous," in *Problems of Paleoclimatology* (A. Nairn, Ed.), Interscience, John Wiley & Sons, London.

Maier-Reimer, E., T. J. Crowley, and U. Mikolajewicz, 1989, "Ocean GCM experiments with an open Central American isthmus," EOS (Transactions American Geophysical Union) **70**, 367–368.

Manabe, S., and A. J. Broccoli, 1985, "The influence of continental ice sheets on the climate of an ice age," *Journal of Geophysical Research* **90**, 2167–2190.

Manabe, S., and K. Bryan, 1985, "$CO_2$-induced change in a coupled ocean-atmosphere model and its paleoclimatic implications," *Journal of Geophysical Research* **90**, 11,689–11,708.

Markov, K. K., and A. A. Velichko, 1967, *Quaternary Period, Vol. 3*, Nedra, Moscow, 440 pp.

Mein, S. V., 1987, *Paleobotanics*, Nedra, Moscow, 401 pp.

Mitchell, J. F. B., N. S. Grahame, and K. H. Needham, 1988, "Climate simulations for 9000 years before present: Seasonal variations and the effect of the Laurentide Ice Sheet," *Journal of Geophysical Research* **93**, 8283–8303.

Molfino, B. E., N. G. Kipp, and J. J. Morley, 1982, "Comparison of foraminiferal, coccolithophorid, and radiolarian paleotemperature equations: Assemblage coherency and estimate concordancy," *Quaternary Research* **17**, 279–313.

Neftel, A., H. Oeschger, T. Staffelbach, and B. Stauffer, 1988, "$CO_2$ record in the Byrd ice core 50,000–5,000 years B. P.," *Nature* **331**, 609–611.

Nikolskaya, M. V., 1982, "Paleobotanic and paleoclimatic reconstructions of Taimyr in Holocene," in *Antropogene of Taimyr*, Nauka, Moscow, pp. 148–157.

Porter, S., and H. Wright (Eds.), 1984, *Late Quaternary Environments of the United States*, University of Minnesota Press, Minneapolis, Minnesota, *Vol. 1*, 408 pp., *Vol. 2*, 277 pp.

Prell, W. L., and J. E. Kutzbach, 1987, "Monsoon variability over the past 150,000 years," *Journal of Geophysical Research* **92**, 8411–8425.

Raynaud, D., J. Chappellaz, J. M. Barnola, Y. S. Korotkevich, and C. Lorius, 1988, "Vostok ice core reveals glacial-interglacial $CH_4$ change about 140 kyr ago: Climatic and $CH_4$ cycle implications," *Nature* **333**, 655–657.

Renault-Miskovsky, J., 1976, "Contribution a la paleoclimatologie du midi Mediterranean pendant la derniere glaciation et de la postglaciaire,"*Bul. Mus. d'Anthrop. et Histoire Monako* **18**, 145–210.

Ronov, A. B., and A. N. Balukhovsky, 1981, "Climatic zonality and climatic trend during the Mesozoic and Cenozoic,"*Lithology and Mineral Fossil* **5**, 118–126 (R).

Rossignol-Strick, M., 1985, "Mediterranean Quaternary sapropels, an immediate response of the African monsoon to variation of insolation,"*Palaeogeography, Palaeoclimatology, Palaeoecology* **49**, 237–265.

Ruddiman, W. F., and J. E. Kutzbach, 1989, "Forcing of Late Cenozoic Northern Hemisphere climate by plateau uplift in southern Asia and the American West," *Journal of Geophysical Research* **94**, 18409–18427.

Savin, S. M., 1982, "Stable isotopes in climatic reconstructions," in *Climate in Earth History*, National Academy Press, Washington, DC, pp. 164–171.

Savin, S. M., R. C. Douglas, and F. G. Stheli, 1975, "Tertiary marine paleotemperatures," *Geological Society Bulletin* **86**, 1499.

Sinitsyn, V. M., 1967, *Introduction to Paleoclimatology*, Nedra, Leningrad, 232 pp. (R).

Stauffer, B., E. Lochbronner, H. Oeschger, and J. Schwander, 1988, "Methane concentration in the glacial atmosphere was only half that of the preindustrial Holocene," *Nature* **332**, 812–814.

Street-Perrott, F. A., and S. P. Harrison, 1985, "Lake levels and climate reconstruction," in *Paleoclimate Analysis and Modelling* (A. D. Hecht, Ed.), John Wiley & Sons, New York, pp. 291–340.

Sundquist, E. T., 1985, "Geological perspectives on carbon dioxide and the carbon cycle," in *The Carbon Cycle and Atmospheric $CO_2$: Natural Variations, Archean to Present, Geophysical Monograph No. 32*, American Geophysical Union, Washington, DC pp. 5–60.

Sundquist, E. T., and W. S. Broecker (Eds.), 1985, *The Carbon Cycle and Atmospheric $CO_2$: Natural Variations, Archean to Present*, *Geophysical Monograph No. 32*, American Geophysical Union, Washington, DC, 627 pp.

US/USSR, 1977, "Soviet-American meeting on climate-nature changes in the Pleistocene and Holocene," *Meteorologiya i Gidrologiya* **5**, 121–123.

Vakhrameev, V. A., 1988, "Climates and Floras during the Jurassic and Cretaceous Time, Nauka, Moscow," Tr. GIN Acad. Sc. USSR, pp. 1–212 (R).

Van Zeist, W., and S. Bottema, 1982, "Vegetational history of the eastern Mediterranean and the Near East during the last 20,000 years," in *Paleoclimates, Paleoenvironments and Human Communities in the Eastern Mediterranean Region in Later Prehistory* (J. L. Bintliff and W. Van Zeist, Eds.), British Archaeological Report, International Series, No. 133, Oxford, pp. 277–321.

Van Zeist, W., H. Woldring, and D. Stapert, 1975, "Late Quaternary vegetation and climate of southwestern Turkey," in *Paleohistoria* **17**, 53–143.

Velichko, A. A., 1982, "Major features of the last climatic microcycle and contemporary environment," in *Paleogeography of Europe for the Last 100,000 Years*. (I. P. Guerassimov and A. A. Velichko, Eds.), Nauka, Moscow, pp. 131–142 (R).

Velichko, A. A. (Ed.), 1984, *Late Quaternary Environments of the Soviet Union*, University of Minnesota, Minneapolis, Minnesota (H. E. Wright and C. W. Barnosky, Eds. of English translation), 327 pp.

Velichko, A. A., 1985, "Empirical paleoclimatology (principles and the degree of accuracy)," in *Methods of Paleoclimatic Reconstructions*, Nauka, Moscow, pp. 7–20.

Velichko, A. A., 1987, "The structure of paleoclimatic Mezo-Cenozoic thermal variations based on data from eastern Europe," in *Earth's Climates in the Geologic Past*, Nauka, Moscow, pp. 5–41 (R).

Velichko, A. A., 1989, "Evolutionary analysis of the contemporary landscape sphere of the Earth and prognosis," *Quaternary International* **1**, 1–8.

Velichko, A. A., M. S. Barash, V. P. Grichuk, Ye. Ye. Gurtovaya, and E. M. Zelikson, 1984, "Climate of the Northern Hemisphere in the last interglacial (Mikulino)," *Izvestiya, AN SSSR, Geography Series*, **1**, 5–18.

Velichko, A. A., V. P. Grichuk, Ye. Ye. Gurtovaya, and E. M. Zelikson, 1983, "Paleoclimate of the USSR during the optimum of the last (Mikulinsky) interglacial," *Izvestiya AN SSSS, Geography Series* **6**, 30–45 (R).

Velichko, A. A., V. P. Grichuk, Ye. Ye. Gurtovaya, E. M. Zelikson, and O. K. Borisova, 1982, "Paleoclimatic reconstructions for the optimum of the Mikulinsky interglacial in Europe," *Izvestiya, AN SSSR, Geography Series* **1**, 15 (R).

Webb III, T. E., 1985, "Holocene palynology and climate," in *Paleoclimate Analysis and Modelling* (A. D. Hecht, Ed.), John Wiley & Sons, New York, pp. 163–196.

Webb III, T. E., J. Cushing, and H. E. Wright, 1983, "Holocene changes in the vegetation of the Midwest," in *Late Quaternary Environments of the United States, Vol. 2*, University of Minnesota, Minneapolis, Minnesota, pp. 142–165.

Weinstein, M., 1976, "The Late Quaternary vegetation of the northern Golan," *Pollen et Spores* **18**, 553–562.

Wijmstra, T. A., 1969, "Palynology of the first 30 metres of a 120 m deep section in northern Greece," *Acta Botanica Neerlandica*, **18**, 511–527.

Winkler, M. G., A. M. Swan, and J. E. Kutzbach, 1986, "Middle Holocene dry period in the northern United States: Lake levels and pollen stratigraphy," *Quaternary Research* **25**, 235–250.

Wolfe, J., 1981, "Distribution of major vegetational types during the Tertiary," in *The Carbon Cycle and Atmospheric $CO_2$: Natural Variations Archaen to Present* (E. T. Sundquist and W. S. Broecker, Eds.) American Geophysical Union Geophysical Monograph, No. 32, pp. 351–375.

Yasamanov, N. A., 1985, *Ancient Climates of the Earth*, Gidrometeoizdat, Leningrad, 292 pp. (R).

Yefimova, N. A., 1987, "The change of moisture conditions on a part of Eurasian territory in the case of global warming of climate," *Meteorologiya i Gidrologiya* **11**, 54–59 (R).

Yelina, G. A., 1981, *Principles and Methods of Reconstruction and Mapping of the Holocene Vegetation*, Nauka, Leningrad, p. 159 (R).

Zagwijn, W. H. 1975, "Variations in climate as shown by pollen analysis, especially in the lower Pleistocene in Europe," in *Ice Ages: Ancient and Modern*, special issue: *The Geographical Journal* **6**, 137–152.

Zubakov, V. A., 1986, *Global Climatic Events of the Pleistocene*, Gidrometeoizdat, Leningrad, 287 pp. (R).

Zubakov, V. A., 1990, *Global Climatic Events during the Neogene*, Gidrometeoizdat, Leningrad (in press) (R).

Zubakov, V. A., and I. I. Borzenkova, 1988, "Pliocene paleoclimates: Past climates as possible analogues of mid-twenty-first century climate," *Palaeogeography, Palaeoclimatology, Palaeoecology* **65**, 35–49.

Zubakov, V. A., and I. I. Borzenkova, 1990, *Global Climate during the Cenozoic*, Elsevier, New York-Oxford-Tokyo, 456 pp.

Zubenok, L. I., 1976, *Evaporation and Nature*, Gidrometeoizdat, Leningrad, 265 pp. (R).

## CHAPTER 3

Angell, J. K., and J. Korshover, 1985, "Surface temperature changes following six major volcanic episodes between 1780 and 1980," *Journal of Climate and Applied Meteorology* **24**, 937–951.

Apasova, Ye. G., and G. V. Gruza, 1982, *Data on Climate Structure and Variability of Precipitation in the Northern Hemisphere*, VNIIGMI-MCD, Obninsk, 212 pp. (R).

Barnett, T. P., 1983, "Recent changes in sea level and their possible causes," *Climate Change* **5**, 15–38.

Barnett, T. P., 1984, "The estimates of 'global' sea-level change: A problem of uniqueness," *Journal of Geophysical Research* **89**, 7980–7988.

Barnett, T. P., 1988, *Global Sea Level Change Revisited*, Special report to the US National Climate Program Office, Washington, DC.

Borzenkova, I. I., K. Ya. Vinnikov, L. P. Spirina, and D. I. Stekhnovsky, 1976, "Air temperature changes in the Northern Hemisphere over 1881–1975," *Meteorologiya i Gidrologiya* **7**, 27–35 (R).

Bradley, R. S., 1988, "The explosive volcanic eruption signal in Northern Hemisphere continental temperature records," *Climatic Change* **12**, 221–242.

Bradley, R. S., H. F. Diaz, J. K. Eischeid, P. D. Jones, P. M. Kelly, and C. M. Goodess, 1987, "Precipitation fluctuations over Northern Hemisphere land areas since the mid-19th century," *Science* **237**, 171–175.

Bradley, R. S., and P. Ya. Groisman, 1989, "Continental scale precipitation variations in the 20th century," in *Proceedings of the International Conference on Precipitation Measurements*, World Meteorological Organization, Geneva (in press).

Bradley, R. S., P. M. Kelly, P. D. Jones, H. F. Diaz, and C. M. Goodess, 1985, *A Climatic Data Bank for the Northern Hemisphere Land Areas, 1851–1980*, report TR017, US Department of Energy, Carbon Dioxide Research Division, Washington, DC, 335 pp.

Budyko, M. I., 1969, "The effect of solar radiation variations on the climate of the Earth," *Tellus* **21**, 611–619.

Budyko, M. I., 1974, *Climatic Changes*, Gidrometeoizdat, Leningrad, 280 pp. (R) (English translation: American Geophysical Union, 1977, 261 pp.).

Callendar, G. S., 1961, "Temperature fluctuations and trends over the Earth," *Quarterly Journal of the Royal Meteorological Society* **87**, 1–12.

Diaz, H. F., R. S. Bradley, and J. K. Eischeid, 1989, "Precipitation over global land areas since the late 1800s," *Journal of Geophysical Research* **94**, 1195–1210.

Ellsaesser, H. W., M. C. MacCracken, J. J. Walton, and S. L. Grotch, 1986, "Global climatic trends as revealed by the recorded data," *Reviews of Geophysics and Space Physics* **24**, 745–792.

Emery, K. O., 1980, "Relative sea-level trends from tide-gauge records," in *Proceedings of the National Academy of Sciences* **77**, 6968–6970.

Etkins, R., and E. S. Epstein, 1982, "The rise of global mean sea level as an indication of climate change," *Science* **215**, 287–289.

Fairbridge, R. W., and O. A. Krebs, 1962, "Sea-level and the Southern Oscillation," *Geophysical Journal of the Royal Astronomical Society* **6**, 532–545.

Folland, C. K., 1989, United Kingdom Meteorological Office, personal communication.

Folland, C. K., and D. E. Parker, 1989, "Observed variations of sea surface temperature," in *Climate-Ocean Interaction* (M. E. Schlesinger, Ed.), NATO Workshop, Kluwer Academic Publishers, Dordrecht, The Netherlands, pp. 21–52.

Frohlich, C., 1988, "Variability of the solar 'constant'," in *Long and Short-term Variability of Climate* (H. Wanner and U. Siegenthaler, Eds.) Springer-Verlag, New York, pp. 6–17.

Gandin, L. S., and R. L. Kagan, 1976, *Statistical Methods for Interpreting Meteorological Data*, Gidrometeoizdat, Leningrad, 359 pp. (R).

Gornitz, V., and S. Lebedeff, 1987, "Global sea level changes during the past century," in *Sea Level Change and Coastal Evolution* (D. Nummedal, O. H. Pilkey, and J. D. Howard, Eds.), Society of Economic Paleontologists and Mineralologists, Special Publication No. 41, Tulsa, Oklahoma, pp. 3–16.

Gornitz, V., S. Lebedeff, and J. Hansen, 1982, "Global sea-level trend in the past century," *Science* **215**, 1611–1614.

Groisman, P. Ya., 1990, "Data on present-day precipitation changes in the extratropical part of the Northern Hemisphere," in *Proceedings of the USSR Academy of Sciences Geographical Series* (in press).

Groisman, P. Ya., and V. V. Koknayeva, 1989, *The influence of urbanization on the estimate of the mean air temperature change in the 20th century over the USSR and USA territories obtained by the State Hydrological Institute network*, State Hydrological Institute, Leningrad, unpublished manuscript (R).

Hansen, J. E., D. Johnson, A. Lacis, S. Lebedeff, and P. Lee, 1981, "Climate impact of increasing atmospheric carbon dioxide," *Science* **213**, 957–966.

Hansen, J. E., and S. Lebedeff, 1987, "Global trends of measured surface air temperature," *Journal of Geophysical Research* **92**, 13,345–13,372.

Hansen, J. E., and S. Lebedeff, 1988, "Global surface air temperatures: Update through 1987," *Geophysical Research Letters* **15**, 323–326.

Humphreys, W. J., 1929, *Physics of the Air*, McGraw Hill Book Company, New York, 654 pp.

Jones, P. D., 1988, "Hemispheric surface air temperature variations: Recent trends and an update to 1987," *Journal of Climate* **1**, 654–660.

Jones, P. D., R. S. Bradley, H. F. Diaz, P. M. Kelly, and T. M. L. Wigley, 1986a, "Northern Hemisphere surface air temperature variations: 1851–1984," *Journal of Climate and Applied Meteorology* **25**, 161–179.

Jones, P. D., and P. M. Kelly, 1988, "Hemispheric and global temperature data," in *Long and Short Term Variability of Climate* (H. Wanner and U. Siegenthaler, Eds.) Springer-Verlag, New York, pp. 18–34.

Jones, P. D., P. M. Kelly, C. M. Goodess, and T. R. Karl, 1989, "The effect of urban warming on the Northern Hemisphere temperature average," *Journal of Climate* **2**, 285–290.

Jones, P. D., S. C. B. Raper, B. Santer, B. S. G. Cherry, C. Goodess, P. M. Kelly, and T. M. L. Wigley, 1985, *A Grid Point Surface Air Temperature Data Set for the Southern Hemisphere*, report TRO22(27), US Department of Energy, Washington, DC.

Jones, P. D., S. C. B. Raper, and T. M. L. Wigley, 1986b, "Southern Hemisphere surface air temperature variations: 1851–1984," *Journal of Climate and Applied Meteorology* **25**, 1213–1230.

Jones, P. D., T. M. L. Wigley, and P. M. Kelly, 1982, "Variations in surface air temperature, Pt. 1, Northern Hemisphere, 1881–1980," *Monthly Weather Review* **110**, 59–70.

Jones, P. D., T. M. L. Wigley, and P. B. Wright, 1986c, "Global temperature variations between 1861 and 1984," *Nature* **322**, 340–434.

Kagan, R. L., 1979, *Averaging Meteorological Fields*, Gidrometeoizdat, Leningrad, 212 pp. (R).

Karl, T. R., 1988, "Multi-year climate fluctuations: The gray area of climate change," *Climatic Change* **12**, 179–197.

Karl, T. R., H. F. Diaz, and G. Kukla, 1988, "Urbanization: Its detection in the United States climate record," *Journal of Climate* **1**, 1099–1123.

Karl, T. R., and P. D. Jones, 1989, "Urban bias in area-averaged surface temperature trends," *Bulletin of the American Meteorological Society* **70**, 265–270.

Karl, T. R., and P. D. Jones, 1990, "Reply to comments on 'Urban bias in area-averaged surface temperature trends'," *Bulletin of the American Meteorological Society* **71**, pp. 572–574.

Karl, T. R., and R. G. Quayle, 1988, "Climate change in fact and in theory: Are we collecting the facts?" *Climate Change* **13**, 5–17.

Karl, T. R., D. Tarpley, R. G. Quayle, H. F. Diaz, D. A. Robinson, and R. S. Bradley, 1989, "The recent climate record: What it can and cannot tell us," *Reviews of Geophysics* **27**, 405–430.

Karl, T. R., and C. N. Williams, Jr., 1987, "An approach to adjusting climatological time series for discontinuous inhomogeneities," *Journal of Climate and Applied Meteorology* **26**, 1744–1763.

Karl, T. R., C. N. Williams, Jr., P. J. Young, and W. M. Wendland, 1986, "A model to estimate the time of observation bias associated with monthly mean maximum, minimum, and mean temperatures for the United States," *Journal of Climate and Applied Meteorology* **25**, 145–160.

Khmelevtsov, S. S. (Ed.), 1986, *Volcanism, Stratospheric Aerosol and Earth's Climate*, Gidrometeoizdat, Leningrad, 256 pp. (R).

Khromov, S. P., 1973, "Solar cycles and climate," *Meteorologiya i Gidrologiya* **9**, 93–110 (R).

Köppen W., 1883, "Über mehrjarige Perioden der Witterung (insbesondere Über die II jarige Periods der temperature)," *der Usterreichischen Gesselschaft für Meteor*. **8**, 241–248, 257–267.

Lisitzin, E., 1974, *Sea Level Changes*, Elsevier, New York.

Lorenz, E. N., 1986, "The index cycle is alive and well," in *Namias Symposium*, Reference Series 86-17, Scripps Institute of Oceanography, La Jolla, California (US Library of Congress number 86-50752).

Mass, C. F., and D. A. Portman, 1989, "Major volcanic eruptions and climate: A critical evaluation," *Journal of Climate* **2**, 566–593.

Meier, M., 1990, "Reduced rise in sea level," *Nature* **343**, 115–116.

Mitchell, Jr., J. M., 1961, "Recent secular changes of global temperature," *Annuals of the New York Academy of Sciences* **95**, 235–250.

Mitchell, Jr., J. M., 1963, "On the world-wide patterns of secular temperature change," in *Changes of Climate, Arid Zone Research* **20**, 161–181.

Parker, D. E., 1989, "Observed climatic change and the greenhouse effect," *Meteorological Magazine* **118**, 128–131.

Peltier, W. R., 1986, "Deglaciation-induced vertical motion of the North American continent and transient lower mantle rheology," *Journal of Geophysical Research* **91**, 9099–9123.

Peltier, W. R., and A. M. Tushingham, 1989, "Global sea level and the greenhouse effect: Might they be connected?" *Science* **244**, 806–810.

Pittock, A. B., 1978, "A critical look at long-term Sun-weather relationships," *Reviews of Geophysics and Space Physics* **16**, 400–420.

Pittock, A. B., 1979, "Solar variability, weather and climate: An update," *Quarterly Journal of the Royal Meteorological Society* **109**, 23–55.

Polar Research Board, 1986, *Glaciers, Ice Sheets, and Sea Level: Effects of a $CO_2$-Induced Climate Change*, Report of the ad hoc Committee on the Relationship between Land Ice and Sea Level, Committee on Glaciology, Commission on Physical Sciences, Mathematics, and Resources, National Research Council, National Academy Press, Washington, DC.

Quinlan, F. T., T. R. Karl, and C. N. Williams, Jr., 1987, *United States Historical Climatology Network (HCN) Serial Temperature and Precipitation Data*, NDP-010, Carbon Dioxide Information and Analysis Center, Oak Ridge National Laboratory, Oak Ridge, Tennessee, 33 pp. (plus appendices).

Robock, A., 1979, "The Little Ice Age: Northern Hemisphere average observations and model calculations," *Science* **206**, 1402–1404.

Sear, C. B., P. M. Kelley, P. D. Jones, and C. M. Goodess, 1987, "Global surface-temperature responses to major volcanic eruptions," *Nature* **330**, 365–367.

Sevruk, B. (Ed.), 1989, *Precipitation Measurement*, Swiss Federal Institute of Technology, Zurich, 584 pp.

Slutz, R. J., S. J. Lubker, J. D. Huscox, S. D. Woodruff, R. L. Jenne, D. H. Joseph, P. M. Steurer, and J. D. Elms, 1985, *Comprehensive Ocean-Atmosphere Data Set (COADS)*, Release I, Environmental Research Laboratory, National Center for Atmospheric Research, National Climate Data Center, Boulder, Colorado, 267 pp.

Toon, O. B., and J. B. Pollack, 1980, "Atmospheric aerosols and climate," *American Scientist* **68**, 268–278.

US-USSR Meeting Report, 1982, "Climatic effects of increased atmospheric carbon dioxide," in *Proceedings of the Soviet-American Meeting on Studying Climatic Effects of Increased Atmospheric Carbon Dioxide*, 15–20 June, 1981, Gidrometeoizdat, Leningrad, 56 pp. (R).

Vannari, P. I., 1911, "Meteorological networks in Russia and other countries," *Notes of the Emperor Russian Geographical Society on General Geography, Collection of Articles on meteorology* **XLVII**, 51–64 (R).

Vinnikov, K. Ya., 1986, *Climate Sensitivity*, Gidrometeoizdat, Leningrad, 224 pp. (R).

Vinnikov, K. Ya., P. Ya. Groisman, and K. M. Lugina, 1990, "Empirical data on contemporary global climate changes (temperature and precipitation)," *Journal of Climate* **3**, 662–677.

Vinnikov, K. Ya., P. Ya. Groisman, K. M. Lugina, and A. A. Golubev, 1987, "The mean surface air temperature variations in the Northern Hemisphere over 1841–1985," *Meteorologiya i Gidrologiya* **1**, 45–55. (R).

Vinnikov, K. Ya., G. V. Gruza, V. F. Zakharov, N. P. Kovyneva, and E. Ya Ran'kova, 1980, "Modern climate in the Northern Hemisphere," *Meteorologiya i Gidrologiya* **6**, 5–17.

Wigley, T. M. L., 1988, "The climate of the past 10,000 years and the role of the Sun," in *Secular Solar and Geomagnetic Variations in the Last 10,000 Years*, Kluwer Academic Publishers, Dordrecht, The Netherlands, pp. 209–224.

Wood, F. B., 1990, "Monitoring global climate change: The case of greenhouse warming," *Bulletin of the American Meteorological Society* **71**, 42–52.

## CHAPTER 4

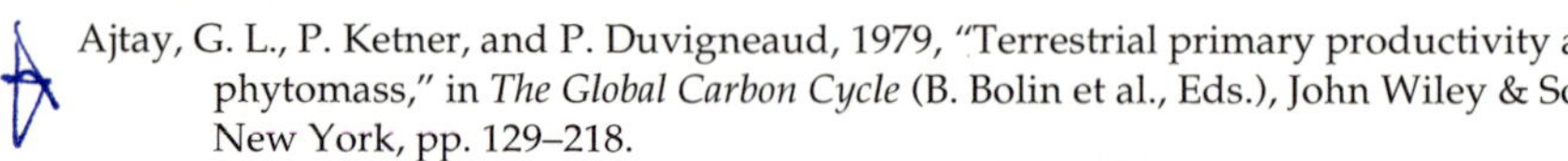

Ajtay, G. L., P. Ketner, and P. Duvigneaud, 1979, "Terrestrial primary productivity and phytomass," in *The Global Carbon Cycle* (B. Bolin et al., Eds.), John Wiley & Sons, New York, pp. 129–218.

Alternative Fluorocarbon Environmental Acceptability Study (AFEAS), 1989, in *Scientific Assessment of Stratospheric Ozone: 1989 Vol. II Appendix: AFEAS Report*, Global Ozone Research and Monitoring Project–Report No. 20, World Meteorological Organization, Geneva.

Andronova, N. G., and I. L. Karol, 1986, "Atmospheric methane and nitrous oxide: Evolution of certain anthropogenic sources," in *Atmospheric Radiative-photochemical Models* (I. Karol, Ed.), Gidrometeoizdat, Leningrad, pp. 5–17 (R).

Bacastow, R. B., and C. D. Keeling, 1973, "Atmospheric carbon dioxide and radiocarbon in the natural carbon cycle," in *Carbon and the Biosphere* (G. W. Woodwell and E. V. Pecan, Eds.), report CONF-720510, US Atomic Energy Commission, Washington, DC, pp. 86–135 (available from US National Technical Information Service, Springfield, Virginia).

Bacastow, R. B., and E. Maier-Reimer, 1990, "Ocean circulation model of the carbon cycle," *Climate Dynamics* **4**, 95–125.

Baes, Jr., C. F., 1982, "Effects of ocean chemistry and biology on atmospheric carbon dioxide," in *Carbon Dioxide Review: 1982* (W. C. Clark, Ed.), Oxford University Press, New York, pp. 187–204.

Barnola, J. M., D. Raynaud, Y. S. Korotkevich, and C. Lorius, 1987, "Vostok ice core provides 160,000-year record of atmospheric $CO_2$," *Nature* **329**, 408–414.

Barron, E. J., and W. M. Washington, 1985, "Warm Cretaceous climates; High atmospheric $CO_2$ as a plausible mechanism," in *The Carbon Cycle and Atmospheric $CO_2$: Natural Variations Archean to Present* (E. T. Sundquist and W. S. Broecker, Eds.), *Geophysical Monograph 32*, American Geophysical Union, Washington, DC, pp. 546–553.

Bates, T. S., J. D. Cline, R. H. Gammon, and S. R. Kelly-Hansen, 1987, "Regional and seasonal variations in the flux of oceanic dimethylsulfide to the atmosphere," *Journal of Geophysical Research* **92**, 2930–2938.

Berner, R. A., A. C. Lasaga, and R. M. Garrels, 1983, "The carbonate-silicate geochemical cycle and its effect on atmospheric carbon dioxide over the past 100 million years," *American Journal of Science* **283**, 641–683.

Bjorkstrom, A., 1979, "A model of $CO_2$ interaction between atmosphere, oceans and land biota," in *The Changing Carbon Cycle, SCOPE 13* (B. Bolin, E. T. Degens, S. Kempe, and P. Ketner, Eds), John Wiley & Sons, New York, pp. 403–457.

Blake, D. R., and F. S. Rowland, 1988, "Continuing worldwide increase in tropospheric methane, 1978 to 1987," *Science* **239**, 1129–1131.

Bolin, B., B. R. Döös, J. Jaeger, and R. A. Warrick (Eds.), 1986, *Greenhouse Effect, Climatic Changes and Ecosystems, SCOPE-29*, John Wiley & Sons, Chichester, UK.

Boyle, E. A., 1988, "The role of vertical chemical fractionation in controlling Late Quaternary atmospheric carbon dioxide," *Journal of Geophysical Research* **93**, 15,701–15,714.

Bretherton, F. P., K. Bryan, J. Hansen, M. I. Hoffert, X. Jiang, S. Manabe, J. Meehl, D. Rind, M. E. Schlesinger, R. Staufer, and T. Volk, 1990, "Time-dependent greenhouse-induced climate change," in *Intergovernmental Panel on Climate Change (IPCC) Report*, (in press) Cambridge University Press, Cambridge, UK.

Broccoli, A. J., and S. Manabe, 1987, "The influence of continental ice, atmospheric $CO_2$, and land albedo on the climate of the last glacial maximum," *Climate Dynamics* **1**, 87–99.

Broecker, W. S., 1982, "Terminations," in *Milankovitch and Climate: Understanding the Response to Astronomical Forcing, Part II* (A. Berger, J. Hays, G. Kukla, and B. Saltzman, Eds.), Reidel, Dordrecht, Netherlands.

Broecker, W. S., 1987, "Unpleasant surprises in the greenhouse?" *Nature* **328**, 123–126.

Broecker, W. S., and T. -H. Peng, 1982, *Tracers in the Sea*, Eldigeo Press, Lamont-Doherty Geological Observatory, Palisades, New York.

Broecker W. S., and T. -H. Peng, 1986, "Carbon cycle, 1985: Glacial to interglacial changes in the operation of the global carbon cycle," *Radiocarbon* **28**, 309–327.

Broecker W. S., and T. -H. Peng, 1989, "The cause of the glacial to interglacial atmospheric $CO_2$ change," *Global Biogeochemical Cycles* **3**, 215–239.

Budyko, M. I., and Yu. A. Izrael (Eds.), 1987, *Anthropogenic Climate Changes*, Gidrometeoizdat, Leningrad, 406 pp. (R) (English translation: Arizona University Press, 1990).

Budyko, M. I., A. B. Ronov, and A. L. Yanshin, 1985, *The History of the Earth's Atmosphere*, Gidrometeoizdat, Leningrad, 209 pp. (R) (English translation: Springer-Verlag, Berlin, 1987, 139 pp).

Byutner, E. K., 1986, *Planetary Gas-exchange of $O_2$ and $CO_2$*, Gidrometeoizdat, Leningrad (R).

Byutner, E. K., and O. K. Zakharova, 1983, "Modeling the carbon cycle by using the fraction derivative model," *Meteorologiya i Gidrologiya* **4**, 14–20 (R).

Byutner, E. K., and O. K. Zakharova, 1989, State Hydrological Institute, personal communication.

Charlson, R. J., J. E. Lovelock, M. O. Andreae, and S. G. Warren, 1987, "Oceanic phytoplankton, atmospheric sulphur, cloud albedo and climate," *Nature* **326**, 655–661.

Chichagova, O. A., 1985, *Radiocarbon Dating of Soil Aging*, Nauka, Moscow, 158 pp. (R).

Cicerone, R. J., 1988, "How has the atmospheric concentration of CO changed?" in *The Changing Atmosphere* (F. S. Rowland and I. S. A. Isaksen, Eds.), John Wiley & Sons, New York.

Cicerone, R. J., and R. S. Oremland, 1988, "Biogeochemical aspects of atmospheric methane," *Global Biogeochemical Cycles* **2**, 299–327.

Connell, P. S., and D. J. Wuebbles, 1989, "Evaluating CFC alternatives from the atmospheric viewpoint," *Air and Waste Management Association paper 89–57*; also report UCRL-99927 Rev. 1, 1989, Lawrence Livermore National Laboratory, Livermore, California.

Craig, H., 1969, "Abyssal radiocarbon in the Pacific," *Journal of Geophysical Research* **74**, 5491–5501.

Craig, H., and C. C. Chou, 1982, "Methane: The record in polar ice cores," *Geophysics Research Letters* **9**, 1221–1224.

DeLuisi, J. J., D. U. Longenecker, C. L. Mateer, and D. J. Wuebbles, 1989, "An analysis of northern middle-latitude Umkehr measurements corrected for stratospheric aerosols for 1979–1986," *Journal of Geophysical Research* **94**, 9837–9846.

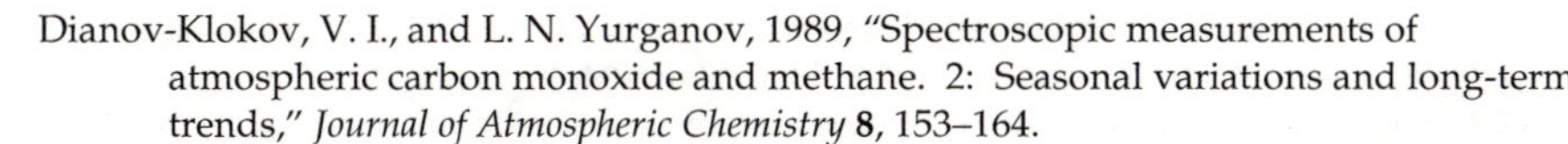


Dianov-Klokov, V. I., and L. N. Yurganov, 1989, "Spectroscopic measurements of atmospheric carbon monoxide and methane. 2: Seasonal variations and long-term trends," *Journal of Atmospheric Chemistry* **8**, 153–164.

Dianov-Klokov, V. I., L. N. Yurganov, E. I. Grechko, and A. V. Dzhola, 1989, "Spectroscopic measurements of atmospheric carbon monoxide and methane. 1: Latitudinal distribution," *Journal of Atmospheric Chemistry* **8**, 139–151.

Druffel, E. R. M., P. M. Williams, and Y. Suzuki, 1989, "Concentrations and radiocarbon signatures of dissolved organic matter in the Pacific Ocean," *Geophysical Research Letters* **16**, 991–994.

Edmonds, J. A., J. M. Reilly, R. H. Gardner, and A. Brenkert, 1986, *Uncertainty in Future Global Energy Use and Fossil Fuel $CO_2$ Emissions: 1975 to 2075*, report DOE/NBB-0081, US Department of Energy, Washington, DC.

Emanuel, W., 1989, Oak Ridge National Laboratory, personal communication.

Fioletov, V. E., 1989, "Global ozone trends," *Meteorologiya i Gidrologiya* **4** (in press) (R).

Friedli, H., H. Lotscher, H. Oeschger, U. Siegenthaler, and B. Stauffer, 1986, "Ice core record of $^{13}C/^{12}C$ ratio of atmospheric carbon dioxide in the past two centuries," *Nature* **324**, 237–238.

Garcia, R. R., and S. Solomon, 1987, "A possible relationship between interannual variability in Antarctic ozone and the quasi-biennial oscillation," *Geophysical Research Letters* **14**, 848–851.

Hansen, J., I. Fung, A. Lacis, D. Rind, S. Lebedeff, R. Ruedy, G. Russell, and P. Stone, 1988, "Global climate changes as forecast by Goddard Institute for Space Studies three-dimensional model," *Journal of Geophysical Research* **93**, 9341–9364.

Hao, W. M., S. C. Wofsy, M. B. McElroy, J. M. Beer, and M. A. Toqan, 1987, "Sources of atmospheric nitrous oxide from combustion," *Journal of Geophysical Research* **92**, 3098–3104.

Hays, J. D., J. Imbrie, and N. J. Shackleton, 1976, "Variations in the Earth's orbit: Pacemaker of the ice ages," *Science* **194**, 1121–1132.

Hoffert, M., 1974, "Global distributions of atmospheric carbon dioxide: A projection," *Atmospheric Environment* **8**, 1255–1249.

Hoffert, M. I., A. J. Callegari, and C. -T. Hseih, 1981, "A box-diffusion carbon cycle model with upwelling, polar bottom water formation and a marine biosphere," in *Carbon Cycle Modelling, SCOPE 16* (B. Bolin, Ed.), John Wiley & Sons, New York, pp. 287–305.

Houghton, R. A., 1986, "Estimating changes in the carbon content of terrestrial ecosystems from historical data," in *The Changing Carbon Cycle* (J. R. Trabalka and D. E. Reichle, Eds.), Springer-Verlag, New York, pp. 175–191.

Houghton, R. A., 1989, "The long-term flux of carbon to the atmosphere from changes in land use," presented at the Third International Conference on Analysis and Evaluation of Atmospheric $CO_2$ Data Present and Past, Hinterzarten, FRG, October 16–20, 1989; abstracted in *Environmental Pollution Monitoring and Research Program Publication No. 59*, World Meteorological Organization, Geneva.

Houghton, R. A., R. D. Boone, J. R. Fruci, J. E. Hobbie, J. M. Melillo, C. Palm, B. J. Peterson, G. R. Shaver, G. M. Woodwell, B. Moore, D. L. Skole, M. Myers, 1987, "The flux of carbon from terrestrial ecosystems to the atmosphere in 1980 due to changes in land use: Geographical distribution of the global flux," *Tellus* **39B**(1–2), 122–139.

Houghton, R. A., J. E. Hobbie, J. M. Melillo, B. Moore, B. J. Peterson, G. R. Shaver, and G. M. Woodwell, 1983, "Changes in the carbon content of terrestrial biota and soils between 1860 and 1980: Net release of $CO_2$ to the atmosphere," *Ecological Monographs* **53**, 235–262.

Imbrie, J., and K. P. Imbrie, 1979, *Ice Ages: Solving the Mystery*, Enslow, Hillside, New Jersey, 224 pp.

Intergovernmental Panel on Climate Change (IPCC), 1990, *Scientific Assessment of Climate Change, Report to IPCC from Working Group 1*, Cambridge University Press, Cambridge, UK (in press).

Isaksen, I. S. A., and O. Hov, 1987, "Calculations of trends in the tropospheric concentrations of $O_3$, OH, CO, $CH_4$, and NO," *Tellus* **398**, 271–285.

Isidorov, V. A., and B. V. Ioffe, 1986, "Non-anthropogenic sources of halocarbons in the terrestrial atmosphere," *Doklady Akademy Nauk USSR* **287**, 86–90 (R).

Jadin, E. A., and E. V. Lysenko, 1988, "The long-term variations in stratospheric dynamics and ozone depletion in Antarctica," *Meteorologiya i Gidrologiya* **10**, 136–140.

Kagan, B. A., and W. A. Ryabchenko, 1981, "Non-linear model of carbon cycle in the ocean," *Doklady Academy Nauk USSR* **285**, 212–215 (R).

Karol, I. L., 1986, "Concerning the possible anthropogenic changes of the gaseous content and temperature of the atmosphere until 2000 AD," *Meteorologiya i Gidrologiya* **9**, 102–107 (R).

Karol, I. L., and V. V. Babanova, 1987, "Global spreading of halocarbons in the atmosphere," *Meteorologiya i Gidrologiya* **3**, 39–46 (R).

Karol, I. L., and A. A. Kiselyev, 1984, "Transient 1-D model of the photochemical transformation of gaseous pollutants of the stratosphere," USSR Academy of Science Izvestia, *Atmospheric and Oceanic Physics* **20**, 957–968 (R).

Karol, I. L., and A. A. Kiselyev, 1988, "Estimates of the influences of the accuracy of photochemical reaction rates on computations of the trace gas concentrations in the atmosphere," USSR Academy of Science Izvestiya, *Atmospheric and Oceanic Physics* **24**, 663–667 (R).

Karol, I. L., A. A. Kiselyev, and E. V. Rozanov, 1986, "Study of the climatic effects of changes in atmospheric trace gases content with the radiation-photochemical model," US/USSR Meeting of Experts on Causes of Recent Climate Change, Leningrad.

Karol, I. L., L. P. Klyagina, A. D. Frolov, and A. M. Shalamyansky, 1987, "Fields of ozone and temperature within the boundaries of air masses," *Meteorologiya i Gidrologiya* **10**, 47–52, (R).

Karol, I. L., L. P. Klyagina, and A. M. Shalamyansky, 1990b, "Total ozone and stratospheric temperature trends in the principal air masses of the Northern Hemisphere," *Meteorologiya i Gidrologiya* (in press) (R).

Karol, I. L., L. P. Klyagina, A. M. Shalamyansky, A. D. Frolov, N. V. Volkova, and N. D. Ionina, 1990a, "Seasonal and interannual variations of total ozone and temperature in the arctic air-mass," *Meteorologiya i Gidrologiya* (in press) (R).

Kasting, J. F., and O. B. Toon, 1989, "Climate evolution on the terrestrial planets," in *Origin and Evolution of Planetary and Satellite Atmospheres* (S. K. Atreya, J. B. Pollack, and M. S. Matthews, Eds.), University of Arizona Press, Tucson, Arizona, pp. 423–449.

Kasting, J. F., and J. C. G. Walker, 1989, "Long term effects of fossil fuel burning, submitted to *Science*.

Keeling, C. D., 1973, "The carbon dioxide cycle: Reservoir models to depict the exchange of atmospheric carbon dioxide with the oceans and land plants," in *Chemistry of the Lower Atmosphere* (S. I. Rasool, Ed.), Plenum Press, New York, pp. 251–329.

Keeling, C. D., R. B. Bacastow, A. E. Bainbridge, C. A. Ekdahl, P. R. Guenther, L. S. Waterman, and J. F. Chin, 1976, "Atmospheric carbon dioxide variations at Mauna Loa Observatory, Hawaii," *Tellus* **28**, 538–551.

Keeling, C. D., R. B. Bacastow, J. Lancaster, T. P. Whorf, and W. G. Mook, 1989, "Evidence for accelerated release of carbon dioxide to the atmosphere, inferred from direct measurements of concentration and $^{13}C/^{12}C$ ratio," presented at the June 1989 Annual Meeting of the Air and Waste Management Association, Anaheim, California; submitted to *Science*.

Khalil, M. A. K., and R. A. Rasmussen, 1982, "Secular trends of methane," *Chemosphere* **12**, 877–883.

Khalil, M. A. K., and R. A. Rasmussen, 1987, "Atmospheric methane: Trends over the last 10,000 years," *Atmospheric Environment* **21**, 2445–2452.

Khalil, M. A. K., and R. A. Rasmussen, 1988a, "Carbon monoxide in the Earth' atmosphere: Indications of a global increase," *Nature* **332**, 242–245.

Khalil, M. A. K., and R. A. Rasmussen, 1988b, "Nitrous oxide: Trends and global mass balance over the last 3000 years," *Annals of Glaciology* **10**, 73–79.

Khalil, M. A. K., and R. A. Rasmussen, 1990, "Constraints imposed by the ice core data on the budgets of nitrous oxide and methane," *Tellus* **42B**, 229–239.

Kheshgi, H. S., 1989, "The sensitivity of $CO_2$ projections to ocean processes," presented at the Third International Conference on Analysis and Evaluation of Atmospheric $CO_2$ Data Present and Past, Hinterzarten, FRG, October 16–20, 1989; abstracted in *Environmental Pollution Monitoring and Research Program Publication No. 59*, World Meteorological Organization, Geneva.

Kobak, K. I., 1988, *Biotical Components of the Carbon Cycle*, Gidrometeoizdat, Leningrad, 246 pp. (R).

Komhyr, W. D., R. H. Gammon, T. B. Harris, L. S. Waterman, T. J. Conway, W. R. Taylor, and K. W. Thoning, 1985, "Global atmospheric $CO_2$ distribution and variations from 1968–1982 NOAA/GMCC $CO_2$ flask sample data," *Journal of Geophysical Research* **90**, 5567–5596.

Kraus, F., W. Bach, and J. Koomry, 1989, *Energy Policy in the Greenhouse, Vol. 1*, International Project for Sustainable Energy Paths (IPSEP), El Cerrito, California.

Lacis, A. A., J. Hansen, P. Lee, T. Mitchell, and S. Lebedeff, 1981, "Greenhouse effect of trace gases, 1970–1980," *Geophysical Research Letters* **8**, 1035–1038.

Lacis, A. A., D. J. Wuebbles, and J. A. Logan, 1990, "Radiative forcing of global climate by vertical distribution change of atmospheric ozone," *Journal of Geophysical Research* **95**, 9971–9981.

Lapenis, A. G., and A. I. Kolomietsev, 1987, "Effects of ocean circulation on marine biota production," *Meteorologiya i Hydrologiya* **1**, 77–84 (R).

Lasaga, A. C., R. A. Berner, and R. M. Garrels, 1985, "An improved geochemical model of atmospheric $CO_2$ fluctuations over the past 100 million years," in *The Carbon Cycle and Atmospheric $CO_2$: Natural Variations Archean to Present* (E. T. Sundquist and W. S. Broecker, Eds.), *Geophysical Monograph* 32, American Geophysical Union, Washington, DC, pp. 397–411.

Linak, W. P., J. A. McSorley, R. E. Hall, J. V. Ryan, R. K. Srivastava, J. O. L. Wendt, and J. B. Mereb, 1989, *$N_2O$ Emissions from Fossil Fuel Combustion*, Air and Waste Management Association paper 89-4.6.

Logan, J. A., 1985, "Tropospheric ozone: Seasonal behavior, trends, and anthropogenic influence," *Journal of Geophysical Research* **90**, 10,463–10,482.

Lorius, C., J. Jouzel, C. Ritz, L. Merlivat, N. I. Barkov, Y. S. Korotkevich, and V. M. Kotlyakov, 1985, "A 150,000-year climatic record from Antarctic ice, *Nature* **316**, 591–596.

MacCracken, M. C., and F. M. Luther (Eds.), 1985, *Projecting the Climatic Effects of Increasing Carbon Dioxide*, report DOE/ER-0237, US Department of Energy, Washington, DC.

Maier-Reimer, E., and K. Hasselmann, 1987, "Transport and storage of $CO_2$ in the Ocean," *Climate Dynamics* **2**, 63–90.

Marland, G., T. A. Boden, R. C. Griffin, S. F. Huang, P. Kanciruk, and T. R. Nelson, 1989, *Estimates of $CO_2$ Emissions from Fossil Fuel Burning and Cement Manufacturing, Based on the United States Energy Statistics and the US Bureau of Mines Cement Manufacturing Data*, report ORNL/CDIAC-25, Oak Ridge National Laboratory, Oak Ridge, Tennessee.

Milankovich, M., 1920, *Theorie Mathematique des Phenomenes Thermiques Produits par la Radiation Solaire*, Gauthier-Villars, Paris, 339 pp.

Mintzer, I. M., 1987, *A matter of degrees: The potential for controlling the greenhouse effects*, Research Report 5, World Resources Institute, Washington, DC, 60 pp.

Mitchell, J. F. B., 1989, "The 'greenhouse' effect and climate change," *Reviews of Geophysics* **27**, 115–139.

Moore III, B., and A. Bjorkstrom, 1986, "Calibrated ocean models by constrained inverse method," in *The Changing Carbon Cycle* (J. R. Trabalka and D. E. Reichle, Eds.), Springer-Verlag, New York, pp. 295–328.

Munk, W. H., 1966, "Abyssal recipes," *Deep Sea Research* **13**, 707–736.

Muzio, L. J., and J. C. Kramlich, 1988, "An artifact in the measurement of $N_2O$ from combustion sources," *Geophysical Research Letters* **15**, 1369–1372.

Neftel, A., E. Moor, H. Oeschger, and B. Stauffer, 1985, "Evidence from polar ice cores for the increase in atmospheric $CO_2$ in the past two centuries," *Nature* **315**, 45–47.

North, G. R., and J. A. Coakley, Jr., 1979, "Differences between seasonal and mean energy balance model calculations of climate and climate sensitivity," *Journal of Atmospheric Science* **36**, 1189–1204.

Oeschger, H., U. Siegenthaler, and A. Gugelman, 1975, "A box diffusion model to study the carbon dioxide exchange in nature," *Tellus* **27**, 168–192.

Pearman, G. I., D. M. Etheridge, F. deSilva, and P. J. Fraser, 1986, "Evidence of changing concentrations of atmospheric $CO_2$, $N_2O$ and $CH_4$ from air bubbles in Antarctic ice," *Nature* **320**, 248–250.

Penner, J. E., 1989, "Cloud albedo, greenhouse effects, atmospheric chemistry and climate change," paper for the 82nd Annual Meeting of the Air and Waste Management Society; also report UCRL-99928, Lawrence Livermore National Laboratory, Livermore, California.

Peterson, J., 1988, National Oceanic and Atmospheric Administration, personal communication.

Plass, G. M., 1972, "Relationship between atmospheric $CO_2$ amount and properties of the sea," *Environmental Science and Technology* **6**, 736–740.

Prinn, R. G., D. Cunnold, R. A. Rasmussen, P. Simmonds, F. Alyea, A. J. Crawford, P. Fraser, and R. Rosen, 1987, "Atmospheric trends in methyl chloroform and the global average for the hydroxyl radical," *Science* **238**, 945–950.

Prospero, J. M., R. J. Charlson, V. Mohnen, R. Jaenicke, A. C. Delany, J. Moyers, W. Zoller, and K. Rahn, 1983, "The atmospheric aerosol system: An overview," *Reviews of Geophysics* **21**, 1607–1629.

Ramanathan, V., L. Callis, R. Cess, J. Hansen, I. Isaksen, W. Kuhn, A. Lacis, F. Luther, J. Mahlman, R. Reck, and S. Schlesinger, 1987, "Climate-chemical interactions and effects of changing atmospheric trace gases," *Reviews of Geophysics* **25**, 1441–1482.

Ramanathan, V., R. J. Cicerone, H. B. Singh, and J. T. Kiehl, 1985, "Trace gas trends and their potential role in climate change," *Journal of Geophysical Research* **90**, 5547–5566.

Rasmussen, R. A., and M. A. K. Khalil, 1984, "Atmospheric methane in the recent and ancient atmospheres: Concentrations, trends, and interhemispheric gradient," *Journal of Geophysical Research* **89**, 11,599–11,605.

Rasmussen, R. A., and M. A. K. Khalil, 1986, "Atmospheric trace gases: Trends and distributions over the last decade," *Science* **232**, 1623–1624.

Raynaud, D., J. Chappellaz, J. M. Barnola, Y. S. Korotkevich, and C. Lorius, 1988, "Climatic and $CH_4$ cycle implications of glacial-interglacial $CH_4$ change in the Vostok ice core," *Nature* **333**, 655–657.

Reinsel, G. C., G. C. Tiao, S. K. Ahn, M. Pugh, S. Basu, J. J. DeLuisi, C. L. Mateer, A. J. Miller, P. S. Connell, and D. J. Wuebbles, 1988, "An analysis of the 7-year record of SBUV satellite ozone data: Global profile features and trends in total ozone," *Journal of Geophysical Research* **93**, 1689–1703.

Reinsel, G. C., G. C. Tiao, A. J. Miller, D. J. Wuebbles, P. S. Connell, C. L. Mateer, and J. J. DeLuisi, 1987, "Statistical analysis of total ozone and stratospheric Umkehr data for trends and solar cycle relationship," *Journal of Geophysical Research* **92**, 2201–2209.

Revelle, R. R., 1983, "Methane hydrates in continental slope sediments and increasing atmospheric carbon dioxide," in *Changing Climate*, National Research Council, National Academy Press, Washington, DC, pp. 252–261.

Revelle, R. R., and H. E. Suess, 1957, "Carbon dioxide exchange between atmosphere and ocean and the question of an increase of atmospheric $CO_2$ during past and present decades," *Tellus* **9**, 18–27.

Robbins, R. C., L. A. Cavanagh, and L. J. Salas, 1973, "Analysis of ancient atmospheres," *Journal of Geophysical Research* **78**, 5341–5344.

Sarmiento, J. L., and J. R. Toggweiller, 1984, "A new model for the role of oceans in determining atmospheric $pCO_2$," *Nature* **308**, 621–624.

Sarmiento, J. L., J. R. Toggweiller, and R. Najaar, 1988, "Ocean carbon cycle dynamics and atmospheric $pCO_2$," *Philosophical Transactions of the Royal Society of London A* **325**, 3–21.

Shaffer, G., 1989a, "A model of biogeochemical cycling of phosporous, nitrogen, oxygen and sulfur in the ocean: One step toward a global climate model," *Journal of Geophysical Research* **94**, 1979–2004.

Shaffer, G., 1989b, "Study of the oceanic carbon cycle and atmospheric $CO_2$ with a high-latitude exchange/diffusion-advection (HILDA) model," presented at the Third International Conference on Analysis and Evaluation of Atmospheric $CO_2$ Data Present and Past, Hinterzarten, FRG, October 16–20, 1989; abstracted in *Environmental Pollution Monitoring and Research Program Publication No. 59*, World Meteorological Organization, Geneva.

Siegenthaler, U., 1984, "Uptake of excess $CO_2$ between the ocean and atmosphere," *Journal of Geophysical Research* **88**, 3599–3608.

Siegenthaler, U, and H. Oeschger, 1978, "Predicting future atmospheric $CO_2$ levels," *Science* **199**, 388–395.

Siegenthaler, U, and H. Oeschger, 1987, "Biospheric $CO_2$ emissions during the past 200 years reconstructed by deconvolution of ice core data," *Tellus* **39B**, 140–154.

Solomon, S., 1988, "The mystery of the Antarctic ozone 'hole'," *Reviews of Geophysics* **26**, 131–148.

Stauffer, B., G. Fischer, A. Neftel, and H. Oeschger, 1985, "Increase of atmospheric methane recorded in Antarctic ice core," *Science* **229**, 1386–1388.

Sundquist, E. T., 1985, "Geological perspectives on carbon dioxide and the carbon cycle," in *The Carbon Cycle and Atmospheric $CO_2$: Natural Variations Archean to Present* (E. T. Sundquist and W. S. Broecker, Eds.), *Geophysical Monograph Series 32*, American Geophysical Union, Washington, DC, pp. 5–59.

Sundquist, E. T., 1986, "Geologic analogs: Their value and limitations in carbon dioxide research," in *The Changing Carbon Cycle* (J. R. Trabalka and D. E. Reichle, Eds.), Springer-Verlag, New York, pp. 371–402.

Sundquist, E. T., and W. S. Broecker (Eds.), 1985, *The Carbon Cycle and Atmospheric $CO_2$: Natural Variations Archean to Present, Geophysical Monograph Series 32*, American Geophysical Union, Washington, DC, 1080 pp.

Tans, P. P., I. Y. Fung, and T. Takahashi, 1990, "Observational constraints on the global atmospheric $CO_2$ budget," *Science* **247**, 1431–1438.

Thompson, A. M., M. A. Huntley, and R. W. Stewart, 1989a, "Perturbations to tropospheric oxidants, 1985–2035: Calculations of ozone and OH in chemically coherent regions," submitted to *Journal of Geophysical Research*.

Thompson, A. M., R. W. Stewart, M. A. Owen, and J. A. Herwehe, 1989b, "Sensitivity of tropospheric oxidants to global chemical and climate change," *Atmospheric Environment* **23**, 519–532.

Tiao, G. C., G. C. Reinsel, J. H. Pedrick, G. M. Allenby, C. L. Mateer, A. J. Miller, and J. J. DeLuisi, 1986, "A statistical trend analysis of ozonesonde data," *Journal of Geophysical Research* **91**, 13,121–13,136.

Toggweiler, J. R., 1990, "Diving into the organic soup," *Nature* **345**, 203–204.

Trabalka, J. (Ed.), 1985, *Atmospheric $CO_2$ and the Global Carbon Cycle*, US Department of Energy, Washington, DC, report DOE/ER-0239.

Verbitsky, M. Ya., and A. G. Lapenis, 1987, "Glaciations, carbon dioxide and calcium in the world ocean," Izvestia, *AN SSSR, Geography Series* **6**, 16–24.

Volk, T., 1987, "Feedbacks between weathering and atmospheric $CO_2$ over the last 100 million years," *American Journal of Science* **287**, 763–779.

Volk, T., and M. I. Hoffert, 1985, "Ocean carbon pumps: Analysis of relative strengths and efficiencies in ocean-driven atmospheric $CO_2$ changes," in *The Carbon Cycle and Atmospheric $CO_2$: Natural Variations Archean to Present* (E. T. Sundquist and W. S. Broecker, Eds.), *Geophysical Monograph 32*, American Geophysical Union, Washington, DC, pp. 91–110.

Volk, R., and Z. Liu, 1988, "Controls of $CO_2$ sources and sinks in the Earth scale ocean," *Global Biogeochemical Cycles* **2**, 73–89.

Wahlen, M., N. Tanaka, R. Henry, B. Deck, J. Zeglen, J. S. Vogel, J. Southon, A. Shemesh, R. Fairbanks, and W. Broecker, 1989, "Carbon-14 in methane sources and in atmospheric methane: The contribution from fossil carbon," *Science* **245**, 286–290.

Walker, J. C. G., P. B. Hays, and J. F. Kasting, 1981, "A negative feedback mechanism for the long-term stabilization of Earth's surface temperature," *Journal of Geophysical Research* **86**, 9776–9782.

Wang, W. -C., D. J. Wuebbles, W. M. Washington, R. G. Isaacs, and G. Molnar, 1986, "Trace gases and other potential perturbations to global climate," *Reviews of Geophysics* **24**, 110–140.

Watson, R. T., and Ozone Trends Panel, M. J. Prather and Ad Hoc Theory Panel, and M. J. Kurylo and NASA Panel for Data Evaluation, 1988, *Present State of Knowledge of the Upper Atmosphere 1988: An Assessment Report*, NASA Reference Publication 1208.

Wigley, T. M. L., 1987, "The effect of model structure on projections of greenhouse-gas-induced climatic change," *Geophysical Research Letters* **14**, 1135–1139.

World Meteorological Organization (WMO), 1985, *Atmospheric Ozone: 1985*, Report No. 16, Global Ozone Research and Monitoring Project, Geneva.

World Meteorological Organization (WMO), 1989, *Scientific Assessment of Stratospheric Ozone 1989*, Geneva.

World Resources Institute (WRI), 1990, *Water Resources 1990–1991*, Oxford University Press, Oxford, UK, 383 pp.

Wuebbles, D. J., 1989, "Beyond $CO_2$—the other greenhouse gases," report UCRL-99883, Lawrence Livermore National Laboratory, Livermore, California; also Paper 89-119.4, Air and Waste Management Association.

Wuebbles, D. J., and J. Edmonds, 1988, *A Primer on Greenhouse Gases*, report DOE/NBB0083, US Department of Energy, Carbon Dioxide Research Division, Washington, DC.

Wuebbles, D. J., K. E. Grant, P. S. Connell, and J. E. Penner, 1989, "The role of atmospheric chemistry in climate change," *Journal of the Air Pollution Control Association* **39**, 22–28.

Wuebbles, D. J., and D. E. Kinnison, 1988, "A two-dimensional model study of past trends in global ozone," in *Proceedings of the Quadrennial Ozone Symposium*, University of Gottingen, FRG.

Wuebbles, D. J., M. C. MacCracken, and F. M. Luther, 1984, *A Proposed Reference Set of Scenarios for Radiatively Active Atmospheric Constituents*, report DOE/NBB-0066, US Department of Energy, Carbon Dioxide Research Division, Washington, DC.

Wyrtki, K., 1962, "The oxygen minima in relation to ocean circulation," *Deep Sea Research* **9**, 11–23.

## CHAPTER 5

Barnett, T. P., 1990, Scripps Institute of Oceanography, personal communication.

Barnett, T., and M. Schlesinger, 1987, "Detecting changes in global climate induced by greenhouse gases," *Journal of Geophysical Research* **92**, 14,772–14,780.

Bryan, K., 1984, "Accelerating the convergence to equilibrium of ocean-climate models," *Journal of Physical Oceanography* **14**, 666–673.

Bryan, K., F. G. Komro, S. Manabe, and M. J. Spelman, 1982, "Transient climate response to increasing atmospheric carbon dioxide," *Science* **215**, 56–58.

Bryan, K., F. G. Komro, and C. Rooth, 1984, "The ocean's transient response to global surface temperature anomalies," in *Climate Processes and Climate Sensitivity, Maurice Ewing Series, Vol. 5* (J. E. Hansen and T. Takahashi, Eds.), American Geophysical Union, Washington, DC, pp. 29–38.

Bryan, K., and M. J. Spelman, 1985, "The ocean's response to a $CO_2$-induced warming," *Journal of Geophysical Research* **90**, 11,679–11,688.

Budyko, M. I., 1968, "On the causes of climate variations," *Sveriges Meteorologiska och Hydrologiska Institut Meddelanden*, ser. B., No. 28.

Budyko, M. I., 1969, "The effect of solar radiation variations on the climate of the Earth," *Tellus* **21**, 611–619.

Budyko, M. I., 1974, *Climatic Changes*, Gidrometeoizdat, Leningrad, 280 pp. (R) (English translation: American Geophysical Union, 1977, 261 pp.)

Cess, R. D., and G. L. Potter, 1988, "A methodology for understanding and intercomparing atmospheric climate feedback processes within general circulation models," *Journal of Geophysical Research* **93**, 8305–8314.

Cess, R. D., G. L. Potter, J. P. Blanchet, G. J. Boer, S. J. Ghan, J. T. Kiehl, H. Le Treut, Z. -X. Li, X. -Z. Liang, J. F. B. Mitchell, J. -J. Morcrette, D. A. Randall, M. R. Riches, E. Roeckner, U. Schlese, A. Slingo, K. E. Taylor, W. M. Washington, R. T. Wetherald, and I. Yagai, 1989, "Interpretation of cloud-climate feedback as produced by 14 atmospheric general circulation models," *Science* **245**, 513–516.

Gates, W. L., 1979, "The Physical Basis of Climate," in *Proceedings of the World Climate Conference,* WMO No. 537, World Meteorological Organization, Geneva, pp. 112–131.

Hansen, J. A., I. Fung, A. Lacis, D. Rind, S. Lebedeff, R. Ruedy, G. Russell, and P. Stone, 1988, "Global climate changes as forecast by Goddard Institute for Space Studies three-dimensional model," *Journal of Geophysical Research* **93**, 9341–9364.

Hansen, J. A., A. Lacis, D. Rind, G. Russell, P. Stone, I. Fung, R. Ruedy, and J. Lerner, 1984, "Climate sensitivity: Analysis of feedback mechanisms," in *Climate Processes and Climate Sensitivity, Maurice Ewing Series, Vol. 5* (J. E. Hansen and T. Takahashi, Eds.), American Geophysical Union, Washington, DC, pp. 130–163.

Hansen, J., and S. Lebedeff, 1987, "Global trends of measured surface air temperature," *Journal of Geophysical Research* **92**, 13,345–13,372.

Hansen, J., and S. Lebedeff, 1988, "Global surface air temperatures: Update through 1987," *Geophysical Research Letters* **15**, 323–326.

Hoffert, M. I., A. J. Callegari, and C. -T. Hsieh, 1980, "The role of deep sea heat storage in the secular response to climate forcing," *Journal of Geophysical Research* **85**, 6667–6679.

Jenne, R., 1975, *Data sets for meteorological research*, NCAR Technical Note TN/1A-III, National Center for Atmospheric Research, Boulder, Colorado, 194 pp.

Jones, P. D., T. M. L. Wigley, and P. B. Wright, 1986, "Global temperature variations between 1861 and 1984," *Nature* **322**, 430–434.

Karol, I. L., and E. V. Rozanov, 1982, "Radiative-convective climate models," *Izvestia—Atmospheric Ocean Physics* **18**, 1179–1191 (R).

Lorenz, E. N., 1984, "Formation of a low-order model of a moist general circulation," *Journal of the Atmospheric Sciences* **41**, 1933–1945.

Manabe, S., and R. J. Stouffer, 1979, "A $CO_2$-climate sensitivity study with a mathematical model of the global climate," *Nature* **282**, 491–493.

Manabe, S., and R. J. Stouffer, 1980, "Sensitivity of a global climate model to an increase of $CO_2$ concentration in the atmosphere," *Journal of Geophysical Research* **85**, 5529–5554.

Manabe, S., and R. T. Wetherald, 1967, "Thermal equilibrium of the atmosphere with a given distribution of relative humidity," *Journal of Atmospheric Sciences* **24**, 241–259.

Manabe, S., and R. T. Wetherald, 1975, "The effects of doubling the $CO_2$ concentration on the climate of a general circulation model," *Journal of Atmospheric Sciences* **32**, 3–15.

Mitchell, J. F. B., C. A. Senior, and W. J. Ingram, 1989, "$CO_2$ and climate: A missing feedback?" *Nature* **341**, 132–134.

Mokhov, I. I., 1981, "On the $CO_2$ influence on the thermal regime of the Earth's climate system," *Meteorologiya i Gidrologiya* **4**, 24–34 (R).

Schlesinger, M. E., 1983, "A review of climate models and their simulation of $CO_2$-induced warming," *The International Journal of Environmental Studies* **20**, 103–114.

Schlesinger, M. E., 1984, "Climate model simulation of $CO_2$-induced climatic change," in *Advances in Geophysics, Vol. 26* (B. Saltzman, Ed.), Academic Press, New York, pp. 141–235.

Schlesinger, M. E., 1985, "Feedback analysis of results from energy balance and radiative-convective models," in *The Potential Climatic Effects of Increasing Carbon Dioxide* (M. C. MacCracken and F. M. Luther, Eds.), report DOE/ER-0237, 280-319, US Department of Energy, Washington, DC, pp. 281–319 (available from NTIS, Springfield, Virginia).

Schlesinger, M. E., 1988, "Quantitative analysis of feedbacks in climate model simulations of $CO_2$-induced warming," in *Physically-Based Modelling and Simulation of Climate and Climatic Change*, (M. E. Schlesinger, Ed.), NATO Advanced Study Institute Series, Kluwer, Dordrecht, The Netherlands, pp. 653–736.

Schlesinger, M. E., and W. L. Gates, 1980, "The January and July performance of the OSU two-level atmospheric general circulation model," *Journal of the Atmospheric Sciences* **37**, 1914–1943.

Schlesinger, M. E., and W. L. Gates, 1981, *Preliminary analysis of four general circulation model experiments on the role of the ocean in climate*, report No. 25, Climatic Research Institute, Oregon State University, Corvallis, Oregon, 56 pp.

Schlesinger, M. E., W. L. Gates and Y. -J. Han, 1985, "The role of the ocean in $CO_2$-induced climate warming: Preliminary results from the OSU coupled atmosphere–ocean GCM," in *Coupled Ocean-Atmosphere Models* (J. C. J. Nihoul, Ed.), Elsevier, Amsterdam, pp. 447–478.

Schlesinger, M. E., and X. Jiang, 1990, "The timescale of climate change induced by increased carbon dioxide," *Journal of Climate* **3** (in press).

Schlesinger, M. E., and J. F. B. Mitchell, 1985, "Model projections of the equilibrium climatic response to increased $CO_2$," in *The Potential Climatic Effects of Increasing Carbon Dioxide*, (M. C. MacCracken and F. M. Luther, Eds.), report DOE/ER-0237, US Department of Energy, Washington, DC, pp. 81–147 (available from NTIS, Springfield, Virginia).

Schlesinger, M. E., and J. F. B. Mitchell, 1987, "Climate model simulations of the equilibrium climatic response to increased carbon dioxide," *Review of Geophysics* **25**, 760–798.

Schlesinger, M. E., and Z. -C. Zhao, 1989, "Seasonal climatic changes induced by doubled $CO_2$ as simulated by the OSU atmospheric GCM/mixed-layer ocean model," *Journal of Climate* **2**, 459–495.

Schlesinger, M. E., Z. -C. Zhao, and D. Vickers, 1989, "Design and critical appraisal of an accelerated integration procedure for atmospheric GCM/mixed-layer ocean models," *Journal of Climate* **2**, 641–655.

Sellers, W. D., 1969, "A global climate model based on the energy balance of the Earth-atmosphere system," *Journal of Applied Meteorology* **8**, 392–400.

Stouffer, R. J., S. Manabe, and K. Bryan, 1989, "Interhemispheric asymmetry in climate response to a gradual increase of atmospheric $CO_2$," *Nature* **342**, 660–662.

Vinnikov, K. Ya., P. Ya. Groisman, and K. M. Lugina, 1990, "The empirical data on modern global climate changes (temperature and precipitation)," *Journal of Climate* **3**, 662—677.

Vinnikov, K. Ya., and I. B. Yeserkepova, 1989, "Empirical data and results of modeling of soil moisture regime," *Meteorologiya i Gidrologiya* **11**, 64–72.

Washington, W. M., and G. A. Meehl, 1984, "Seasonal cycle experiment on the climate sensitivity due to a doubling of $CO_2$ with an atmospheric general circulation model coupled to a simple mixed-layer ocean model," *Journal of Geophysical Research* **89**, 9475–9503.

Washington, W. M., and G. A. Meehl, 1989, "Climate sensitivity due to increased $CO_2$: Experiments with a coupled atmosphere and ocean general circulation model," *Climate Dynamics* **4**, 1–38.

Wetherald, R. T., and S. Manabe, 1986, "An investigation of cloud cover change in response to thermal forcing," *Climatic Change* **8**, 5–23.

Wilson, C. A., and J. F. B. Mitchell, 1987, "A doubled $CO_2$ climate sensitivity experiment with a global climate model including a simple ocean," *Journal of Geophysical Research* **92**, 13,315–13,343.

## *CHAPTER 6*

Barnola, J. M., D. Raynaud, Y. S. Korotkevich, and C. Lorius, 1987, "Vostok ice core provides 160,000-year record of atmospheric $CO_2$," *Nature* **329**, 408–414.

Bolin, B., B. R. Döös, J. Jaeger, and R. A. Warrick, Eds., 1986, *Greenhouse Effect, Climatic Change and Ecosystems, SCOPE-29*, John Wiley & Sons, Chichester, UK.

Budyko, M. I., 1971, *Climate and Life*, Gidrometeoizdat, Leningrad, 472 pp. (R) (English translation: 1974, Academic Press, London, 470 pp.).

Budyko, M. I., 1972, *Man's Impact on Climate*, Gidrometeoizdat, Leningrad (R).

Budyko, M. I., 1974, *Climatic Changes*, Gidrometeoizdat, Leningrad, 280 pp. (R) (English translation: American Geophysical Union, 1977, 261 pp.).

Budyko, M. I., 1979, "Semi-empirical model of atmosphere's thermal regime and real climate," *Meteorologiya i Gidrologiya* **4**, 5–17 (R).

Budyko, M. I., 1980, *The Earth's Climate: Past and Future*, Gidrometeoizdat, Leningrad, 352 pp. (R) (English translation: Academic Press, 1982, 307 pp.).

Budyko, M. I., 1988, "Climate of the late 20th century," *Meteorologiya i Gidrologiya* **10**, 5–24 (R).

Budyko, M. I., K. Ya. Vinnikov, O. A. Drozdrov, N. A. Yefimova, 1978, "Future Climatic Changes," *Izvestia, AN SSSR, Geography Series* **6**, 5–20.

Budyko, M. I., and P. Ya. Groisman, 1989, "The 1980s warming," *Meteorologiya i Gidrologiya* **3**, 5–10 (R).

Budyko, M. I., and Yu. A. Izrael (Eds.), 1987, *Anthropogenic Climatic Changes*, Gidrometeoizdat, Leningrad, 406 pp. (R) (English translation: Arizona University Press, 1990).

Budyko, M. I., A. B. Ronov, and A. L. Yanshin, 1985, *The History of the Earth's Atmosphere*, Gidrometeoizdat, Leningrad, 209 pp. (R) (English translation: Springer-Verlag, Berlin, 1987, 139 pp.).

Budyko, M. I., and M. A. Vasishcheva, 1971, "The influence of astronomical factors on the Quaternary glaciations," *Meteorologiya i Gidrologiya* **6**, pp. 37–47 (R).

Budyko, M. I., and N. A. Yefimova, 1984, "Annual variations in meteorological elements as a model of climatic change," *Meteorologiya i Gidrologiya* **1**, 5–10 (R).

Budyko, M. I., N. A. Yefimova, and I. Yu. Lokshina, 1989, "Expected anthropogenic changes of the global climate," *Izvestia, AN SSSR, Geography Series* **5**, 45–55.

Drozdov, O. A., 1981, "Soil moisture during the climate fluctuations," *Meteorologiya i Gidrologiya* **4**, 12–28.

Flohn, H., 1981, *Major Climatic Events Associated with a Prolonged $CO_2$-induced Warming*, report ORAU-IEA 81-8(M), Oak Ridge, Tennessee: Oak Ridge Associated Universities.

Grotch, S. L., 1988, *Regional Intercomparisons of General Circulation Model Predictions and Historical Climate Data*, report DOE/NBB-0084, US Department of Energy, Washington, DC.

Hansen, J. A., A. Lacis, D. Rind, G. Russell, P. Stone, I. Fung, R. Ruedy, and J. Lerner, 1984, "Climate sensitivity: Analysis of feedback mechanisms," in *Climate Processes and Climate Sensitivity, Geophysical Monograph No. 29, Vol. 5*, pp. 130–163.

Kelchevskaya, L. S, 1983, *Soil Moisture in the European USSR*, Gidrometeoizdat, Leningrad, 183 pp. (R).

Kellogg, W. W., 1977–1978, "Effects of human activities on global climate," Part I, *WMO Bulletin* **26**, 2229–2240 (1977); Part II, *WMO Bulletin* **27**, 3–10 (1978).

Korzun, V. I. (Ed.), 1974, *World Water Balance and Earth's Resources*, Gidrometeoizdat, Leningrad, 640 pp. (R).

Manabe, S., and R. T. Wetherald, 1980, "On the distribution of climate change resulting from an increase in $CO_2$-content of the atmosphere," *Journal of the Atmospheric Sciences* **37**, 99–118.

Manabe, S., and R. T. Wetherald, 1987, "Large-scale changes in soil wetness induced by an increase in atmospheric carbon dioxide," *Journal of Atmospheric Sciences* **44**, 1211–1236.

Meshcherskaya, A. V., N. A. Boldyreva, and N. D. Shapayeva, 1982, *Mean Regional Stores of Soil Available Moisture and Snow Cover Thickness: Statistical Analysis and Examples of Usage*, Gidrometeoizdat, Leningrad, 243 pp. (R).

Mitchell, J. F. B., 1990, "Greenhouse warming: Is the Holocene a good analogue?" *Journal of Climate* (in press).

Mitchell, J. F. B., N. S. Grahame, and K. H. Needham, 1988, "Climate simulations for 9000 years before present: Seasonal variations and the effect of the Laurentide Ice Sheet," *Journal of Geophysical Research* **93**, 8283—8303.

National Research Council (NRC), 1982, *Carbon Dioxide and Climate*, National Academy of Sciences Press, Washington, DC, 72 pp.

Schlesinger, M. E., and J. F. B. Mitchell, 1987, "Climate model simulations of the equilibrium climatic response to increased carbon dioxide," *Review of Geophysics* **25**, 760–798.

US-USSR Meeting Report, 1982, "Climatic effects of increased atmospheric carbon dioxide," *Proceedings of the Soviet-American Meeting on Studying Climatic Effects of Increased Atmospheric Carbon Dioxide*, 15–20 June, 1981, Gidrometeoizdat, Leningrad, 56 pp. (R).

Verigo, S. A., and L. A. Razumova, 1973, *Soil Moisture*, Gidrometeoizdat, Leningrad, 328 pp. (R).

Vinnikov, K. Ya., and I. B. Yeserkepova, 1989, "Empirical data and the results of modeling soil moisture," *Meteorologiya i Gidrologiya* **11**, 64–72.

Vinnikov, K. Ya., and N. A. Lemeshko, 1987, "Soil moisture content and runoff for the USSR territory with global warming," *Meteorologiya i Gidrologiya* **12**, 96–103 (R).

Vinnikov, K. Ya., N. A. Lemeshko, and N. A. Speranskaya, 1989, "Soil moisture and runoff on the extratropical part of the Northern Hemisphere," *Meteorologiya i Gidrologiya* (in press) (R).

Washington, W. M., and G. A. Meehl, 1984, "Seasonal cycle experiment on the climate sensitivity due to doubling of $CO_2$ with an atmospheric general circulation model coupled to a simple mixed layer ocean model," *Journal of Geophysical Research* **89**, 9475–9503.

Zhukov, V. (Ed.), 1986, "Mean multi-year stores of available moisture under winter crops and spring crops over regions, territories, republics and economic regions," in *Vol. 1, European Part of the USSR*, Gidrometeoizdat, Leningrad (R).

Zubenok, L. I., 1976, *Evaporation on the Continents*, Gidrometeoizdat, Leningrad, 264 pp. (R).

## *CHAPTER 7*

Acock, B., and L. H. Allen, Jr., 1985, "Crop responses to elevated carbon dioxide concentrations," in *Direct Effects of Increasing Carbon Dioxide on Vegetation* (B. R. Strain and J. D. Cure, Eds.), US Department of Energy, Washington, DC, pp. 33–97.

Adams, R. M., C. Rosenzweig, R. M. Peart, J. T. Ritchie, B. A. McCarl, J. D. Glyer, R. B. Curry, J. W. Jones, K. J. Boote, and L. H. Allen, Jr., 1990, "Global climate change and US agriculture," *Nature* **345**, 219–224.

American Association for the Advancement of Sciences (AAAS), 1990, *Climate Change and US Water Resources*, (P. Waggoner, Ed.), John Wiley & Sons, New York.

Baker, J. T., L. H. Allen, Jr., K. J. Boote, P. Jones, and J. W. Jones, 1989, "Response of soybean to air temperature and carbon dioxide concentration," *Crop Science* **29**, 98–105.

Beran, M., 1986, "The water resource impact of future climate change and variability," in *Effects of Changes in Stratospheric Ozone and Global Climate. Vol. 1: Overview* (J. G. Titus, Ed.), US Environmental Protection Agency, Washington, DC, pp. 299–330.

Beran, M., 1987, *The Water Resource Impacts of Future Climatic Change and Variability*, World Climate Program WMO 17, No. 129, World Meteorological Organization, Geneva, pp. 1–30.

Budyko, M. I., and Yu. A. Izrael (Eds.), 1987, *Anthropogenic Climatic Changes*, Gidrometeoizdat, Leningrad, 406 pp. (R) (English translation: Arizona University Press, 1990).

Budyko, M. I., K. Ya. Vinnikov, O. A. Drozdrov, N. A. Yefimova, 1978, "Future climatic changes," *Izvestia, AN SSSR, Geography Series* **6**, 5–20.

Bultot, F., A. Coppens, G. L. Dupriez, D. Gellens, and F. Meulenberghs, 1988, "Repercussions of a $CO_2$ doubling on the water cycle and on the water balance, a case study for Belgium," *Journal of Hydrology* **99**, 319–347.

Byron, E. R., A. Jacoby, and C. R. Goldman, 1989, "The effects of global climate change on the water quality of mountain lakes and streams," in *The Potential Effects of Global Climate Change on the United States, Vol. E*, US Environmental Protection Agency, Report to Congress, Washington, DC.

Changnon, Jr., S. A., 1987, "An assessment of climate change, water resources, and policy research," *Water International* **12**, 67–76.

Cohen, S. J., 1986, "Impacts of $CO_2$-induced climatic change on water resources in the Great Lakes Basin," *Climatic Change* **8**, 135–153.

Cure, J. D., 1985, "Carbon dioxide doubling responses: A crop survey," in *Direct Effects of Increasing Carbon Dioxide on Vegetation* (B. R. Strain and J. D. Cure, Eds.), US Department of Energy, Washington, DC, pp. 99–116.

Dooge, J. C. I., 1986, "Effects of $CO_2$ increases on hydrology and water resources," presented at *European Economic Community Symposium on $CO_2$ and Other Greenhouse Gases: Climatic and Associated Impacts*, Brussels, November 4.

Dudek, D. J., 1989, "Climate change impacts upon agriculture and resources: A case study of California," in J. B. Smith and D. A. Tirpak, *The Potential Effects of Global Climate Change on the United States, Vol. C*, US Environmental Protection Agency, Draft Report to Congress, Washington, DC.

Food and Agriculture Organization (FAO), 1985, *World Agricultural Statistics*, United Nations, Rome.

Flaschka, I. M., C. W. Stockton, and W. R. Boggess, 1987, "Climatic variation and surface water resources in the Great Basin region," *Water Resources Bulletin* **23**, 47–57.

Gifford, R. M., 1979, "Growth and yield of $CO_2$-enriched wheat under water-limited conditions," *Australian Journal of Plant Physiology* **6**, 367–378.

Glantz, M. H. (Ed.), 1988, *Societal Responses to Regional Climatic Change: Forecasting by Analogy*, Westview Press, Boulder, Colorado, 428 pp.

Glantz, M. H., and J. H. Ausubel, 1988, "Impact assessment by analogy: Comparing the impacts of the Ogallala aquifer depletion and $CO_2$-induced climate change," in *Societal Responses to Regional Climatic Change: Forecasting by Analogy* (M. H. Glantz, Ed.), Westview Press, Boulder, Colorado, 428 pp.

Glantz, M. H., and T. M. Wigley, 1987, "Climatic variations and their effects on water resources," in *Resources and World Development: Report of Dahlem Workshop, Part B: Water and Lands*, April 27 to May 2, 1986, Berlin, pp. 625–641.

Gleick, P. H., 1986a, *Regional Water Availability and Global Climatic Change: The Hydrologic Consequences of Increases in Atmospheric $CO_2$ and Other Trace Gases*, Ph.D. Thesis, report ERG-DS-86-1, Energy and Resources Group, University of California, Berkeley, California, 688 pp.

Gleick, P. H., 1986b, "Methods for evaluating the regional hydrologic impacts of global climatic changes," *Journal of Hydrology* **88**, 99–116.

Gleick, P. H., 1987a, "The development and testing of a water balance model for climate impacts assessment: Modeling the Sacramento Basin," *Water Resources Research* **23**(6), 1049–1061.

Gleick, P. H., 1987b, "Regional hydrologic consequences of increases in atmospheric $CO_2$ and other trace gases," *Climatic Change* **10**, 137–161.

Gleick, P. H., 1988a, "The effects of future climatic changes on international water resources: The Colorado River, the United States, and Mexico," *Policy Sciences* **21**, 23.

Gleick, P. H., 1988b, "Climate change and California: Past, present, and future vulnerabilities," in *Societal Responses to Regional Climatic Change: Forecasting by Analogy* (M. H. Glantz, Ed.), Westview Press, Boulder, Colorado, pp. 307–327.

Gleick, P. H., 1989, "Climate Change, Hydrology, and Water Resources," *Review of Geophysics* **27**(3), 329-344.

Gleick, P. H., 1990, "Vulnerabilities of water supply, " in *Climate Change and US Water Resources* (P. Waggoner, Ed.), John Wiley & Sons, New York, 223-240.

Jacoby, H., 1989, "Water quality," in *Climate Change and US Water Resources*, (P. Waggoner, Ed.), John Wiley & Sons, New York, 307-328.

Karl, T. R., and W. Reibsame, 1989, "The impact of decadal fluctuations in mean precipitation and temperature on runoff: A sensitivity study over the United States," *Climatic Change* **15**(3), 423-447.

Kassam, A. H., 1977, "Net biomass production and yield of crops," in *Present and Potential Land Use by Agro-ecological Zones Project*, Food and Agriculture Organization of the United Nations, Rome.

Keerberg, O. F., and Yu. A. Vijl, 1982, "Regulation systems and energetics of the pentozophosphate reduction cycle," in *Physiology of Photosynthesis*, Nauka, Moscow, pp. 104–108 (R).

Kellogg, W. W., and Z. -C. Zhao, 1988, "Sensitivity of soil moisture to doubling of carbon dioxide in climate model experiments. Part I: North America," *Journal of Climate* **1**, 348–366.

Kimball, B. A., J. R. Mauney, J. W. Radin, F. S. Nakayama, S. B. Idso, D. L. Hendrix, D. H. Akey, S. G. Allen, M. G. Anderson, and W. Hartung, 1986, *Effects of Increasing Atmospheric $CO_2$ on the Yield and Water Use of Crops*, 039 in series *Response of Vegetation to Carbon Dioxide*, US Departments of Energy and Agriculture, Washington, DC, 125 pp.

Klemes, V., 1985, *Sensitivity of water resource systems to climate variations*, report WCP-98, World Climate Applications Programme, World Meteorological Organization.

Langbein et al., 1949, *Annual Runoff in the United States*, US Geological Survey Circular 5, US Department of the Interior, Washington, DC.

Laysk, A. H., 1977, *Photosynthesis and Respiration Kinetics of C3 Species*, Nauka, Moscow, 196 pp.

Manabe, S., and R. T. Wetherald, 1986, "Reduction in summer soil wetness induced by an increase in atmospheric carbon dioxide," *Science* **232**, 626–628.

Manabe, S., and R. T. Wetherald, 1987, "Large-scale changes in soil wetness induced by an increase in atmospheric carbon dioxide," *Journal of Atmospheric Sciences* **44**, 1211–1235.

Martinec, J., and A. Rango, 1989, "Effects of climate change on snowmelt runoff patterns," in *Remote Sensing and Large-Scale Global Processes, Proceedings of the IAHS Third International Assembly*, Baltimore, Maryland, IAHS Publication No. 186.

Mather, J. R., and J. Feddema, 1986, "Hydrologic consequences of increases in trace gases and $CO_2$ in the atmosphere," in *Effects of Changes in Stratospheric Ozone and Global Climate, Vol. 3*, US Environmental Protection Agency, Washington, DC, pp. 251–271.

Mauney, J. R., K. E. Fry, and G. Guinn, 1978, "Relationship of photosynthetic rate to growth and fruiting of cotton, soybean, sorghum, and sunflower," *Crop Science* **18**, 259–263.

Mearns, L. O., R. W. Katz, and S. H. Schneider, 1984, "Extreme high-temperature events: Changes in their probabilities with changes in mean temperature," *Journal of Climate and Applied Meteorology* **23**, 1601–1613.

Mearns, L. O., S. H. Schneider, S. L. Thompson, and L. R. McDaniel, 1990, "Analysis of climate variability in general circulation models: Comparison with observations and changes in variability in $2 \times CO_2$ experiments," *Journal of Geophysical Research* (in press).

Menzhulin, G. V., and M. V. Nickolaev, 1987, "Methods for study year-to-year variability of cereals crops yields," *Transactions of the State Hydrological Institute* **327**, 113–131 (R).

Menzhulin, G. V., M. V. Nickolaev, and S. P. Savvateev, 1983, "Estimates of technological and climate components of cereal yields," *Transaction State Hydrological Institute* **280**, 111–119.

Miller, K. A., 1990, "Water, electricity, and institutional innovation," in *Climate and Water: Climatic Variability, Climate Change, and the Planning and Management of U.S. Water Resources*, American Association for the Advancement of Sciences (AAAS), John Wiley & Sons, New York.

Mitchell, J. F. B., and D. A. Warrilow, 1987, "Summer dryness in northern mid-latitudes due to increased $CO_2$," *Nature* **330**, 238–240.

Nemec, J., and J. Schaake, 1982, "Sensitivity of water resource systems to climate variation," *Hydrological Sciences* **27**, 327–343

Parry, M. L., T. R. Carter, and N. T. Konijn (Eds.), 1988, *The Impact of Climatic Variations on Agriculture. Vol. I: Assessments in Cool Temperate and Cold Regions*, Kluwer Academic Publishers, Dordrecht, The Netherlands, 876 pp.

Pavlova, V. N., and O. G. Sirotenko, 1985, "On the use of dynamic models for estimating the effect of possible changes and fluctuations of climate on crop capacity," *Transactions All-Union Institute Agricultural Meteorology* **10**, 81–90 (R).

Peart, R. M., J. W. Jones, R. B. Curry, K. Boote, and L. H. Allen, Jr., 1989, "Impact of climate change on crop yield in the southeastern USA: A simulation study," in J. B. Smith and D. A. Tirpak, *The Potential Effects of Global Climate Change on the United States, Vol. C*, US Environmental Protection Agency, Draft Report to Congress. Washington, DC.

Pitovranov, S. E., V. Iakimets, V. I. Kiselev, and O. D. Sirotenko, 1988, "The effects of climatic variations on agriculture in the subarctic zone of the USSR," in *The Impact of Climatic Variations on Agriculture. Vol. I: Assessments in Cool Temperate and Cold Regions* (M. L. Parry, T. R. Carter, and N. T. Konijn, Eds.), Kluwer Academic Publishers, Dordrecht, The Netherlands, pp. 615–722.

Revelle, R. R., and P. E. Waggoner, 1983, "Effects of a carbon dioxide-induced climatic change on water supplies in the western United States," in *Changing Climate*, National Academy of Sciences, National Academy Press, Washington, DC.

Rind, D., R. Goldberg, J. Hansen, C. Rosenzweig, and R. Ruedy, 1990, "Potential evapotranspiration and the likelihood of future drought," *Journal of Geophysical Research* (in press).

Rind, D., R. Goldberg, and R. Ruedy, 1989, "Change in climate variability in the 21st century," *Climatic Change* **14**, 5–38.

Ritchie, J. T., D. B. Baer, and T. Y. Chou., 1989, "Effect of global climate change on agriculture: Great Lakes region," in J. B. Smith and D. A. Tirpak, *The Potential Effects of Global Climate Change on the United States, Vol. C*, US Environmental Protection Agency, Draft Report to Congress, Washington, DC.

Rosenzweig, C., 1985, "Potential $CO_2$-induced climate effects on North American wheat-producing regions,"*Climate Change* **7**, 367–389.

Rosenzweig, C., 1989, "Potential effects of climate change on agricultural production in the Great Plains: A simulation study," in J. B. Smith and D. A. Tirpak, *The Potential Effects of Global Climate Change on the United States, Vol. C*, US Environmental Protection Agency, Draft Report to Congress, Washington, DC.

Sanderson, M., and L. Wong, 1987, "Climatic change and Great Lakes water levels," in *The Influence of Climate Change and Climatic Variability on the Hydrologic Regime and Water Resources*, Proceedings of the Vancouver Symposium, August 1987 (S. I. Solomon, M. Beran, and W. Hogg, Eds.), IAHS Publication No. 168, pp. 477–487.

Schwarz, H. E., 1977, "Climatic change and water supply: How sensitive is the Northeast?" in *Climate, Climatic Change, and Water Supply*, National Academy of Sciences, Washington, DC.

Shiklomanov, I. A., 1987, "Effects of climate changes on Soviet rivers," presented at the *International Symposium of the XIXth General Assembly of the International Union of Geodesy and Geophysics (IUGG)*, August 9–22, Vancouver, British Columbia, Canada.

Shiklomanov, I. A., 1988, *Studies of Land Water Resources: Results, Problems, and Prospects*, Gidrometeoizdat, Leningrad, 170 pp. (R).

Shiklomanov, I. A., 1989, "Anthropogenic changes of climate, water resources, and water management problems," *Conference on Climate and Water, Vol. 2*, Helsinki, Finland, pp. 310–347.

Shiklomanov, I. A., and O. L. Markova, 1987, *Problems of Water Availability and Water Transfers in the World*, Gidrometeoizdat, Leningrad, 293 pp. (R).

Singh, B., 1987, "The impacts of $CO_2$-induced climate change on hydroelectric generation potential in the James Bay Territory of Quebec," in *The Influence of Climate Change and Climatic Variability on the Hydrologic Regime and Water Resources*, Proceedings of the Vancouver Symposium, August 1987 (S. I. Solomon, M. Beran, and W. Hogg, Eds.), IAHS Publication No. 168, pp. 403–418.

Sirotenko, O. D., and A. P. Boiko, 1980, "The use of the imitation model of soil-plant-atmosphere system to estimate the influence of $CO_2$ concentration fluctuations on agrocenoses productivity," in *Transactions of the Soviet-American Symposium on Problems of Atmospheric $CO_2$*, Hydrometeoizdat, Leningrad, pp. 243–251 (R).

Smith, J. B., and D. A. Tirpak, 1989, *The Potential Effects of Global Climate Change on the United States*, US Environmental Protection Agency, Draft Report to Congress, Washington, DC.

Stockton, C. W., and W. R. Boggess, 1979, *Geohydrological Implications of Climate Change on Water Resource Development*, US Army Coastal Engineering Research Center, Fort Belvoir, Virginia.

United States Environmental Protection Agency, 1984, *Potential Climatic Impacts of Increasing Atmospheric $CO_2$ with Emphasis on Water Availability and Hydrology in the United States*, Strategic Studies Staff, Office of Policy Analysis, Office of Policy, Planning and Evaluation, Washington, DC.

United States Environmental Protection Agency, 1988, "The potential effects of global climate change on the United States," *Vol. 1: Regional Studies*, Draft Report to Congress, Office of Policy, Planning, and Evaluation, Washington, DC.

Vinnikov, K. Ya., N. A. Lemeshko, and N. A. Speranskaya, 1989, "Soil moisture and runoff in the extratropical part of the Northern Hemisphere," *Meteorologia and Gidrologia* (in press) (R).

Warrick, R. A., 1984, "The possible impacts on wheat production of a recurrence of the 1930s drought in the US Great Plains," *Climatic Change* **6**, 5–26.

Wigley, T. M. L., and P. D. Jones, 1985, "Influences of precipitation changes and direct $CO_2$ effects on streamflow," *Nature* **314**, 149–152.

Wilhite, D. A., 1988, "The Ogallala aquifer and carbon dioxide: Are policy responses applicable?" in *Societal Responses to Regional Climatic Change: Forecasting by Analogy* (M. H. Glantz, Ed.), Westview Press, Boulder, Colorado, 428 pp.

Wilson, C. A., and J. F. B. Mitchell, 1987, "A doubled $CO_2$ climate sensitivity experiment with a global climate model including a simple ocean," *Journal of Geophysical Research* **92**, 13,315–13,343.

Wong, S. C., 1979, "Elevated atmospheric partial pressure of $CO_2$ and plant growth: I. Interactions of nitrogen nutrition and photosynthetic capacity in $CO_3$ and $CO_4$ species," *Oecologia* (Berlin) **44**, 68–74.

World Meteorological Organization (WMO), 1987, *Water Resources and Climatic Change: Sensitivity of Water-Resources Systems to Climate Change and Variability*, WCAP-4, WMO/TD-No. 247, World Meteorological Organization, Geneva, 50 pp.

# *APPENDIX. Acronyms*

| | |
|---|---|
| AAAS | American Association for the Advancement of Science |
| AGC | Atmospheric general circulation |
| BUV | Backscattered ultraviolet |
| CCN | Cloud condensation nuclei |
| CERES | Crop Estimation through Resource and Environment Synthesis |
| CFC | Chlorofluorocarbon |
| COADS | Comprehensive Ocean-Atmosphere Data Set |
| COS | Carbonyl sulfide |
| DJF | December-January-February |
| DMS | Dimethylsulfide |
| ENSO | El Niño/Southern Oscillation |
| EPA | (US) Environmental Protection Agency |
| FAO | (UN) Food and Agricultural Organization |
| GCM | General circulation model |
| GFDL | Geophysical Fluid Dynamics Laboratory |
| GISS | Goddard Institute for Space Studies |
| GMCC | Geophysical Monitoring for Climate Change |
| GMCC | Global Monitoring for Climate Change |
| GWP | Greenhouse Warming Potential |
| HCFC | Halogenated chlorofluorocarbon |
| HCN | (US) Historical Climate Network |
| HFC | Halogenated fluorocarbon |
| IIASA | International Institute for Applied Systems Analysis |
| IPCC | Intergovernmental Panel on Climate Change |
| JJA | June-July-August |
| MLO | Mixed layer ocean |
| NADW | North Atlantic Deep Water. |
| NASA | (US) National Space and Aeronautics Administration |
| NCAR | National Center for Atmospheric Research |
| NH | Northern Hemisphere |
| NMHC | Nonmethane hydrocarbons |
| NOAA | (US) National Oceanographic and Atmospheric Administration |
| NPP | Net primary productivity |
| OGCM | Oceanic general circulation model |
| OSU | Oregon State University |
| PgC | Petagrams ($10^{15}$ g) of carbon; 1 billion tonnes of carbon |
| ppbv | Parts per billion by volume |
| ppmv | Parts per million by volume |
| SBUV | Solar backscattered ultraviolet |

| | |
|---|---|
| SDDI | Supply Demand Drought Index |
| SH | Southern Hemisphere |
| SOYGRO | An agricultural crop model |
| SST | Sea surface temperature |
| SWV | Stratospheric water vapor |
| TgN | Teragrams ($10^{12}$ g) of nitrogen |
| UKMO | United Kingdom Meteorological Office |
| UNEP | United Nations Environment Program |
| UV | Ultraviolet |
| WMO | World Meteorological Organization |